S0-BJM-845

SOCIOECONOMIC AND ENVIRONMENTAL IMPACTS OF BIOFUELS

Evidence from Developing Nations

Biofuels are currently in the middle of a heated academic and public policy debate. Biofuel production has increased fivefold in the past decade and is expected to double by 2020. Most of this expansion will happen in developing nations. This book is the first of its kind, providing a comprehensive overview of the biofuel debate in developing countries. The chapters are written by a multidisciplinary team of experts who expose the key drivers and impacts of biofuel production and use. The book covers impacts as diverse as air pollution, biodiversity loss, deforestation, energy security, food security, greenhouse gas emissions, land use change, rural development, water consumption, and other socioeconomic issues. It has a wide geographical focus accommodating examples from countries in Africa, America, and Asia. As such, this book will become an indispensable companion to academics, practitioners, and policy makers who wish to know more about biofuel issues in the developing world.

ALEXANDROS GASPARATOS is a James Martin Research Fellow at the Biodiversity Institute, University of Oxford. He has published on a wide range of topics, including biofuels, food security, energy security, ecosystem services, urban biodiversity, and sustainability assessment. He has been involved in several major research projects during his time at Oxford University, the United Nations University (Yokohama, Japan), and the University of Dundee. Dr. Gasparatos is committed to policy-relevant research and has contributed to policy reports that were launched during the Tenth Conference of the Parties of the Convention for Biological Diversity (CBD-COP10). He has a background in ecological economics (PhD, University of Dundee), environmental science (MSc, Imperial College, London), and chemistry (BSc, University of Patras).

PER STROMBERG is a Visiting Research Fellow at the United Nations University (Yokohama, Japan) and an environmental economist at the Swedish Environmental Protection Agency. He is a Cambridge European Trust Fellow and holds a PhD and MSc in environmental economics (University of Cambridge and University College, London, respectively) and a BSc in economics (Stockholm University). He was awarded the James Claydon Prize in Economics from the University of Cambridge. He has lectured at universities in Japan and Peru and was responsible for the MSc module on environmental economics at the United Nations University. Currently his research focuses on economics of climate change, ecosystem services, and biofuel production, and he has published widely on economic development and the environment. He has been a researcher for the United Nations (UN) Development Programme, the UN Economic Commission for Latin America, the Organisation for Economic Cooperation and Development, the economics unit of the Delegation of the European Commission in Mexico, and the International Institute for Environment and Development. He previously headed the Sustainable Development Governance unit at the United Nations University's Institute of Advanced Studies.

SOCIOECONOMIC AND ENVIRONMENTAL IMPACTS OF BIOFUELS

Evidence from Developing Nations

Edited by

ALEXANDROS GASPARATOS
University of Oxford

PER STROMBERG
United Nations University

CAMBRIDGE UNIVERSITY PRESS

CAMBRIDGE UNIVERSITY PRESS
Cambridge, New York, Melbourne, Madrid, Cape Town,
Singapore, São Paulo, Delhi, Mexico City

Cambridge University Press
32 Avenue of the Americas, New York, NY 10013-2473, USA

www.cambridge.org
Information on this title: www.cambridge.org/9781107009356

© Cambridge University Press 2012

This publication is in copyright. Subject to statutory exception
and to the provisions of relevant collective licensing agreements,
no reproduction of any part may take place without the written
permission of Cambridge University Press.

First published 2012

Printed in the United States of America

A catalog record for this publication is available from the British Library.

Library of Congress Cataloging in Publication Data
Socioeconomic and environmental impacts of biofuels : evidence from developing nations / [edited by]
Alexandros Gasparatos, Per Stromberg.
p. cm.
Includes bibliographical references and index.
ISBN 978-1-107-00935-6 (hardback)
1. Biomass energy – Economic aspects – Developing countries.   2. Biomass energy – Social aspects –
Developing countries.   3. Biomass energy – Environmental aspects – Developing countries.
I. Gasparatos, Alexandros.   II. Stromberg, Per.
HD9502.5.B543D448   2012
333.95'39091724–dc23   2012000284

ISBN 978-1-107-00935-6 Hardback

Cambridge University Press has no responsibility for the persistence or accuracy of URLs for external or third-party
Internet Web sites referred to in this publication and does not guarantee that any content on such Web sites is, or will
remain, accurate or appropriate.

Contents

Contributors

Ricardo Abramovay is full professor in the Department of Economics, University of Sao Paulo, Brazil.

Bothwell Batidzirai is junior researcher in the Department of Science, Technology, and Society, Copernicus Institute, Utrecht University, Netherlands.

Ryan Blanchard is a PhD student in the Department of Botany and Zoology, Stellenbosch University, South Africa.

Gareth D. Borman is an MSc student at the School of Animal, Plant, and Environmental Sciences, University of the Witwatersrand, Johannesburg, South Africa.

Matteo Borzoni is research Fellow at the Istituto di Management of Scuola Superiore Sant'Anna, Pisa, Italy.

Miguel Carriquiry is associate scientist at the Center for Agricultural and Rural Development, Iowa State University, Ames, Iowa, United States.

Mark Elder is principal researcher and manager of the Policy and Governance Team, Institute for Global Environmental Strategies, Hayama, Japan.

Amani Elobeid is associate scientist at the Center for Agricultural and Rural Development, Iowa State University, Ames, Iowa, United States.

Karl-Heinz Erb is associate professor of land use and global change at the Institute of Social Ecology, Alpen-Adria University, Vienna, Austria.

Colin Everson is research Fellow in the School of Agriculture, Earth, and Environmental Sciences, University of KwaZulu-Natal, Pietermaritzburg, and Extraordinary Professor in the Department of Plant Production and Soil Science, University of Pretoria, South Africa.

Jacinto F. Fabiosa is codirector of the Center for Agricultural and Rural Development, Iowa State University, Ames, Iowa, United States.

Yan Gao is postdoctoral Fellow at the Graduate School of Environmental Science, Hokkaido University, Japan.

John Garcia-Ulloa is a PhD student in the Department of Environmental Sciences, Swiss Federal Institute of Technology Zurich, Switzerland.

Alexandros Gasparatos is a James Martin Research Fellow at the Biodiversity Institute, Oxford University, and a visiting research Fellow at the Institute of Advanced Studies of the United Nations University, Yokohama, Japan.

P. Winnie Gerbens-Leenes is assistant professor at the University of Twente, Netherlands.

Mark B. Gush is senior scientist at Natural Resources and the Environment, South African Council for Scientific and Industrial Research, Stellenbosch, South Africa.

Helmut Haberl is associate professor at the Institute of Social Ecology, Alpen-Adria University, Vienna, Austria.

Jason Hill is assistant professor in the Department of Bioproducts and Biosystems Engineering, University of Minnesota, Minneapolis, Minnesota, United States.

Arjen Y. Hoekstra is scientific director of the Water Footprint Network and professor in Multidisciplinary Water Management, University of Twente, Netherlands.

Francis X. Johnson is senior research Fellow at the Climate and Energy Programme, Stockholm Environment Institute, Sweden.

Lian Pin Koh is assistant professor of applied ecology and conservation in the Department of Environmental Sciences, Swiss Federal Institute of Technology Zurich, Switzerland.

Fridolin Krausmann is professor of sustainable resource use and deputy director of the Institute of Social Ecology, Alpen-Adria University, Vienna, Austria.

Christian Lauk is research Fellow at the Institute of Social Ecology, Alpen-Adria University, Vienna, Austria.

Janice S. H. Lee is a PhD student in the Department of Environmental Sciences, Swiss Federal Institute of Technology Zurich, Switzerland.

Markku Lehtonen is research Fellow at the Science and Technology Policy Research, University of Sussex, Brighton, United Kingdom.

Omar Masera is professor at the Centro de Investigaciones en Ecosistemas of the National Autonomous University of Mexico, Mexico City, Mexico.

Andreas Mayer is research Fellow at the Institute of Social Ecology, Alpen-Adria University, Vienna, Austria.

Siwa Msangi is senior research Fellow at the Environment and Production Technology Division, International Food Policy Research Institute, Washington, D.C., United States.

Christoph Plutzar is research Fellow at the Institute of Social Ecology, Alpen-Adria University, Vienna, Austria.

Stephen Polasky is the Fesler-Lampert Professor of Ecological/Environmental Economics, University of Minnesota, St. Paul, Minnesota, United States.

Jane Romero is policy researcher, Institute for Global Environmental Strategies, Hayama, Japan.

Daisuke Sano is policy researcher and submanager of the Policy and Governance Team, Institute for Global Environmental Strategies, Hayama, Japan.

Margaret Skutsch is associate professor in environmental geography, Autonomous University of Mexico, Mexico City, Mexico, and senior researcher at the Centre for Studies in Technology and Sustainable Development, University of Twente, Netherlands.

Julia Steinberger is lecturer in ecological economics at the Sustainability Research Institute, University of Leeds, United Kingdom.

Per Stromberg is visiting research Fellow at the Institute of Advanced Studies of the United Nations University, Yokohama, Japan, and economist at the Environmental Economics Unit, Swedish Environmental Protection Agency, Stockholm, Sweden.

Anne Sugrue is a South African National Energy Research Institute PhD Fellow, University of Johannesburg, South Africa.

Theo H. van der Meer is chair of thermal engineering at the Faculty of Engineering Technology, University of Twente, Netherlands.

Graham P. Von Maltitz is senior researcher at Natural Resources and the Environment, Council for Scientific and Industrial Research, Pretoria, and a PhD student at the Sustainability Research Unit, Nelson Mandela Metropolitan University, Port Elizabeth, South Africa.

Kristina Wagstrom is postdoctoral researcher in the Department of Civil Engineering, University of Minnesota, Minneapolis, Minnesota, United States.

Preface

Energy security, economic development, and environmental protection have become three recurrent and closely intertwined policy themes in national and international policy arenas. Presently, fossil fuels are by far the predominant energy carriers driving the world economy. However, their scarcity and uneven geographical distribution can severely affect national economies and international markets. At the same time, fossil fuel combustion is singled out as the most important driver of human-induced climate change, a phenomenon with potentially catastrophic effects in the medium and long term. It is no wonder that the development of copious amounts of cheap, renewable, evenly distributed, and environmentally friendly energy has started featuring prominently in the energy strategies of developed and developing countries alike.

Perhaps the most controversial among the different types of renewable fuel options currently pursued are biofuels, a type of liquid fuel derived from biomass. Biofuels have been identified as potentially viable substitutes for conventional transport fuels. Currently twenty-four countries have enacted the mandatory blending of biofuels with conventional transport fuel (e.g., Brazil, China, the European Union, India, and the United States). Several other countries are designing other types of biofuel-related policies (e.g., Indonesia, the Philippines, and sub-Saharan African nations).

Although certain biofuel practices were initially viewed as environmentally friendly, awareness is emerging about the complexity of biofuel chains and their impacts on the environment and society. Studies have confirmed that first-generation biofuels[1] can have negative impacts on biodiversity, ecosystem functioning, the climate, food security, and the inclusion of the poor. Conversely, certain first-generation biofuel practices can be net-energy suppliers, can be economically and socially beneficial, and may emit fewer greenhouse gases and other atmospheric pollutants during their life cycles when compared to conventional fossil fuels.

---

[1] Biofuels from "sugar, starch and oil bearing crops or animal fats that in most cases can also be used as food and feed" (IEA, 2010a: 22).

In recent years, the biofuel polemic has started featuring very prominently in academic and policy discussions. The sometimes contradictory and controversial findings of the different studies have further fueled the debate. However, conflicting findings regarding the overall impact of biofuels are unsurprising given that the drivers, impacts, and trade-offs of biofuel production and use vary greatly, depending on the environmental and socioeconomic contexts within which biofuels are produced and consumed.

Although literature on the topic is growing, a consistent and cohesive overview is lacking. Considering the preceding, the present volume critically discusses the main drivers, policies, and, especially, impacts that first-generation biofuels can have in different developing nations. The geographical focus reflects the editors' conviction that developing nations will be the biggest winners (or losers) from a shift toward greater biofuel production. The focus on first-generation biofuels reflects the fact that these biofuel practices will make up the bulk of biofuel expansion in developing countries in the coming decade. Despite some discussion about the production of second-generation biofuels (i.e., lignocellulosic biofuels) in developing nations, the fact remains that first-generation biofuels and their impacts will remain highly relevant in these parts of the world in the foreseeable future.

The book is divided into five parts. Part One introduces the key socioeconomic and environmental drivers and impacts surrounding biofuel production and places them within a wider global context (Chapter 1). The major impacts discussed in Part One include energy provision (Chapter 1), rural development (Chapter 1), food security (Chapters 1 and 2), land use change (Chapters 2 and 5), greenhouse gas emissions (Chapter 3), air pollution (Chapter 3), water consumption (Chapter 4), and deforestation (Chapter 5).

The subsequent three parts (Parts Two to Four) provide a rigorous analysis of the preceding (and other context-specific) drivers, impacts, and associated trade-offs in key biofuel-producing developing regions such as Brazil (Chapters 6–8), Southeast Asia (Chapter 9), China (Chapter 10), and sub-Saharan Africa (Chapters 11–13). Contributions span different spatial scales (from the local to the subnational, national, regional, and global scale) and cover a broad range of biofuel production practices, including sugarcane bioethanol (Chapters 6–8 and 12), soybean biodiesel (Chapter 6), palm oil biodiesel (Chapter 9), and jatropha biodiesel (Chapters 10 and 13).

In more detail, Chapter 6 provides a comprehensive overview of the drivers, policies, and impacts of the Brazilian bioethanol and biodiesel programs, while Chapter 7 focuses on the distinct socioeconomic impacts of the bioethanol program (and the power relations that have emerged) in the northeast of Brazil. Chapter 8 explores how global ethanol demand will affect regional land use in Brazil. Chapter 9 shifts the focus to palm oil biodiesel and its impact on biodiversity in Southeast Asia and makes concrete proposals on how to minimize such negative effects. Chapter 10 identifies the main sustainability impacts of jatropha cultivation (for biodiesel) in the Yunnan region of China and proposes solutions that can enhance the viability of jatropha biodiesel. Chapter 11 discusses how biofuel expansion in developed countries might affect African households and proposes a stylized model of household economic behavior to better understand the welfare impacts that are transmitted through biofuel markets to the household level. Chapter 12 looks at the intersection between energy security, agroindustrial development, and international trade in southern Africa and identifies how a regionally integrated expansion of the sugarcane agroindustry (for bioethanol) offers opportunities for improving energy security and competitiveness in the region. Chapter 13 provides a comprehensive overview of the environmental and socioeconomic impacts of jatropha biodiesel across southern Africa.

Finally, in Part Five, the main findings are synthesized. We identify the key lessons learned from the biofuel experience of the countries studied throughout this book (Chapter 14). In our effort to make this work useful to a broad range of readers, we conclude by providing a number of proposals to academics, practitioners, and policy makers that can promote the sustainability of the biofuel economy.

The different chapters adopt highly diverse methodologies to assess and explain the diverse environmental and socioeconomic impacts of biofuels. Methods used in this edited volume range from material balances (Chapter 2) to life cycle assessments (Chapters 3 and 6), water footprint analysis (Chapter 4), remote sensing (Chapter 5), sociological research (Chapter 7), partial equilibrium models (Chapter 8), local interviews and surveys (Chapter 10), econometric models (Chapters 8 and 11), and a number of other field techniques (Chapters 9 and 13).

It is our belief that the wide focus and multiple academic perspectives employed in this book provide a sober, balanced, and cohesive overview of the true potential and real impacts of biofuel expansion in developing countries.

Acknowledgments

First, we would like to thank all the contributors to this book, whose expertise and diverse academic viewpoints acted as a major source of inspiration during the development of this truly multidisciplinary work.

This edited book has benefited from our continuous interaction and fruitful discussions with several academics. We are particularly grateful for the constant input we received during our involvement in the Biofuel for Sustainable Development program, which was funded by the Japanese Ministry of the Environment (Global Environmental Research Fund, project ID E-0802) and conducted in collaboration with several Japanese academic institutions. We also benefited significantly from the feedback we received during two special sessions dedicated to the impacts of biofuels in developing nations during the 2010 Biannual Conference of the International Society for Ecological Economics.

A major component of this book was conceived and undertaken during the editors' fellowships at the Institute of Advanced Studies of the United Nations University (UNU-IAS). We would like to acknowledge the constant support of our UNU-IAS colleagues and particularly of the institute's assistant director, Jose Puppim de Oliveira.

We are also thankful to the Japan Society for the Promotion of Science for providing research funding that enabled this book. Alexandros Gasparatos would further like to acknowledge the support of the Oxford Martin School through a James Martin Fellowship.

Finally, we would like to thank Paloma White for her significant help during the editorial process and our editor at Cambridge University Press, Matt Lloyd, for his understanding and encouragement during the production of this volume.

Foreword

One of the major challenges of the twenty-first century is how to meet growing energy demand in a sustainable manner. Energy demand worldwide has increased with the growth in population and in per capita energy use. Over the past several decades, energy demand has increased most rapidly in developing countries. Even so, per capita energy use in developing countries remains far lower than in developed countries. Energy demand is projected to continue to grow in the coming decades, due to continued population increases and the continued need for economic development.

Where will the supply come from to meet this growing energy demand? Fossil fuels currently supply approximately 80 percent of world energy demand, but overwhelming reliance on fossil fuels is not a sustainable energy strategy. The fossil fuel supply is finite. While peak oil may or may not be reached anytime soon, fossil fuels are an exhaustible resource and cannot be relied on indefinitely. Even if fossil fuels were not an exhaustible resource, continued reliance on fossil fuels causes major environmental problems. Combustion of fossil fuels has been the primary driver of increases in the greenhouse gas concentrations in the atmosphere that intensify global warming. Fossil fuels are also a principal contributor to local and regional air pollution and other environmental problems.

The world needs alternative energy supplies that can replace a substantial portion of fossil fuel use. To be a viable alternative, however, an energy supply source must satisfy three criteria:

- *Energy supply*: be producible in large quantities
- *Economy*: be cost competitive
- *Environment*: have relatively low environmental impact

Renewable energy will undoubtedly be an increasing part of the energy supply picture in the future. Renewable energy in the form of sunlight, wind, and tides is more than sufficient to supply human needs and can be produced in an environmentally sound manner. The main challenge, however, will be finding sufficient amounts of low-cost renewable energy.

Biofuels are a potentially attractive source of renewable energy. Biofuels can reduce the reliance on fossil fuels, especially for liquid transportation fuel, and can be produced in an environmentally sustainable manner. Plants absorb carbon dioxide during growth so that biofuels should offer carbon savings relative to burning fossil fuels. In addition, second-generation biofuels derived from perennial grasses and grown on lower-quality soils could reduce environmental impacts and lessen the competition with food production.

Whether biofuels are better for the environment, and whether they are cost competitive, have been the subjects of heated debate. The contentious high-stakes nature of the biofuels versus fossil fuels debate has often generated far more heat than light. Some critics of biofuels have made overly broad claims that biofuels starve the poor, only survive because of government subsidies, and are environmentally harmful because of impacts on water supplies (both quantity and quality) as well as causing habitat loss and carbon release from land conversion. Some proponents have made overly rosy statements about the pace of technological improvement, cost competitiveness, and environmental friendliness of biofuels.

Of course, not all biofuels are created equal. For example, the carbon footprint of biofuels generated from residual biomass from lands already in agricultural production differs greatly from that of biofuels generated from biomass grown on land newly converted from native forests. Similarly, the economics of sugarcane ethanol, corn-grain ethanol, and cellulosic ethanol grown from various biomass feedstocks are all different. Furthermore, there are several unanswered questions regarding the potential positive or negative social impacts of biofuels. Whether biofuels are an attractive proposition may depend on the manner in which biomass is grown and converted into biofuels.

This book provides a much-needed balanced and evidence-based treatment of the relative merits of biofuels. The focus on biofuel production in developing countries is particularly needed. The tropics offer the most favorable conditions for growing biomass. Developing countries are also likely to have the fastest growth in energy demand in the coming decades. The book provides a wealth of detailed evidence on the specific impacts of biomass production, conversion into biofuels, and subsequent use in different regions. This book is a welcome addition to the literature and one that promises to add much-needed light to the subject.

Stephen Polasky
University of Minnesota

Part One

Global overview

1

Biofuels at the confluence of energy security, rural development, and food security: A developing country perspective

PER STROMBERG[1]

Institute of Advanced Studies, United Nations University, Yokohama, Japan

ALEXANDROS GASPARATOS

Oxford University, Oxford, United Kingdom

Abstract

First-generation biofuels, such as bioethanol and biodiesel from food crops, have been identified as potentially viable substitutes for conventional transport fuels, with several countries mandating the blending of biofuel with conventional transport fuel. A number of policy concerns ranging from energy security to climate mitigation and economic development have been identified as driving forces behind biofuel expansion. Even though energy security is usually the overarching policy driver of biofuel expansion, depending on the local context, other policy concerns, such as rural development, are becoming important shapers of biofuel policies, particularly in developing nations. This local context is also a major determinant of the extent to which biofuel production and use are beneficial to the environment and human well-being. This chapter initially identifies the main drivers of biofuel expansion in developing nations and proceeds to discuss three highly interlinked socioeconomic impacts associated with biofuel production and use: energy security, rural development, and food security.

Keywords: biofuel expansion, energy security, food security, rural development

1. Introduction

Biofuels are a class of liquid fuels derived from biomass through diverse chemical processes (e.g., transesterification of vegetal oils, fermentation of sugar and starch-rich crops). Different biofuel classifications have been proposed in the past decade, with the most enduring being first-generation and second-generation biofuels.[2] Owing

---

[1] Per Stromberg also works for the Swedish Environmental Protection Agency, Stockholm, Sweden.
[2] An emerging classification distinguishes *conventional* from *advanced* biofuels (IEA, 2010a).

to the wide range of feedstocks and technologies used for biofuel production, as well as their varied impacts (e.g., greenhouse gas (GHG) emissions, competition with food), there is sometimes disagreement as to what constitutes a first-generation or a second-generation biofuel (IEA, 2010a). For the purposes of this edited volume, we define *first-generation biofuels* as those biofuels that are produced from "sugar, starch and oil bearing crops or animal fats that in most cases can also be used as food and feed" (IEA, 2010a: 22). *Second-generation biofuels*, conversely, are those that are produced from cellulose, hemicellulose, or lignin (IEA, 2010a).

In this classification, the likelihood that a biofuel will directly compete with food competition constitutes the major criterion for determining whether it is a first- or second-generation biofuel. We adopt this classification because the "food versus fuel" debate has been at the core of the biofuel polemic, especially in developing nations.[3]

The most common types of first-generation biofuels are bioethanol and biodiesel, which are usually used as substitutes for conventional transport fuel (IEA, 2004a). Other uses of first-generation liquid biofuels include cooking and rural electrification (FAO, 2009a).

Energy security, climate change mitigation, foreign exchange savings, and rural development are usually perceived as the principal driving forces behind past and present biofuel expansion (Yan and Lin, 2009). Energy security is in most cases the overarching policy concern for biofuel expansion in places as diverse as Brazil (Chapter 6), the United States (U.S. House of Representatives, 2007), the European Union (EU, 2009), China (Chapter 10), India (MNRE, undated; Zhou and Thomson, 2009), and Indonesia (Zhou and Thomson, 2009). For some countries, particularly those located in sub-Saharan Africa and Southeast Asia, trade balance and rural development have played more significant roles as motivators for biofuel development (Jumbe et al., 2009; Bekunda et al., 2009; Zhou and Thomson, 2009). Conversely, climate change mitigation has influenced marginally, if at all, developing countries to pursue biofuel production (e.g., Zhou and Thomson, 2009).[4] In fact, with the exception of the EU member states, no country has ranked climate change highly as a justification for its biofuel policies.

Brazil, with its Proálcool program, has been since the mid-1970s a pioneer of large-scale biofuel production and use for transport purposes. Thanks to several interconnected factors, the Brazilian biofuel experiment is generally seen as a success (Abramovay, 2008; Fischer et al., 2009) (see Chapter 6) and consequently a number of countries across the world are trying to imitate the Brazilian experience (see Chapter 12). To achieve this, various policy instruments are being put in place to promote biofuel expansion, with the most common being blending mandates (Section 2.3).

---

[3] Oil from *Jatropha curcas* is a common biodiesel feedstock, particularly in China, India, and sub-Saharan Africa. While it is not edible, it is derived from an oil crop and is therefore technically a first-generation biofuel. This is why it is discussed in this volume (see Chapters 10 and 13).

[4] Developing nations have been included in Annex A of the United Nations Framework Convention on Climate Change (UNFCCC). As a result, they are not currently legally obligated to reduce their GHG emissions by the Kyoto Protocol.

This steady demand for biofuels has resulted in global biofuel production increasing more than fivefold in the last decade, with significantly greater biofuel expansion expected to take place in the next decade (OECD-FAO, 2010). OECD-FAO (2010) predicts a doubling of biofuel production by 2020, mainly through the expansion of first-generation biofuels in developing nations. According to different scenarios, biofuel consumption in 2035 will increase by 162.9 to 505.7 percent for Organisation for Economic Co-operation and Development (OECD) countries and by 302.4 to 678.0 percent for non-OECD countries when compared to 2009 consumption levels (IEA, 2010a). For developing nations, the largest increases in biofuel demand are expected to take place in Brazil, China, and India.

At the same time, biofuel expansion has been associated with a number of positive and negative socioeconomic and environmental impacts, which are discussed throughout this edited volume. Some of the impacts include GHG emissions (Chapter 3), air quality (Chapter 3), water consumption (Chapters 4 and 13), deforestation (Chapter 5), land use change (Chapter 8), biodiversity loss (Chapters 9 and 13), and several other socioeconomic issues (Chapters 2, 6, 7, 10, 11, 12, and 13). However, perhaps none of these impacts is more emblematic in the biofuel debate than the potential impacts on food prices and food security (Chapters 2, 6, 10, 11, and 13).

By drawing on significant evidence coming from developing nations, this chapter discusses three highly interlinked socioeconomic impacts associated with biofuel production and use: energy security (Section 2), rural development (Section 3), and food security (Section 4). Given the overwhelming literature on the topic, the aim of this chapter is not to provide a comprehensive review but rather to highlight the interconnected nature of these three impacts and how it complicates efforts to understand the net impact of biofuel expansion on human well-being.

2. Energy provision

2.1. Biofuel and feedstock types

2.1.1. First-generation bioethanol

First-generation bioethanol can be derived through fermentation of sugar-rich crops, such as sugarcane (*Saccharum officianarum*), sugar beet (*Beta vulgaris*), or sweet sorghum (*Sorghum* spp.), or starch-rich crops, such as maize (*Zea mays*), wheat (*Triticum* spp.), or cassava (*Manihot esculenta*) (Fischer et al., 2009). After fermentation and distillation, ethanol can be directly blended with conventional gasoline in different proportions. For example, a mix of 5 percent ethanol and 95 percent gasoline is denoted as "E5."

Bioethanol is by far the most widely produced biofuel on a global scale, with its production having increased by 351.2 percent between 2000 and 2008 (Figure 1.1). Currently the largest bioethanol producers are the United States (from maize), Brazil (from sugarcane), the EU (from sugar beet and wheat), China (from maize), and India

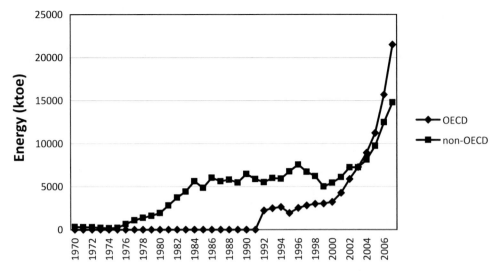

Figure 1.1. Global bioethanol production (1970–2008). *Source:* IEA (2011a).

(from molasses) (IEA, 2010a) (Figure 1.2). Less popular feedstocks include cassava (Southeast Asia and China), sweet sorghum (China), and sweet potato (China). In addition to these countries, several other developing nations around the world (e.g., in sub-Saharan Africa; see Chapter 12) are promoting bioethanol policies, including

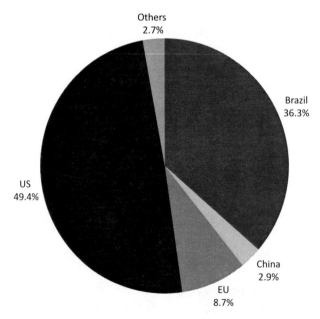

Figure 1.2. Bioethanol production by country (2008). *Source:* IEA (2011a).

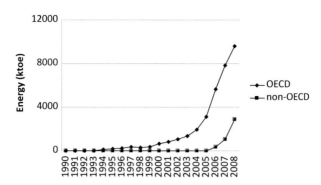

Figure 1.3. Global biodiesel production (1990–2008). *Source:* IEA (2011a).

blending mandates (refer to Section 2.3). This edited volume contains chapters specifically addressing the impacts of sugarcane bioethanol in Brazil (Chapters 6–8) and southern Africa (Chapter 12).

2.1.2. First-generation biodiesel

First-generation biodiesel is produced through the transesterification of animal fats and vegetable oils with the most common feedstocks, including oil from rapeseed (*Brassica napus*), soybean (*Glycine max*), sunflower (*Helianthus annuus*), palm (*Elaeis guineensis*), and jatropha (*Jatropha curcas*) (Fischer et al., 2009). Less popular biodiesel feedstocks include oil from coconut (*Cocos nucifera*) and castor bean (*Ricinus communis*) as well as numerous other oil-bearing crops. Following initial processing that varies depending on the type of feedstock, the derived fatty acid methyl esters are blended with conventional diesel in different proportions, for example, B5 (5% biodiesel, 95% diesel). In some cases, pure plant oil, derived from plants such as jatropha, can be used directly as a fuel for transport, cooking, and power generation (IEA, 2010a).

Global biodiesel production increased by 1,829.6 percent between 2001 and 2009 (Figure 1.3). The largest producers and consumers of biodiesel are the EU (mainly from rapeseed) and the United States (mainly from soybean). Emerging producers include Brazil and Argentina (from soybeans) and Malaysia and Indonesia (from palm oil) (Figure 1.4). India, China, and several sub-Saharan and Southeast Asian nations are showing considerable interest in the production of biodiesel from jatropha.

This edited volume contains chapters dedicated specifically to the impacts of soybean biodiesel in Brazil (Chapter 6), palm oil biodiesel in Southeast Asia (Chapter 9) and jatropha biodiesel in China (Chapter 10), and southern Africa (Chapter 13).

2.1.3. Second-generation biofuels

Second-generation biofuels are produced from cellulose, hemicelluloses and lignin (IEA, 2010a: 22). In contrast to first-generation biofuels, second-generation biofuels

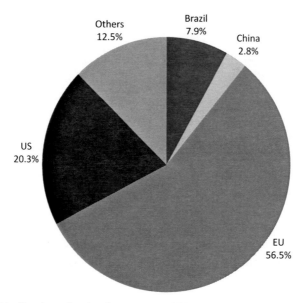

Figure 1.4. Biodiesel production by country (2008). *Source:* IEA (2011a).

can be derived from nonedible plants or from nonedible parts of food crops. Several potential feedstocks for second-generation biofuels have been identified, including the following:

- Short rotation coppice: poplar (*Populus spec.*), willow (*Salix spec.*), eucalyptus (*Eucalyptus spec.*)
- Perennial grasses: miscanthus (*Miscanthus sinensis*), switchgrass (*Panicum virgatum*), reed canary grass (*Phalaris arundinacea*)
- Agricultural by-products: straw, stover, shells, husks, cobs, bagasse, pulp and fruit bunches from different food crops such as corn, rice, sugarcane, sweet sorghum, sugar beet, oil palm, and jatropha
- Forestry by-products: treetops, branches, woodchips, sawdust, bark

For converting lignocellulosic material into liquid biofuel, two main conversion routes are currently pursued: the biochemical route and the thermochemical route. The biochemical route entails the hydrolysis of the lignocellulosic material into sugars and the subsequent fermentation of these sugars into ethanol (Gupta and Demirbas, 2010). The thermochemical route requires heating the feedstock to high temperatures (e.g., through pyrolysis or gasification) and the subsequent transformation of the gas into liquid fuel through different biomass-to-liquids processes such as the Fischer-Tropsch process (Gupta and Demirbas, 2010).

Despite its large global potential (e.g., IEA, 2010a; Chapter 2), there is currently no commercial second-generation biofuel production anywhere in the world. Furthermore, even though there are considerable relevant research activities going on in developed countries, such as the United States and the EU, with the exception of Brazil and China, plans to produce second-generation biofuels in other developing countries are almost nonexistent. This is to a large extent because of the lack of appropriate infrastructure and the shortage of skilled labor and R&D capabilities (IEA, 2010a).

Despite this lack of interest in second-generation biofuel production from developing nations, the fact remains that such biofuel practices can provide numerous economic, environmental, and social benefits. Apart from high net-energy provision and GHG savings, second-generation biofuels avoid direct competition with food production (see Section 4 and Chapters 2, 6, 10, 12, and 13), while they can be an important agent of rural development (IEA, 2010a). In this sense, second-generation biofuels can be very promising energy options for developing countries, such as Brazil, China, and India, that have the infrastructure and R&D potential to pursue their production (IEA, 2010a). Other countries, particularly those located in sub-Saharan Africa, that have much fewer resources to pursue this energy option domestically can still profit from the growing global biomass market by producing and exporting the feedstock to developed countries such as the United States and the EU. In this sense, the biofuel mandates in developed countries can be an important driver for second-generation biofuel production in developing nations (IEA, 2010a).

2.2. Energy balances

When assessing a biofuel's viability, a key consideration is the degree to which the biofuel provides a net energy gain compared to conventional fossil fuels. Two indicators that can facilitate such comparisons are the energy return on investment (EROI) and the percentage energy improvement over conventional fuel (percentage fossil energy improvement). The EROI is defined as the ratio of the total energy supplied by biofuel combustion to the total energy used in the biofuel production process. For a given biofuel practice, an EROI higher than 1 means that the practice is a net energy provider. The percentage fossil energy improvement, in contrast, provides a measure of the amount of nonrenewable energy used during a biofuel's whole life cycle. Generally speaking, biofuel practices with high EROIs and high positive percentage fossil energy improvements provide the largest energy gains and are to be preferred if net energy gain is the sole criterion for determining biofuel viability.

Life cycle analysis (LCA) has been identified as the most appropriate tool for calculating EROI and percentage fossil energy improvement (Menichetti and Otto, 2009; Hill et al., 2006; Zah et al., 2007). Biofuel life cycles are rather complicated and include an array of different stages that range from feedstock production (essentially an agricultural activity) to feedstock transport, feedstock processing, biofuel production and biofuel distribution, storage, dispensing, and combustion (Hess et al., 2009; Delucchi, 2006). The processes adopted during these stages can have a series of environmental and socioeconomic impacts (discussed throughout this edited volume) affecting the different biofuels' sustainability performance.

Tables 1.1 and 1.2 contrast the energy content and the EROIs of different biofuel production practices from all over the world. In most cases, the energy contents and EROIs were directly taken from the cited sources. When they were not provided by the original source, we derived them using standard fuel properties.

The results of our meta-analysis show relatively high EROIs for sugarcane bioethanol, wheat bioethanol, palm oil biodiesel, and jatropha biodiesel. Conversely, corn bioethanol and certain soybean biodiesel practices demonstrate low EROIs – some cases lower than 1. The results of our meta-analysis are consistent with the results of the meta-analysis conducted by de Vries et al. (2010), according to whom sugarcane bioethanol, sweet sorghum bioethanol, and oil palm biodiesel provide the highest EROIs. Sugar beet bioethanol, cassava bioethanol, rapeseed biodiesel, and soybean biodiesel have the next highest EROIs. Maize bioethanol and wheat bioethanol exhibited the lowest EROIs.

An LCA meta-analysis conducted by Menichetti and Otto (2009) concludes that most current first-generation biofuel production practices provide – albeit to different degrees – positive percentage fossil energy improvements. Sugarcane bioethanol provides by far the highest and most consistent percentage fossil energy improvements (in the range of 80%–90%). Other biofuel production practices, such as maize–sugar beet–wheat bioethanol and rapeseed–soybean–sunflower–palm oil biodiesel, provide mostly positive, but highly variable, percentage fossil energy improvements. Finally, a comparative LCA study ranked different biodiesel production chains according to their use of nonrenewable energy as follows (in decreasing order of energy consumption): soybean (Argentina), soybean (Brazil), rapeseed (EU), rapeseed (Switzerland), palm oil (Malaysia), and soybean (United States) (Panichelli et al., 2009).[5]

Several LCAs have shown that jatropha biodiesel is generally a net energy provider, with the biodiesel production stage (transesterification) being the most energy-intensive stage (Achten et al., 2008; Reinhardt et al., 2007a). LCAs on the production and use of straight jatropha oil as a biofuel have shown significant net energy gains when compared to conventional fuels (Gmunder et al., 2010), though using straight jatropha oil as a fuel without any prior processing is not as energy efficient as jatropha biodiesel, and it can cause the combustion engine to malfunction.

Considering the findings of the preceding meta-analyses, it becomes obvious that several first-generation biofuel practices can provide significant net energy gains. As a result, it is reasonable to conclude that some first-generation biofuels meet the net energy provision criterion suggested by Hill et al. (2006) and therefore constitute feasible energy options in the short to medium term. However, certain authors have given the reminder that the moderately high EROIs of most biofuel practices are

---

[5] All these biodiesel practices had lower cumulative energy demand than conventional diesel.

Table 1.1. *Energy content and EROI for different bioethanol practices*

Feedstock	Country	J/kg Feedstock	J/ha	EROI (Jout/Jin)	Source[a]
Sugarcane	Brazil	1.91E+06	1.32E+11	8.5	Smeets et al. (2008); Macedo et al. (2004)
	Brazil	1.87E+06	1.27E+11	3.1	Smeets et al. (2008); Oliveira et al. (2005)
	Brazil	1.99E+06	1.59E+11	3.9	Smeets et al. (2008); Oliveira et al. (2005)
	Brazil	1.76E+06	1.40E+11	9.3	Boddey et al. (2008)
	Brazil	1.63E+06	1.30E+11	8.2	Pereira and Ortega (2010)
	Brazil	1.75E+06	1.30E+11	NA[b]	de Vries et al. (2010)
	Colombia	1.63E+06	1.85E+11	NA[b]	Quintero et al. (2008)
	Mexico	1.76E+06	1.23E+11	4.7	Garcia et al. (2011)
	Southern Africa	1.22E+06	9.60E+10	8.0	von Maltitz and Brent (2008)
Corn	United States	8.22E+06	7.64E+10	1.5	Hill et al. (2006)
	United States	8.14E+06	6.68E+10	NA[b]	de Vries et al. (2010)
	United States	8.36E+06	7.30E+10	1.7	Shapouri et al. (2004)
	United States	7.85E+06	6.79E+10	0.8	Pimentel and Patzek (2005)
	United States	8.30E+06	NA[b]	0.9	Nielsen and Wenzel (2005)
	United States	8.56E+06	7.46E+10	0.9	Kim and Dale (2008)
	Colombia	9.42E+06	1.41E+11	NA[b]	Quintero et al. (2008)
	China	8.36E+06	5.43E+10	0.9	Xunmin et al. (2009)
	Southern Africa	4.07E+07	3.47E+10	1.3	von Maltitz and Brent (2008)
Wheat	United Kingdom	7.80E+06	6.71E+10	1.6	Royal Society (2008)
	Netherlands	7.64E+06	6.30E+10	1.8	SenterNovem (2005)
	Switzerland	8.67E+06	5.57E+10	2.5	Gnansounou et al. (2009)
	Europe	7.53E+06	6.18E+10	NA[b]	de Vries et al. (2010)
	Australia	5.65E+06	9.89E+09	5.8	Grant et al. (2008)

(continued)

Table 1.1 (continued)

Feedstock	Country	J/kg Feedstock	J/ha	EROI (Jout/Jin)	Source[a]
Sweet sorghum	China	1.78E+06	NA[b]	NA[b]	Li and Chan-Halbrendt (2009)
	China	1.84E+06	8.65E+10	NA[b]	de Vries et al. (2010)
	China	2.51E+07	1.00E+11	NA[b]	Zhang et al. (2010)
	China	1.42E+06	9.18E+10	0.7	Xunmin et al. (2009)
	Southern Africa	1.48E+06	8.86E+10	1.0	von Maltitz and Brent (2008)
Sugar beet	United Kingdom	2.25E+06	1.17E+11	1.2	Royal Society (2008)
	Europe	2.05E+06	1.25E+11	NA[b]	de Vries et al. (2010)
Cassava	China	3.71E+06	NA[b]	NA[b]	Li and Chan-Halbrendt (2009)
	China	4.01E+06	1.33E+11	1.5	Dai et al. (2006)
	China	9.79E+06	NA[b]	1.3	Leng et al. (2008)
	China	8.91E+06	1.19E+11	1.5	Xunmin et al. (2009)
	Thailand	2.89E+06	7.80E+10	NA[b]	de Vries et al. (2010)
	Thailand	2.89E+06	7.82E+10	1.9	Nguyen et al. (2007)
	Thailand	3.40E+06	NA[b]	0.8	Silalertruska and Gheewala (2009)
Molasses	Thailand	4.75E+06	2.45E+09	0.8	Nguyen et al. (2007)
	Thailand	4.59E+06	NA[b]	0.8	Silalertruska and Gheewala (2009)
	Nepal	4.85E+06	8.32E+09	0.6	Khatiwada and Silveira (2009)
Sweet potato	China	2.97E+06	NA[b]	NA[b]	Li and Chan-Halbrendt (2009)

[a] De Vries et al. (2010) reports averages from several case studies. Apart from Leng et al. (2008), all results are reported based on feedstock wet weight.

[b] The information was not readily reported or could not be derived using the information reported in the respective study.

12

Table 1.2. *Energy content and energy ratios for different biodiesel practices*

Feedstock	Country	J/kg Feedstock	J/ha	EROI (Jout/Jin)	Source[a]
Soybean	United States	6.74E+06	1.79E+10	3.2	Hill et al. (2006)
	United States	6.69E+06	1.74E+10	NA[b]	de Vries et al. (2010)
	Brazil	7.00E+06	2.04E+10	1.1	Gasparatos et al. (2011)
	Brazil	4.05E+06	1.15E+10	2.5	Cavalett and Ortega (2010)
	Argentina	5.83E+06	1.51E+10	1.1	Panichelli et al. (2009)
	China	6.48E+06	1.17E+10	1.0	Xunmin et al. (2009)
	Southern Africa	3.52E+06	9.41E+09	3.0	von Maltitz and Brent (2008)
Palm oil	Southeast Asia	9.43E+06	1.70E+11	NA[b]	de Vries et al. (2010)
	Thailand	5.54E+06	9.52E+10	2.4	Pleanjai and Gheewala (2009)
	Thailand	6.39E+06	1.17E+11	2.6	Thamsiriroj and Murphy (2009)
Rapeseed	United Kingdom	1.30E+07	4.03E+10	2.3	Royal Society (2008)
	Netherlands	1.35E+07	4.44E+10	NA[b]	SenterNovem (2005)
	Europe	1.41E+07	4.67E+10	NA[b]	de Vries et al. (2010)
Canola	Southern Africa	8.44E+06	1.24E+10	2.5	von Maltitz and Brent (2008)
Sunflower	Southern Africa	1.81E+07	1.48E+10	3.0	von Maltitz and Brent (2008)
Jatropha	China	1.16E+07	5.79E+10	2.0	Xunmin et al. (2009)
	India	1.02E+07	1.73E+10	1.4	Achten et al. (2010a)
	India	3.30E+07	7.87E+10	1.9	Whitaker and Heath (2009)
	Thailand	8.26E+06	1.03E+11	1.4	Prueksakorn and Gheewala (2008)
	Africa	1.43E+07	5.73E+10	4.7	Ndong et al. (2009)
	Africa	9.83E+06	1.38E+10	1.8	Ndong et al. (2009)

[a] De Vries et al. (2010) report averages from several case studies. Apart from Achten et al. (2010a) and Ndong et al. (2009), all reported results are based on feedstock wet weight.

[b] The information was not readily reported or could not be derived using the information provided in the respective study.

Table 1.3. *Biofuel mandates in selected countries*

Country	Target/Mandate
Argentina	E5 (2010); B7 (2010)
Australia	E6 (2011) in New South Wales; B2 in New South Wales
Brazil	E20–25; B5 (2010)
Canada	E5 (2010); B2 (2012)
China	E10 in nine provinces
Colombia	E10; B10 (2010); B20 (2012)
France	7% by energy content
Germany	6.75% by energy content
India	E10; B5 (2012); 20% biofuels (2017)
Italy	B3.5 (2010)
Japan	500 million L (2010)
Kenya	E10 (2010); B5 (2012)
Paraguay	E24; B5
Peru	E7.8 (2010); B5 (2010)
South Africa	E2; B2
South Korea	B2.5 (2010); B3 (2012)
Spain	5.83% by energy content
Sweden	5.75% by energy content
Thailand	B2; B5 (2012)
United Kingdom	3.6% (2010); 4.2% (2011); 4.7% (2012); 5.3% (thereafter)
United States	49 billion L (2010); 78 billion L (2015); 136 billion L (2022)
Zambia	E5 (2011); B10 (2011)

Source: IEA (2010a).

much lower than the EROIs of conventional fossil fuels (about 15-20) (Cleveland et al., 2006) and are thus not high enough to ensure energy security (Hagens et al., 2006). Additionally, the overreliance on fossil fuel–intensive inputs (e.g., fertilizers and agrochemicals) for feedstock production throws doubt onto biofuels' long-term energy viability as long as the current production practices are employed. Finally, several non-energy-related impacts also need to be considered and are currently far from being adequately managed, as discussed in greater length throughout this edited volume.

2.3. Energy security

As already mentioned, energy security is usually the main motivation behind biofuel expansion. Several countries around the world have promoted biofuel policies as a response to this concern (see Chapters 6, 9, and 12). Blending mandates are the most commonly adopted policy instrument and have already been adopted by several countries around the world (Table 1.3). They legally require that conventional transport fuel be blended with bioethanol and biodiesel and set the proportions of each in the blended fuel. Such programs are usually supported by different types of subsidies, targeting both the fuel's production and consumption (GSI, 2011) (refer to Section 1.3).

First-generation biofuels that are net energy providers (high EROI, positive and high percentage fossil energy improvement) could be a dependable and renewable source of energy capable of improving energy security at the national and local levels.

Brazil offers the only example of a country that has significantly improved its energy security through biofuel production and use.[6] Brazil's success can be attributed to an interconnected set of factors, including sugarcane ethanol's high EROI (refer to Table 1.1 and Chapter 6). However, in special cases, biofuels can promote energy security despite not providing net energy gains. LCA results suggest that while molasses bioethanol in Nepal is not a net energy provider, its nonrenewable energy requirement is only about 8.3 percent (Khatiwada and Silveira, 2009). This implies a significant percentage fossil energy improvement. In this case this biofuel practice is a relatively viable option owing to Nepal's abundant supply of the required renewable resources coupled with the financial burden that fossil fuel import entails on the national economy (Khatiwada and Silveira, 2009).

Biofuel may also improve energy security at the local level. For example, the Indonesian government has committed to promoting energy security in the country through the Energy Self-Sufficient Villages program, which aims to make 1,000 villages capable of meeting their own energy needs from locally available renewable resources, including biofuels, hydropower, and wind energy (Kusdiana and Saptono, 2008). Biofuel-based energy solutions are one of the main avenues explored using several different feedstocks such as jatropha, coconut oil, palm oil, cassava, and sugarcane (Kusdiana and Saptono, 2008).

Rural households' access to readily available and locally produced liquid biofuels can contribute to their energy security. In a number of cases in Africa, Asia, and Latin America, energy security at the local level (usually the village level) was improved through small-scale biofuel projects (FAO, 2009a; Energia, 2009) (see also Section 3.3). In addition, access to locally produced liquid biofuels could help alleviate poverty, for example, by reducing the high shadow cost in terms of time and energy that people invest to collect fuel wood in some of the least developed countries (Ewing and Msangi, 2009). This positive effect of liquid biofuels is particularly significant when transport costs of imported fossil fuel are high (e.g., in landlocked countries) or when road infrastructure is poorly developed (Kojima and Johnson, 2006).

3. Rural development

Rural development is another key driver of biofuel expansion in developing nations and can be promoted through both domestic and export demand. In macroeconomic terms, feedstock and biofuel production can generate employment and income opportunities

---

[6] Sugarcane bioethanol constituted 20.4% of the total energy consumed in the transport sector and 11.1% of the total energy consumed within the country in 2009 (MME, 2011) (see Chapter 6).

as well as enhance trade balances and foreign currency reserves. Of these positive effects, the first two are directly linked to rural development because of the agricultural phase (feedstock production) and therefore contribute significantly to rural poverty alleviation and human well-being.

3.1. Feedstock production and rural employment

In developing nations, the largest labor requirements in the biofuel production chain, and consequently the largest employment opportunities, come from feedstock production (i.e., the agricultural sector).[7] Less labor is needed for the production of the biofuel itself, the commercialization of new market commodities (e.g., bioethanol), in the energy sector, and in other stages of the biofuel production chain.

Brazil, Indonesia, and Malaysia have substantial and highly competitive feedstock production sectors, which are legacies of their plantation history. In Brazil, the bioethanol sector, with a history spanning over 30 years, has contributed significantly to income and employment generation (refer to Chapter 6 for a detailed discussion of the impact of the Brazilian bioethanol and biodiesel programs on rural employment).[8] The case of Indonesia is somewhat different. Oil palm cultivation provides the main biofuel feedstock in the country, palm oil, but has a rather low labor intensity, with only 0.5 jobs/ha, when compared to 2–4 jobs/ha for sugarcane (Winrock, 2009). Despite this relatively low labor intensity, Winrock (2009) estimated that up to 57 percent of Riau's population, and between 10 and 50 percent in 11 other Indonesian regions, derive economic gain in one way or another by the palm oil industry. This figure includes employees and family dependants in downstream processing and associated services.

For certain biofuel production practices the mechanization of agricultural production is important to attain increases in productivity and competitiveness. Mechanization might, however, take a toll on rural employment, considering that a single tractor can displace several workers (Smeets et al., 2008) (see Chapter 6). Finally, it should be noted that downstream activities in the biofuel chain, such as biofuel manufacturing, processing, and distribution, require more skilled labor (often abroad) and, as a general rule, employ far fewer people. For example, although it has been suggested that Indonesian biodiesel production alone could provide 2.5 million jobs (Cassman and Liska, 2007; Sargeant, 2001), Dillon et al. (2008) found that only 1,000 persons were directly employed in downstream biofuel activities in the whole country.

---

[7]   This includes land clearing, which is the first stage of the biofuel production chain that requires labor (and hence employment), particularly in areas where the feedstock has not been cultivated on former agricultural land (Koh and Wilcove, 2007).

[8]   In Brazil and the Philippines, sugarcane bioethanol policies have aimed to use the idle production capacity of sugar distilleries, which have traditionally suffered high unemployment due to harsh competition in international sugar markets (e.g., Javier, 2008).

3.2. Feedstock production and income generation

Often higher prices of agricultural commodities (see Section 4.1) benefit farmers in developing countries by increasing their income (OECD, 2008). Essentially, the same applies for higher biofuel prices and feedstock producer incomes. Moreover, cross-sectoral comparisons in different countries have suggested that biofuel feedstock production can be a more lucrative activity than crop cultivation for food purposes. For example, the average salary of sugarcane cutters in São Paulo State is generally higher than for agricultural workers for other crops (Smeets et al., 2008) (see Chapter 6).[9]

The effect of biofuel prices on income generation can differ depending on the stage of the biofuel production chain. In the case of Indonesia, it has been suggested that farmers, plantation owners, and exporters will gain from the generally higher biofuel prices, whereas processors and net food buyers further down the production chain are expected to be negatively affected (Dillon et al., 2008) (refer to Section 4).

3.3. The role and risks of different feedstock production systems

The effect of biofuel production on employment and income depends, to a great extent, on the type of feedstock production system adopted and its institutional context. There is a significant debate about which of the two production schemes – large scale (i.e., large plantations) or smallholder – can contribute most to employment and income generation in developing countries (e.g., Arndt et al., 2010a; 2010b) (see also Chapters 6 and 13).

3.3.1. Large-scale feedstock production

Large-scale feedstock production can enable economies of scale and often lower transaction costs for accessing markets (Pickett et al., 2008).[10] A comparison of the Brazilian bioethanol and biodiesel programs can provide some insight into this assertion. The bioethanol program has traditionally adopted a plantation model, whereas the biodiesel program initially aimed to boost feedstock production by smallholders. The bioethanol program reached very competitive biofuel production only a few years after its inception, whereas the biodiesel program still struggles to attain competitiveness (see Chapter 6). Furthermore, large-scale plantations can support rural development indirectly by leasing land either from smallholders or from regional and/or municipal governments, who then, ideally, translate these incomes into other welfare-enhancing investments. However as Cotula et al. (2008) and Vermeulen and

---

[9] Even though the average salary is above the minimum wage levels mandated by the Brazilian government, it is still not high enough to allow recipients to escape poverty (Martinelli and Filoso, 2008).

[10] E.g., through integrated production units (integrated feedstock production and processing facilities) linked with good road infrastructure to major transport hubs and/or consumption.

Cotula (2010) discuss, the regulation of such land leases is an important prerequisite for avoiding instances of land grabbing that do not benefit the local population.

3.3.2. Feedstock production by smallholders

Some authors suggest that feedstock production by smallholders can offer significant economic benefits to rural populations (see Chapters 10 and 13). One important reason for this, is the potential for technology transfers that can increase productivity in other economic activities, especially the cultivation of crops for food. In Mozambique, the adoption of a biofuel program based on decentralized, small-contract farming units – outgrowers – could contribute up to 0.6 percent of gross domestic product (GDP) (Arndt et al., 2010a). In fact, the effect of Mozambique's planned biofuel program on GDP would be higher if such an outgrower approach were to be adopted rather than a large-scale plantation model (Arndt et al., 2010a). Even accounting for some food displacement effects, the Mozambique program could reduce poverty by approximately 6 percent over a 12-year start-up period.

The main benefits of outgrower schemes (i.e., contract farming) compared with large plantations are (1) better use of unskilled labor, (2) land rents are appropriated by the smallholders themselves, and (3) higher and more evenly distributed technology spillovers (Arndt et al., 2010b). The outgrower approach may result in a more even income distribution (Arndt et al., 2010a, 2010b).

3.3.3. Small-scale production for local consumption

A growing body of literature suggests that small-scale biofuel initiatives for local consumption can contribute positively to poverty alleviation and rural development (FAO, 2009a; Energia, 2009). Small-scale biofuel schemes are particularly beneficial when alternative local energy carriers are costlier (e.g., in remote areas with high fuel transportation costs) or are associated with other high indirect costs (e.g., the time spent for fuel wood collection that deviates labor from other, more economically productive activities). Such benefits materialize to a large extent because of the local production and utilization of a renewable energy carrier, which also enhances local energy security (refer to Section 2.3).

Examples of small-scale initiatives include rural electrification in Mali, Cambodia, Uganda, and India (from straight jatropha oil); biodiesel water pumping and irrigation in India and Nepal (from jatropha and other local underutilized oil seeds); and biodiesel production in Guatemala, Thailand, and South Africa (from jatropha, sunflower seeds, and soybean oil) (FAO, 2009a; Energia, 2009).

3.3.4. Market-oriented smallholders: Negotiation power and risk exposure

A fact that is not always acknowledged in government expansion programs and seed company campaigns that target smallholders is that shifting to biofuel feedstock production can be a risky endeavor for smallholders.

Smallholders have low negotiation power because of high transaction costs and asymmetric information (e.g., lower access to information, including information about prices in national and international commodity markets), which essentially restrict their market access and market power. Evidence from India shows that middlemen capture the rent from higher commodity prices instead of letting the rents trickle down to the producers (Agoramoorthy et al., 2009). The Brazilian sugarcane sector is notorious for the concentration of wealth and power in the hands of a few actors who manage to control much of the economic rents from biofuel feedstock production (see Chapters 6 and 7). Similar findings have been reported for jatropha plantations (Ariza-Montobbio et al., 2010). In a similar fashion, Richardson (2010) reports that the Zambian ethanol program's aim to benefit smallholders has failed due in part to the strong market power of a local ethanol company that succeeded in influencing the price structure of ethanol to its own benefit, limiting its tax contribution.

High market and production chain uncertainties expose farmers to financial risk. The 2007–2008 food crisis provides an example of this. During this crisis, worldwide agricultural commodity stocks were diminished, rendering food (and as an extension, biofuel feedstock) production more vulnerable to supply and demand shocks (Wright, 2009).

The link between food and energy systems adds additional market uncertainty, as can be seen when observing the price volatility of food commodities over the past decade (Woods, 2006; Robles et al., 2009). High energy prices can have a significant knock-on effect on the price of production inputs, including fuel and energy-intensive commodities such as fertilizers and pesticides, and, as a result, a large negative effect on poverty alleviation (Arndt et al., 2010a).

Other production risks stem from insufficient knowledge about new technologies (Agoramoorthy et al., 2009). This problem is accentuated in boom and bust investment cycles, with an example being the large-scale jatropha expansion in parts of the developing world. Many of these large-scale expansions were implemented despite an insufficient understanding of the links between jatropha cultivation requirements (e.g., water, fertilizer) and expected yields (see Chapter 13).

Additionally, infrastructure and marketing channels may be a source of risk due to the production chain being vulnerable to interruptions. For example, transport interruption and capacity constraints of palm oil processing facilities have in some cases resulted in the harvested crops perishing before being processed, causing subsequent income loss for smallholders in parts of Africa (FAO, 2002b).

Finally, high production costs, including the acquisition of seed, can also affect the viability of feedstock production for smallholders. One relevant example is the case of genetically modified crops. In one case a single company controlled 33 percent of genetically modified soybean seeds, therefore enjoying a large influence on the price formation of such seeds (FAO, 2002a).

3.3.5. International market risks

The largest increase in value added from the production of first-generation biofuel feedstocks is experienced in developed countries (Fischer et al., 2009). It is in this context that barriers to international trade imposed by industrialized countries should be seen (Pickett et al., 2008).

Trade barriers (e.g., subsidies) in industrialized countries reduce the competitiveness of agricultural commodities and biofuel exports from developing countries. An interesting example is the case of the bioethanol industry in Pakistan, which, benefiting from special trade arrangements, was, by 2005, the second largest foreign supplier of bioethanol to the European Union (behind Brazil). The subsequent removal of these arrangements in 2006 dealt a hard blow to the Pakistani bioethanol industry (Khan et al., 2010).

3.3.6. The role of subsidies

Currently most countries' biofuel programs are supported by subsidies (GSI, 2011; OECD, 2008), with Brazil being the only major exception.[11] In 2008 Indonesia subsidized its biofuel sector with USD 40 million, excluding infrastructure investments (Dillon et al., 2008). While subsidized biofuel programs may generate employment and income, it is not obvious to what extent they contribute to net public welfare in general and rural development in particular, especially when compared with other means of poverty alleviation and employment creation. Once the cost of subsidies and the opportunity cost of factor endowments such as labor, financial and natural capital are taken into account, the net contribution of biofuel development to employment is often unclear, as is its relative poverty-alleviation performance when compared to other economic activities with similar labor intensity as biofuel feedstock cultivation (e.g., Rajagopal, 2008). There is also some evidence that biofuel subsidies may worsen a country's income distribution due to rent capture by already privileged groups. For example, in Indonesia, 60 percent of the portion of the population that has the highest incomes received 83 percent of subsidies for fuel (gasoline and biofuels), while the remaining portion of the population received only 17 percent of the total subsidies (Dillon et al., 2008).

While the production of biofuel feedstock is economically viable if corresponding fossil fuel prices are high enough, in developing countries the steps from feedstock to final biofuel are generally not cost competitive with its market substitute, fossil fuels. This is one of the reasons why biofuel-related exports from developing countries consist, to a large extent, of feedstock rather than refined ethanol or biodiesel.

---

[11] Even though the Brazilian bioethanol program benefited from major direct subsidies in the first two decades of its existence, this is no longer the case (see Chapter 6).

4. Food security and access to food

4.1. Competition between biofuels and food production

With the exception of jatropha, all first-generation biofuel feedstocks considered in this edited volume are food crops. Some of them are staple food crops (e.g., corn, wheat), others are key vegetable oils (e.g., palm oil), whereas others are important components of the food industry as a whole (e.g., sugarcane, soybeans). As a result, it has been suggested that biofuel expansion risks competing with food production directly by diverting food crops for biofuel production (see Chapter 2) and indirectly through competition for land (Chapters 6 and 8), agricultural labor (Chapter 10), and other production inputs such as water (Chapter 4).[12]

It has been estimated that in 2007, 1.6 percent of the world's cultivated land was used for biofuel feedstock production (Fischer et al., 2009). Simulations conducted by the same authors suggest that if 2020 global biofuel blending targets are to be met, then biofuel production from cereals will disrupt significantly the production of food and feed, particularly in developing nations. For this reason, countries such as India and China have prohibited the use of edible crops for biofuel production. Instead, molasses and jatropha are targeted as feedstock in India, while jatropha and low-quality corn from stockpiles are targeted in China (Zhou and Thomson, 2009). Other countries, such as Indonesia, have adopted less direct measures to prevent food–biofuel competition such as imposing high export taxes on feedstocks. For example, the Indonesian government's latest attempt to secure a sufficient supply of palm oil for domestic users was to increase the export tax to 25 percent (*Jakarta Post*, 2011).

4.2. Biofuels and food prices

Between 2002 and 2007, food prices increased by approximately 140 percent, a phenomenon most dramatically witnessed during the 2007–2008 food crisis (Fischer et al., 2009). Even though the exact mechanisms of how biofuel expansion affects food prices cannot be easily delineated, it is believed that biofuel subsidies in developed countries, globalized trade, speculation, and high fossil fuel and fertilizer prices play a significant role (Runge and Senauer, 2007; RFA, 2008; Mitchell, 2008).

Various studies have reached highly different conclusions regarding the contribution of biofuel production to food prices. The available evidence suggests that biofuels' contribution to the increase of commodity food prices observed during the 2007–2008 food crisis ranged between 3 and 30 percent (Runge and Senauer, 2007;

---

[12] E.g., in parts of Tripa Province (Indonesia), the expansion of oil palm plantations has led to water shortages in adjacent wetlands. As a consequence, rice paddies, which provide the main staple crop in the region, are no longer established (Mulyoutami et al., 2010).

Mitchell et al., 2008; Fisher et al., 2009; Mueller et al., 2011; Ajanovic, 2011). Biofuel programs in Brazil, the United States, and China are predicted to further affect international corn and sugar prices in the future (Koizumi, 2009; Koizumi and Ohga, 2009). It is also expected that biofuels will continue to affect grain prices in the near future (von Braun et al., 2008; OECD, 2008; Rosegrant et al., 2008; Kretschmer et al., 2009; Fabiosa et al., 2009). Simulations by Fischer et al. (2009) points to that by 2020, global biofuel expansion might increase the price of agricultural commodities by 8 to 34 percent (for cereals), 9 to 27 percent (for other crops), and 1 to 6 percent (for livestock).

4.3. Impact on food security

High international biofuel prices may divert agricultural production originally supplying domestic food markets toward biofuel production (Ewing and Msangi, 2009). As a result, biofuel expansion may affect the food security of millions of people, particularly in net food importing countries (FAO, 2002a). As a result, as many as 131 million people (by 2020) and 136 million people (by 2030) might be at risk of hunger due to biofuel expansion (Fischer et al., 2009). Additionally, caloric intake in sub-Saharan Africa may decrease by 11 percent by 2020 as a consequence of food–biofuel competition in the region (Msangi et al., 2008). Moreover, little is understood about how climatic factors will affect feedstock yields, and as a result food–biofuel competition, in the future (e.g., RFA, 2008; Stromberg et al., 2011).

Hence the effect of biofuel expansion on food security is not so easy to unravel. When tackling the question of how biofuel expansion has affected (or will affect) food security, it is key to consider how the welfare effects of higher food and feedstock prices are distributed among the net producers and net consumers of crops used for biofuel production (and across different regions of the world).

On the one hand, high feedstock prices can benefit feedstock producers (large-scale producers or smallholders) (OECD, 2008) and agricultural workers thanks to higher income opportunities. Higher income can have a positive ripple effect on food security. For example, in the state of São Paulo, workers in the sugarcane sector have benefited from higher incomes, which might have had an indirect positive impact on their food security (Smeets et al., 2008) (see Chapter 6).

On the other hand, high feedstock prices could increase households' living expenditures due to high food prices (OECD, 2008). Poor people in developing, food-deficit countries are considered particularly vulnerable because food constitutes a high share of their total household expenditures (Runge and Senauer, 2007). For example, the average Ghanaian and Rwandan household spends 67 percent and 86 percent of their respective incomes on food (Ahmed et al., 2007; Von Braun, 2008). Evidence suggests that in parts of Africa, the negative effects of high food prices are likely to outweigh the benefits of higher sales prices of agricultural commodities (Wodon and Zaman, 2008; Hoffler and Owour Ochieng, 2008).

In general, developing countries' subsistence farmers, landless persons, and urban poor face higher risks of biofuel-driven food insecurity (Fischer et al., 2009). RFA (2008) suggests that biofuel-induced food price increases can affect the economic well-being of the urban poor in the short term (e.g., a 2% poverty increase is predicted in Nicaragua) but that the effect declines in the long term. In fact, the national economies of low-income, food-deficit countries[13] have already experienced decreases of more than 1 percent of their economic activity with a 10 percent increase in food prices in the aftermath of the 2007–2008 food crisis, to which biofuel expansion might have been a major contributor, as discussed in Section 4.2 (RFA, 2008). Data from 73 developing countries show that for most countries, it would take 0.1 percent of GDP to eliminate that part of urban poverty that has been caused by the rise in food prices since 2005, but for some countries (e.g., Nigeria, Nicaragua, Haiti), it could take more than 3 percent of GDP (Dessus et al., 2008).

5. Discussion

In this chapter we have shown how certain biofuel practices can be net energy providers and directly promote rural development by creating employment and income opportunities. However biofuel production can also compete for land (and other inputs) with food crops, hence negatively affecting food security. Depending on land allocation, biofuel expansion could also negatively affect employment and income opportunities. Apart from these easily identified linkages, many more indirect linkages between biofuels, energy security, rural development, and food security complicate the picture.

For example, locally produced and consumed biofuels can increase local energy security (e.g., through rural electrification) and have a positive effect on income generation, increasing food security as a result (e.g., FAO, 2009a). Conversely, high feedstock prices, despite offering higher income opportunities, might increase households' living expenditures due to high food prices (OECD, 2008), thus outweighing the monetary benefits, as evidenced in parts of Africa (Wodon and Zaman, 2008; Hoffler and Owour Ochieng, 2008).

Given these challenges, policy makers need to delineate policy goals carefully before designing policies meant to promote biofuel expansion. Different production systems (e.g., large-scale plantations, smallholders, and small-scale biofuel schemes) could be more or less suitable, depending on the goal. For example, direct local income generation may call for different types of strategies than national income generation intended to be subsequently distributed to other welfare-enhancing activities.

---

[13] E.g., Armenia, Egypt, Haiti, Honduras, Mongolia, and several sub-Saharan African countries (RFA, 2008).

Similarly, local energy security and national energy security concerns require different production systems to be promoted. Moreover, system-specific risks and other conditions need to be factored in. It is also important to acknowledge that the structure of the existing farming sector (technical capacity of farmers, resource endowments including land titles, etc.) may influence the success of different production systems.

Regarding the link between biofuel expansion and food prices, it is interesting to note how different model assumptions affect the expected effect of biofuel expansion on commodity prices and food security (see Chapter 11). Of particular importance is the scope and time horizon of the modeling technique. Models addressing interactions between feedstock production and other sectors of the economy (i.e., general equilibrium models) show lower price effects than partial equilibrium models that only focus on one economic sector (Timilsina and Shrestha, 2010). In the same fashion, price effects are higher in the short term since in the longer term, the interaction between locally produced crops' substitute and complement commodities causes prices to move toward the initial equilibrium. Additionally, once feedstock by-products are included, the price effect of biofuel targets further declines. Taheripour et al. (2010) show that models that omit by-products actually overstate cropland conversion due to biofuel demand by 27 percent (in the United States and the European Union). However, given the obvious limitations of trying to reallocate food supplies across time (e.g., via stockpiling), such temporal lags in readjustments between supply and demand remain a severe challenge. It appears prudent to acknowledge the absolute constraints presented by both the market mechanism and technological advances in handling shortages in on factor inputs such as land and water. Notably, agricultural commodity prices are predicted to fall in the short term but increase by 20 to 30 percent on average in real terms between 2011 and 2020, a drastic increase when compared to 2001–2010 prices (OECD-FAO, 2010). This highlights the fact that instances of competition for land, including biofuel feedstock production, will continue to have consequences for food security.

An even more difficult task is to delineate the net impact of biofuel expansion on human welfare. The main reason for this is the existence of significant environmental and socioeconomic feedbacks. For example, expanding the agricultural frontier for biofuel production can help to induce rural development. However, there is not an automatic relationship to assure this. An analysis of 286 Brazilian municipalities found a boom and bust pattern related to deforestation (Rodrigues et al., 2009). After an initial improvement of human welfare, as measured by the Human Development Index, the indicator returned to its initial level. One explanation for this might be that the benefits derived from the mechanization of agriculture are soon overridden by the exhaustion of natural resources and increasing population pressure. One distributional effect to consider in the Brazilian (Rodrigues et al., 2009) and the Indonesian (Tata and van Noordwijk, 2010) contexts is that labor used for forest clearing and on large plantations is often provided by internal immigrants. It has been reported that in several cases, the income generated during the forest-clearing phase does not reach the local population (Mulyoutami et al., 2010).

Another source of complexity in assessing the welfare effect of biofuel expansion is that the negative effects of biofuel production on human welfare might not be proportional between genders. Araujo and Quesada-Aguilar (2007) make the case that biofuel initiatives have the potential to benefit women. If gender considerations are left out when planning biofuel policies, the livelihoods of women and their families might be threatened. Rossi and Lambrou (2008) discuss potential risks of first-generation biofuel expansion in developing regions, explaining that in several cases, these risks are gender differentiated, with women being more likely to face the negative socioeconomic and environmental impacts. For example, evidence reveals that when food prices rose in Indonesia (allegedly partly due to biofuel expansion), mothers in poor families decreased their food intake to feed their children (Actionaid, 2010). Nevertheless, there are also cases in which both small- and large-scale biofuel initiatives have contributed significantly to the well-being of women (e.g., Energia, 2009; FAO, 2009a). Actually understanding the net welfare contribution of biofuels across genders can be very complicated. Little research has been conducted to understand the trade-offs of increasing women's participation in biofuel production. Nevertheless, Arndt et al. (2010c) have shown that in Mozambique, significant trade-offs can be expected if women are more actively involved in feedstock production. Increasing women's participation is not expected to affect overall economic growth in the country (Arndt et al., 2010a), but it is expected to curb the negative effects – resulting from higher food prices – that biofuel production has on poverty alleviation (Arndt et al., 2010c).

6. Conclusions

Biofuel production chains are complex and interconnected with other socioeconomic activities and environmental processes. This complexity does not only stem from the different uses of biofuel feedstock (fuel, food, and fiber) and the many processes that biofuel chains are comprised of, but also reflects the multiple policy concerns and interests that lie behind biofuel expansion (e.g., energy security and rural development). Furthermore, biofuel production is tightly interconnected with both food and energy markets. For example, the processes and outcomes of feedstock production affect other economic sectors in ways that can be difficult to foresee and adequately plan for.[14] It has been shown in other contexts that complexity and interconnectedness are particularly difficult to control for (e.g., Acharya and Richardson, 2009; Folke, 2006).

---

[14] Examples from this chapter include the effects of fossil fuel prices on biofuel feedstock and food prices through commodity markets at the macro level as well as boom and bust patterns of ill-planned biofuel expansion programs exposing small farmers to livelihood risks at the micro level.

Despite the fact that certain biofuel practices are net energy providers, their moderately low EROIs raise concerns about their long-term viability in the cases in which they are not combined with high fossil fuel energy savings. Furthermore, biofuel production can also compete for land (and other inputs) with food crops, hence negatively affecting food security. Conversely, biofuel production can directly promote rural development through creating employment and income opportunities.

The net contribution of biofuels to income and employment depends, among other things, on the opportunity cost associated with forgone alternative uses of factor inputs such as land, water, labor, and financial resources[15] (both in the short and long term). This trade-off in the use of productive resources is often not highlighted in the literature. The trade-off is both dynamic (with an impact differentiated across time) and has distributional consequences (with an impact differentiated across countries, economic sectors, and population groups).

Markets do not always allocate resources to the most efficient use, nor is the most efficient use necessarily the most desirable from a societal viewpoint. Owing to their market power, landowners and actors further down the production chain are likely to capture the rents from higher international crop prices, leaving smallholders unable to benefit (Taheripour and Tyner, 2007).

All the preceding suggests that unraveling the potential and the negative impacts of different biofuel production practices at the confluence of energy security, food security, and rural development is a very complicated task. Integrated efforts must be made if biofuel expansion is to have a net benefit for the welfare of all actors involved in biofuel chains.

Acknowledgments

This chapter was written while A.G. was based at the Institute of Advanced Studies of the United Nations University (UNU). A.G. would like to acknowledge the financial support of the Japanese Society for the Promotion of Science (JSPS) through a JSPS-UNU Postdoctoral Fellowship.

---

[15] This includes fiscal resources such as subsidies, smallholder loans with low interest rates, or government ownership of production facilities.

2

The interrelations of future global bioenergy potentials, food demand, and agricultural technology

KARL-HEINZ ERB, ANDREAS MAYER, FRIDOLIN KRAUSMANN,
CHRISTIAN LAUK, CHRISTOPH PLUTZAR
Alpen-Adria University, Klagenfurt-Vienna-Graz, Austria

JULIA STEINBERGER
University of Leeds, Leeds, UK

HELMUT HABERL
Alpen-Adria University, Klagenfurt-Vienna-Graz, Austria

Abstract

Enhancing global bioenergy production intimately relates to food production and food security. Here we present a scoping study that aims to explore the interlinkages of bioenergy and food production at the global scale. On the basis of a food-first approach, that is, calculating bioenergy potentials for 2050 only on land that is not needed for food or feed production, we consistently integrated existing mainstream scenarios of the development of food demand and agricultural technology for 2050 in a biophysical biomass-balance model. This balance model is built from highly detailed consistent biophysical databases and combines four different assumptions on yield developments and two on cropland expansion with four assumptions on diets and two on feeding efficiencies of livestock. Out of the possible 64 scenario combinations, 43 were found to be "feasible," that is, biomass required to match the respective food, fiber, and feed demand could be met with the assumed cropland area and yields. For all feasible scenarios, three components of the global primary bioenergy potential were calculated: (1) bioenergy from crop residues, (2) bioenergy from dedicated crop plantations on land available within the agricultural scenarios, and (3) bioenergy available after assuming an intensification of high-quality grazing land and cultivation of the set-free area. Deforestation and bioenergy from forestry were not considered in this top-down approach. Global primary bioenergy potentials were found to range from 58 to 178 EJ/yr, resulting in a global biofuel potential of 9–53 EJ/yr. The range of results mainly depends on diets; richer diets were associated with lower bioenergy potentials and vice versa. The bioenergy potential grows with growing cropland yields and feeding efficiency. The lion's share of this bioenergy potential is found in developing countries, in particular Latin America and sub-Saharan Africa. The realization of these potentials, however, will be associated with far-reaching transformations of the land systems in these regions, with a high risk of reducing food security.

Keywords: agriculture, bioenergy potential, biomass flows, land use change, livestock

1. Introduction

Enlarging bioenergy production and at the same time securing provision of food of sufficient quantity and quality to meet the needs of the global population is a formidable challenge. Currently an amount of 225 EJ of biomass is used annually by human society for the provision of food, feed, fuel, and fiber (Krausmann et al., 2008). Approximately 46 EJ/yr, about 10 percent of the global primary energy use, comes from bioenergy in the form of solid, liquid, and gaseous fuels from primary (harvest) and secondary (waste utilization) sources (IEA, 2006). Seventy percent of these 46 EJ/yr are used by households in developing countries, whereas a mere 4.2 EJ/yr stems from dedicated bioenergy crops.

Food and purpose-grown bioenergy production are intimately interlinked as both land uses compete for the same resource: high-quality, fertile land (see Chapter 1). Strategies that aim at fostering the production of purpose-grown bioenergy have evoked fears about food security, especially in developing countries (Royal Society, 2008). Food security denotes the – desirable – state when "all people, at all times, have physical, social and economic access to sufficient, safe and nutritious food that meets their dietary needs and food preferences for an active and healthy life" (FAO, 2010d: 8). It is estimated that currently about two billion people do not or only barely meet the daily nutritional intake of 2,100 kcal that is considered the threshold to malnutrition. Every third person in sub-Saharan Africa and every tenth person in Latin America does not reach this threshold (FAO, 2009b; Haggblade et al., 2010; Kidane et al., 2006; Pinstrup-Andersen, 2009). Many factors related to access to food and to the production of sufficient quantities of food influence food insecurity (Godfray et al., 2010). On the production side, it is either biophysical or socioeconomic constraints (soil fertility, the lack of means of production such as water or fertilizers, land tenure, technology, democratic deficits, or lack of subsidies for farmers in developing countries) that relate to food insecurity. On the distribution side, access to food and the quality of diets (e.g., healthy and nutritious food) are the most pertinent issues. These issues relate to market failures, food aid, selective subsidies of cash crops, or high world market prices (Devereux, 2009).

The relationship between food and bioenergy production is complex and character-ized by large knowledge gaps as well as by a limited understanding of the magnitude and geographic distribution of future bioenergy potentials. Recent estimates of future primary bioenergy potentials range from 27 to over 1,000 EJ/yr for 2050 (Table 2.1). The differences between these studies largely result from differing assumptions on the

Table 2.1. *Global estimates of bioenergy potentials*

Global bioenergy potential (EJ/yr)	Description	Reference
27	Energy crops on current abandoned farmland	Field et al. (2008)
5–34	Global energy crop potential in 2025	Sims et al. (2006)
32–41	Energy crops on current abandoned farmland	Campbell et al. (2008)
34–120	Energy crop potential based on LPJmL	WBGU (2008)
160–270	Global primary bioenergy potentials, including residues and forestry	Haberl et al. (2011)
65–300	Energy crops on abandoned farmland, grassland	van Vuuren et al. (2009)
300–650	Energy crops on abandoned farmland, "rest" land	Hoogwijk et al. (2005)
215–1,272	Energy crops on surplus pasture and farmland	Smeets et al. (2007)

many factors that influence the availability of biomass for energy and material use such as the availability and suitability of land, achievable yield levels of energy crops, and technological progress in agriculture and livestock breeding. Furthermore, the relation of bioenergy production to food and feed production, but also land requirements relating to soil protection, biodiversity, and forest conservation, are currently only insufficiently quantified and subject to large knowledge gaps (Haberl et al., 2010).

To explore the scope of the impact of bioenergy production on food provision and security, we here present a first-order study aimed at systematically analyzing the interrelations between global food demand, agricultural technology, and global bioenergy supply in 2050. We use a biophysical model that calculates global biomass balances from combinations of different individual factors affecting food–bioenergy interactions. In particular, we focus on changes in diets, agricultural technology (yields, feeding efficiency of livestock), and cropland expansion.

In a top-down scenario approach, we analyze biophysical constraints of future bioenergy potentials resulting from future food demand. To adequately tackle the many parameters and their associated uncertainties relating to future food and bio-energy production, we combine existing scenarios and forecasts on food demand and agricultural developments. Thus we follow a "food first" approach, which means that we first assess how much land is required to satisfy food demand and then calculate energy potentials on suitable land not needed for food and feed (and fiber) production. Even though we are aware that policies can be implemented that tackle the sovereignty of food over biofuels, the aim of this chapter is not to elaborate on the policy implications of following such paths, nor is it to predict future bioenergy supply in 2050; rather, we intend to provide a sound basis for exploring the implications of bioenergy strategies for food supply and food security. As a consequence, economic and social effects of land competition (e.g., land rents, food and bioenergy prices, land grabbing) are not considered in the quantitative framework. The study focuses on agricultural land uses only; bioenergy potentials in forestry as well as bioenergy

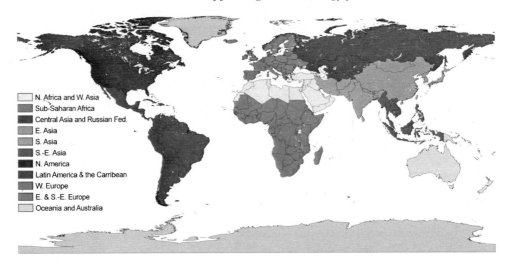

Figure 2.1. World regions used in this study. (See color plate.)

potentials from animal manure (e.g., based on biogas), municipal solid waste, and algae are beyond the scope of this chapter.

2. Methods and data

2.1. The biomass-balance model

The study is carried out on the level of the 11 world regions shown in Figure 2.1. All model calculations are performed at this level of regions, without considering any further subregional details. This is important to note, as individual countries do not necessarily follow the average characteristics of their regions.

The study draws from highly detailed consistent biophysical databases on global socioeconomic biomass flows, land use, and the human appropriation of net primary production available for the year 2000 (Erb et al., 2007; Haberl et al., 2007; Krausmann et al., 2008). These databases fulfill multiple consistency criteria across scales and domains. Biomass flows are traced from the net primary production (NPP) of each land use class to national-level data on final biomass consumption. Spatial scales range from high-resolution data sets (available at 5′ geographic resolution, i.e., about 10×10 km at the equator, covering ~98% of Earth's land, excluding Antarctica) to the country level (~160 countries) and the level of the previously described 11 world regions. On the basis of these data sets, a biophysical biomass-balance model that consistently matches global land demand for biomass products (food, feed, fibers) with gross agricultural production and land use in the year 2000 is developed (Figure 2.2). The model has been documented extensively elsewhere (Erb et al., 2009; Haberl et al., 2011) and will here only be described to the point that is necessary to understand its basic assumptions and their methodological implications for the results obtained.

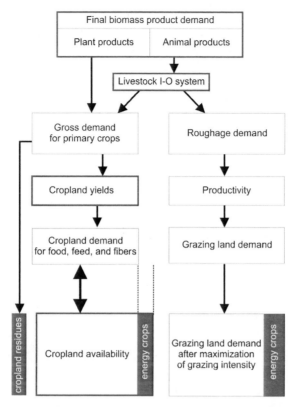

Figure 2.2. Flowchart of the biomass-balance model used to calculate the bioenergy potentials in 2050, based on the combination of assumptions ("scenarios") (boxes with bold frames).

The biomass-balance model calculates the demand for cropland products and roughage for grazing in 2050 based on human diets and livestock feeding efficiencies and contrasts the resulting figures with the amount of cropland production and roughage supply in 2050 (from cropland areas and yields as well as the area, NPP, and grazing intensity on grazing lands). Regional discrepancies of production and consumption of biomass are assumed to be compensated through trade in the bio-physical balance model. For each of the parameters (future diets, livestock feeding efficiencies, cropland yields, and cropland area), widely ranging assumptions were extracted from the literature and the previously mentioned databases (see later). Each scenario is characterized by a unique combination of these possible assumptions. Scenarios in which global cropland area demand exceeds global cropland availability by more than 5 percent are labeled as "nonfeasible" and are not considered further (food first approach). We assume that a difference of 5 percent is not significant given the uncertainties in the biomass-balance model. Roughage supply was not found to be a limiting factor, a consequence of the large extent of area suitable for grazing at the global scale (Erb et al., 2007). Thus the availability of grazing land does not influence the evaluation of the feasibility of the scenarios.

If a scenario was classified as "feasible," a global primary bioenergy potential of energy crops and cropland residues was calculated, distinguishing between three main components of agricultural areas (Haberl et al., 2011):

Bioenergy crops on cropland. On cropland area not required for food, feed, and fiber production according to the respective scenario assumptions, we calculated the potential for primary bioenergy production from dedicated crops. We assumed that bioenergy crops can reach the same aboveground productivity as potential vegetation (Campbell et al., 2008; Field et al., 2008), that is, the vegetation assumed to exist in the absence of human land use. Data on potential NPP (NPP_0) were taken from prior work (Haberl et al., 2007). This component of the bioenergy potential was set to be zero if cropland demand equaled (or surpassed by less than 5%) the cropland area assumed to exist in 2050.

Bioenergy crops on current grazing areas. We assumed that high-quality grazing land (highly productive meadows and pastures) would also be suitable for the production of energy crops such as switchgrass (*Panicum virgatum*), *Miscanthus* sp., short-rotation coppice, or others. The potential for producing such crops on grazing land was calculated by assuming that grazing intensity (the ratio of grazed biomass to the aboveground NPP) on the most suitable grazing lands (i.e., grazing land of quality class 1, according to Erb et al., 2007) can be increased to 67 percent of the actual aboveground NPP for developing regions and to 75 percent for industrialized regions. As a consequence, areas formerly under grazing class 1 are considered to be available for cultivation of bioenergy crops without reducing roughage supply. In these areas, the bioenergy yield potential is assumed to be equal to the current aboveground NPP (Campbell et al., 2008; Field et al., 2008); NPP data were taken from Haberl et al. (2007). In scenarios in which the demand for biomass from cropland for food, feed, and fiber exceeded the production of biomass on cropland by less than 5 percent, we assumed that the missing biomass would also have to be produced on grazing areas; that is, we accordingly reduced the bioenergy potential on current grazing areas.

Energy potential from unused residues on cropland. Crop residues were calculated by applying harvest indices and usage factors derived from Krausmann et al. (2008). We assumed that 50 percent of the amount of unused crop residues are required to maintain soil fertility and can therefore not be used to produce bioenergy (WBGU, 2008).

We assumed that dry-matter biomass (i.e., bone dry biomass) has a gross calorific value of 18.5 MJ/kg to derive primary energy supply potentials from the biomass flow data. These calculations do not take any conversion or production losses into account. The conversion from primary biomass to final energy (such as biofuels) or useful energy (such as mechanical energy) usually entails significant losses (Campbell et al., 2008; Field et al., 2008; WBGU, 2008) and was performed on the basis of conversion factors presented in Chapter 1.

To account for these losses in the particularly relevant case of biofuels, we derived a biofuel potential based on the calculated primary bioenergy potential combined with region-specific conversion ratios between primary biomass and biofuels in terms of

Table 2.2. *Assumed conversion rates between primary energy biomass potentials and derived first-generation biofuels, based on the upper range for biomass/biofuel conversion rates found in case studies and the highest regional harvest index for each cultivar according to Haberl et al. (2007)*

Feedstock	Conversion biofuel/ biomass (GJ/GJ)	Reference
Sugarcane (ethanol)	61.4%	Oliveira et al. (2005)
Sugar beet (ethanol)	34.3%	Royal Society (2008)
Grain sorghum (ethanol)	30.5%	Zhang et al. (2010)
Cassava (ethanol)	29.1%	Oliveira et al. (2005)
Wheat (ethanol)	27.3%	Gnansounou et al. (2009)
Maize (ethanol)	24.2%	Kim and Dale (2008)
Rapeseed (biodiesel)	30.4%	de Vries et al. (2010)
Soybeans (biodiesel)	18.9%	Chapter 6
Palm oil (biodiesel)	15.9%	Thamsiriroj and Murphy (2009)
Jatropha (biodiesel)	10.8%	Ndong et al. (2009)
Considered pathways, based on preceding case studies		
Starch–oil–sugar beet (temperate regions)	30.0%	
Starch–oil–sugarcane (tropical regions)	45.0%	

Note: All conversion rates refer to the total aboveground harvestable plant biomass (corn and straw in the case of cereals) and not only to the feedstock going into the conversion process, such as the grains. For further explanation, see the text.

energy (Table 2.2). We therefore collected conversion rates between biomass feedstocks used for first-generation biofuels from a review of case studies (for references, see Table 2.2). These ratios for first-generation feedstocks were combined with harvest indices for each respective cultivar to derive a ratio between primary biomass in terms of all harvestable biomass (including by-products). We allowed for technological progress until 2050 by using the best cultivar/biofuel conversion rate for each cultivar and the highest regional harvest index for all regions. The regional differences between biofuel cultivars in different regions were considered by assuming a path of starch–oil–sugar beet with a conversion ratio of 30 percent between primary biomass potential and biofuels for temperate regions and a path of starch–oil–sugarcane with a conversion ratio of 45 percent for tropical regions, reflecting in particular the higher conversion ratio in the case of sugarcane.

2.2. Assumptions on future diets

The total food demand in 2050 was derived from the United Nations medium-population forecast (UN, 2007a), assuming changes in regional per capita diets. The four diet scenarios are based on different nutritional levels and proportions of animal protein (Table 2.3). All four scenarios are nutritionally adequate, in a strict average sense, in terms of energy, protein content, and diversity of food sources:

- The *high-meat scenario* assumes a fast acceleration of economic growth and consumption patterns in the coming decades, leading to a global convergence to Western diet patterns and increases in the shares of animal products, sugar, and vegetable oil. All the regions attain diets at or above 3,000 kcal/cap/d, an extreme increase for most regions. Twenty-one percent of the total calorie intake stems from animal sources.
- The *current trend* scenario, in line with the prospects for global diets in 2050 made by the FAO (2006a), maintains current growth trends and strong regional differences in diet levels and compositions. All regions attain diets above 2,700 kcal/cap/d, and the world average is almost 3,000 (compared to 2,788 in 2000). This scenario represents a quantitative and qualitative enrichment of diets for the poorest areas, while the richest areas do not significantly increase or change their diets.
- The *less meat* scenario satisfies growing food demands, both from population growth and better nutritional levels, with a lower-meat diet. The diet levels attain the same levels as in the current trend scenario, but with 30 percent of the protein coming from animal products. Animal products are substituted to a certain extent with cereals, roots, and pulses.
- The *fair and frugal* scenario reduces the fraction of protein from animal sources to 20 percent. Moreover, a universal diet level of 2,800 kcal/cap/d is imposed in every region. This results in a radical change in global dietary patterns, with substantial reductions in caloric intake for most industrialized regions and improvements in regions with the lowest 2000 levels.

2.3. Assumptions on livestock feeding efficiency

Livestock feeding efficiencies allow the estimation of the gross feed demand per unit of output (meat, milk, eggs, etc.). We derived trajectories of the biomass feeding efficiencies of livestock for ruminants and monogastric species for the time period from 1961 to 2000 at the regional level (Krausmann et al., 2008) on the basis of statistical data reported by the Food and Agriculture Organization (FAO). This information was used to calculate feeding efficiencies for different livestock-rearing systems (subsistence, market-oriented extensive, market-oriented intensive) based on assumptions on the future development of livestock systems in each region (see Erb et al., 2009). Two alternative developments were modeled: intensive and organic. For both scenarios, we assumed a reduction of the regional subsistence fractions by 50 percent. Keeping the share of extensive market-oriented livestock production constant, we assumed that this reduction is in favor of intensive livestock systems in the intensive scenario and in favor of organic livestock systems in the organic scenario. On the basis of a literature review documented in Erb et al. (2009), we assumed that producing one metric ton of dry matter of animal product (meat and eggs in the case of nongrazers, meat and milk

Table 2.3. *Diet scenarios for 2050 compared to the situation in the year 2000*

	2000		High Meat		Current Trend		Less Meat		Fair and Frugal	
	kcal/cap	%[a]	kcal/cap	%[a]	kcal/cap	%[a]	kcal/cap	%[a]	kcal/cap	%[a]
Northern Africa and western Asia	2,958	10	3,300	23	3,194	12	3,194	7	2,800	8
Sub-Saharan Africa	2,247	7	3,000	19	2,801	8	2,801	7	2,800	8
Central Asia and Russian Federation	2,784	22	3,300	23	3,075	23	3,075	7	2,800	7
Eastern Asia	2,935	19	3,300	25	3,143	21	3,143	7	2,800	8
Southern Asia	2,425	9	3,000	19	2,751	13	2,751	9	2,800	10
Southeast Asia	2,677	8	3,000	19	2,862	11	2,862	6	2,800	8
North America	3,748	27	3,600	27	3,749	27	3,749	6	2,800	7
Latin America	2,836	20	3,300	23	3,063	21	3,063	8	2,800	8
Western Europe	3,431	31	3,600	27	3,524	32	3,524	7	2,800	7
Eastern and southeastern Europe	3,072	25	3,600	27	3,253	27	3,253	9	2,800	9
Oceania and Australia	3,017	28	3,600	27	3,214	29	3,214	8	2,800	7
World	**2,788**	**16**	**3,171**	**21**	**2,993**	**16**	**2,993**	**8**	**2,800**	**8**

[a] Percentage of animal sources of total caloric intake per capita.

in the case of grazers) requires 20 percent more feed input in organic livestock-rearing systems.

2.4. Assumptions on cropland yields

We derived a baseline scenario on yields and cropland area with information published by the FAO (Bruinsma, 2003; FAO, 2006a), combining these data with a database containing country-level data on production, area harvested, yields for the major crop groups, and the extent of arable land, including fallows, for the period 1960–2000 in decadal steps (see Erb et al., 2009). We constructed a consistent database containing information on harvested area, yield per unit of harvested area and year, cropping intensity, and total production volume in 2050 for each of the seven major crop groups used in the study (cereals, oil-bearing crops, sugar crops, pulses, roots and tubers, vegetables and fruits, and other crops, including fibers, coffee, etc.) in the 11 world regions. The outcomes on crop yield development were cross-checked for plausibility with alternative data on yield developments (Rosegrant et al., 2001) and corrected for maximum achievable yields for individual cultivars (Fischer et al., 2000) in cases where our assessment resulted in implausibly high yields for 2050. These resulting deviations of results from the FAO projections for 2050 were found to be small compared to the other uncertainties involved in such a long-term projection. The yield and area of fodder crops, which are not reported by FAO, were extrapolated from the previously mentioned database (Krausmann et al., 2008).

Taking the FAO trajectory for agricultural yields as a starting point, we derived three additional yield scenarios, as follows:

1. A maximum yield level (MAX) was assumed to be 9 percent higher than the FAO forecast, in line with the TechnoGarden scenarios in the Millennium Ecosystem Assessment (MA, 2005), which represents the highest yield scenarios in MA (Global Orchestration), +9 percent to the MA current trend scenario.
2. A minimum yield level was defined by calculating crop yields in 2050 under wholly organic conditions (ORG), which assumes that 100 percent of the cropland area is planted according to standards of organic agriculture. According to a literature review documented in Erb et al. (2009), yields in organic agriculture were assumed to be about 40 percent lower than those of industrialized intensive agricultural systems, if calculated over the whole crop rotation cycle (i.e., taking the cultivation of N-fixating leguminous plants into account; Connor, 2008; Halberg et al., 2006; von Fragstein und Niemsdorff and Kristiansen, 2006). This yield reduction was applied to high-input cropland systems such as those prevailing in industrialized countries. The deviation of the organic yield scenario from the baseline was based on assumptions on the regional mix between high-input and low-input agriculture (see Table 2.4).
3. We calculated intermediate (INT) yields between FAO and ORG as the arithmetic mean between these two variants. This corresponds to a diversity of mid-range scenarios, for instance, half of the area being managed with organic techniques and the other half with

Table 2.4. *Regional average yields on cropland in 2000 and according to the four yield scenarios*

	2000	MAX	FAO [t dm/ha/yr]	INT	ORG
Northern Africa and western Asia	1.4	2.6	2.4	2.1	1.8
Sub-Saharan Africa	1.0	2.3	2.1	2.0	2.0
Central Asia and Russian Federation	1.1	1.5	1.4	1.3	1.1
Eastern Asia	3.6	5.6	5.1	4.2	3.3
Southern Asia	2.0	4.1	3.7	3.6	3.4
Southeast Asia	2.8	5.6	5.1	4.6	4.1
Northern America	2.7	5.0	4.6	3.7	2.8
Latin America and the Caribbean	2.1	4.6	4.2	3.6	3.0
Western Europe	3.9	6.5	5.9	4.8	3.7
Eastern and southeastern Europe	2.4	3.8	3.5	2.8	2.1
Oceania and Australia	1.6	2.1	1.9	1.8	1.6
World	**2.2**	**4.0**	**3.6**	**3.2**	**2.7**

intensive high-yield systems, a situation in which cropland agroecosystems are not pushed to their very limits due to environmental considerations or a trajectory in which FAO yield expectations cannot be met for economic (lack of investment) or biophysical (physiological limits, soil degradation, etc.) reasons.

2.5. Assumptions on cropland expansion

According to the baseline scenario derived from FAO reports (Bruinsma, 2003; FAO, 2006a), global cropland will be 16.6 million km^2 in 2050, that is, 9 percent larger than in 2000. This estimate is in line with the findings of other studies, which assume that growth in agricultural production will depend mostly on yield increases and only to a smaller extent on the increase in cropland area growth (IAASTD, 2009; Tilman et al., 2001). Cropland expansion in the FAO projection is expected to be the highest in regions with low population density: sub-Saharan Africa, Latin America and the Caribbean, and Oceania/Australia.

In a second scenario, we assumed that cropland expansion will be double that of the FAO projection (massive expansion). In regions where, according to the FAO projection, cropland is assumed to contract, we keep cropland areas constant at the year 2000 level. This second cropland expansion scenario is still lower than the cropland expansion projected by other global scenario studies (IAASTD, 2009) and is in line with assumptions in other studies (IIASA and FAO, 2000; Ramankutty et al., 2002). It is assumed that the additional cropland (massive expansion minus business as usual (BAU) expansion) has a 10 percent lower productivity than the cropland of the BAU expansion because a stronger expansion would occur on more marginal lands.

Changes in urban and infrastructure areas are modeled separately in both cropland expansion scenarios. We calculate the area of rural infrastructure as a percentage of cropland area in each region, using factors derived from prior work (Erb et al., 2007), and the extent of urban areas by combining data on per capita urban area demand in 2000 with the UN medium forecast of urbanization (UN, 2007a; for a detailed description of the approach, see Erb et al., 2009). Accordingly, rural infrastructure areas differ in the two cropland expansion variants. As infrastructure areas are small compared with, for example, existing cropland or grazing land, the errors introduced by this method are assumed to be small.

We assume that the cropland and infrastructure area expansion in both scenarios would consume the most well suited grazing areas, which is in accordance with other studies. For example, regions with much land in the most well suited grazing land (i.e., grazing land of quality class 1, according to Erb et al., 2007) are also those in which the FAO's Global Agroecological Zones and other studies (IIASA and FAO, 2000; Ramankutty et al., 2002) indicate the existence of great potential for cropland expansion. Conversely, regions with little cropland expansion potential according to IIASA and FAO (2000) and Ramankutty et al. (2002) also have small areas of most well suited grazing land (Erb et al., 2007). The most well suited grazing land is large enough to account for this change, with the exception of the region "North Africa and western Sahara" (Erb et al., 2007). In this region, cropland and infrastructure area expansion is assumed to affect grazing land of quality class 2 as well. Reduced roughage production in 2050 as a result of the consumption of grazing areas by cropland was explicitly accounted for in the biomass-balance model in the respective regions. Grazing land expansion through forest clearance was not considered. Despite this restrictive assumption, grazing area was not found to limit roughage supply in any of our scenarios.

3. Results

Figure 2.3 displays the primary bioenergy potential in the 43 scenarios classified as "feasible" (out of a total of 64 scenarios). Note that we calculate the total primary bioenergy production potential on agricultural areas (cropland and grazing land), not the additional potential compared to the present situation. According to Sims et al. (2006), presently, 4.2 EJ/yr of primary bioenergy are produced on croplands, plus an unknown amount of fuel wood or shrub biomass in other agricultural areas (nonforested).

The primary bioenergy potential in the feasible scenarios ranges from 58 to 178 EJ/yr. The highest bioenergy potential (178 EJ/yr) occurs in the highly unlikely combination of maximum yields, intensive livestock systems, massive land use change, and a fair and frugal diet. The lowest bioenergy potential (58 EJ/yr) is found in the only feasible scenario that succeeds in supporting the high-meat diet with FAO yields; the same diet scenarios but with MAX yields are feasible in two cases and reach considerably larger energy potentials (72 and 79 EJ/yr). The scenario that combines all BAU assumptions (current trend diet, intensive livestock systems, FAO yields, and FAO cropland expansion; hereinafter called the *baseline* scenario) resulted in a

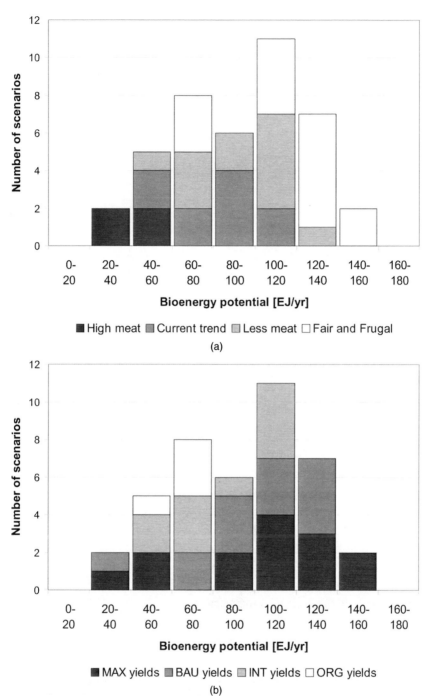

Figure 2.3. Histogram of primary bioenergy potentials of the 43 feasible scenario combinations (a) classified according to the four diet scenarios and (b) classified according to the four yield scenarios.

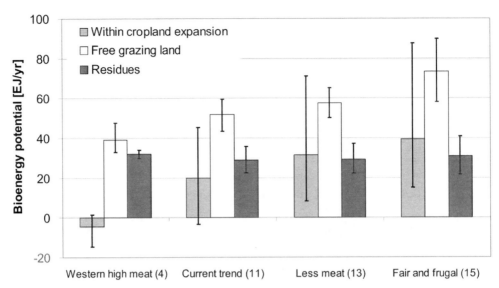

Figure 2.4. Dependency of the gross primary bioenergy potential on assumptions on diets, based on the 43 feasible scenarios.[1]

bioenergy potential of 105 EJ/yr, located between the two extremes. More than 50 percent of all scenarios range between 80 and 140 EJ/yr, with peaks between 120 and 140 EJ/yr (11 combinations) and between 80 and 100 EJ/yr (8 combinations).

Our results suggest that the bioenergy potential grows with crop yield, cropland expansion, and the efficiency of the livestock system and shrinks with the expected quantity and quality of the food supply. According to our calculations, cropland expansion has a relatively low impact: on one hand, it increases the bioenergy potential on cropland, but on the other hand, it reduces the bioenergy potential on current grazing areas because cropland expansion is expected to consume the most well suited grazing areas. The histograms in Figure 2.4 show that diets rich in calories and animal products result in small bioenergy potentials. The largest bioenergy potentials can only be achieved when assuming a fair and frugal diet. Achieving large bioenergy potentials also requires high yields, whereas low yields are only feasible in combination with modest diets. The highest bioenergy potential compatible with the current trend diet was 128 EJ/yr, and the highest bioenergy potential compatible with intermediate yields was 132 EJ/yr.

Figure 2.4 underlines that diets have a strong effect on the total bioenergy potential. It shows the arithmetic mean plus the minimum and maximum level of all scenarios within each of the diets. Bioenergy potentials under the assumption of intensifying grazing in high-quality grazing land and using the set-free area for bioenergy crops

---

[1] The number in parentheses indicates the number of feasible scenarios with respect to each diet. The columns indicate bioenergy potentials (left) according to primary bioenergy potential on cropland within the assumed cropland expansion; (center) on grazing land set free by grazing intensification, and (right) from cropland residues. Whiskers indicate the range of scenarios. Biomass potentials within the assumed cropland expansion can turn negative if a scenario's cropland area demand exceeds supply by less than 5%.

are the largest in all diet scenarios. Without this assumption, bioenergy potentials would be significantly lower; on cropland that is assumed will prevail in 2050 under the basic land use change assumptions, primary bioenergy potentials would range between 13 and 105 EJ/yr and yield 46 EJ/yr in the baseline scenario. The amount of primary bioenergy provided within the land use change assumption is highly variable: whereas within the frugal diet group, a mean of 40 EJ/yr could be produced, this amount decreases with the quantitative increase of diets. In the Western high-meat scenario, it even turns negative, as a result of the methodological assumption that demand can exceed supply by 5 percent (see earlier). Residues play a substantial role, and their contributions are the most constant in the diet breakdown. The generation of residues follows linearly the amount of food and feed produced. The low consumption of animal products in the less meat and fair and frugal diets allows production of substantial amounts of primary bioenergy, even if organic agriculture is adopted (low end of the range), because of the very low grazing intensity and the low demand for crop residues by the livestock sector. In contrast, in the case of the high-meat diet, the high consumption of animal products implies a high grazing intensity and low free residue potential. In these four scenario combinations, bioenergy potentials on dedicated cropland are low or zero.

The current trend diet, with a global average of 3,000 kcal/cap/d and a considerable growth in the global average protein from animal products, can be realized with several different combinations of yields, livestock systems, and land use changes. This diet is feasible over the whole range of assumptions on the conversion efficiencies in the livestock system (intensive and organic), but it requires, at least, yield increases as assumed in the INT yield assumption. Combinations of a current trend diet with organic yields are found to be infeasible in all cases.

The less meat diet assumes the same level of calorific intake as the current trend scenario but assumes a reduced share (–26% globally) of animal products. This demand scenario has a much broader feasibility space than the current trend scenario and is classified as feasible in 13 cases. The fair and frugal diet scenario has the broadest feasibility space and would be feasible for all combinations of land use change and livestock systems with intermediate yields and yields large bioenergy potentials with FAO and MAX crop yields. It was even classified as feasible with ORG yields, with the exception of the combination of FAO land use change and organic livestock-rearing systems.

In the baseline scenario, cropland expands by 9 percent on global average, and in the massive expansion scenario, it expands by 18 percent. Adding the expansion of purpose-grown biofuels to the entire high-quality grazing land results in total cropland area growing by 40 percent, with huge regional differences.[2] Cropland area

---

[2] Note that according to the no-deforestation assumption followed in this study, the two cropland expansion scenarios yield similar values of total cropland expansion according to diets because land expansion is assumed to reduce grazing land and grazing land is assumed to be used at a maximum level by increasing grazing intensity and using the set-free area for biofuels.

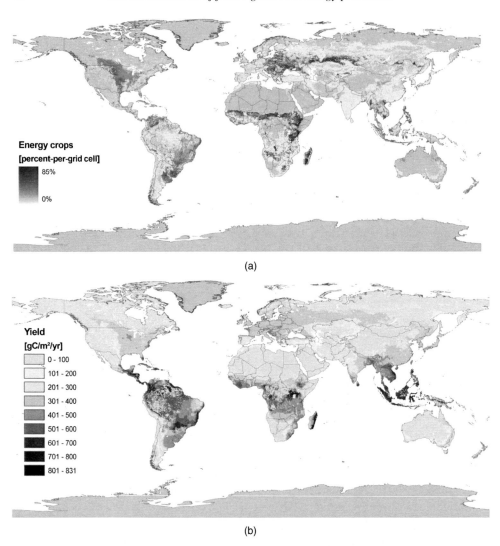

(a)

(b)

Figure 2.5. (a) Location and (b) yield of dedicated energy crops in 2050 in the baseline scenario. (See color plate.)

would more than double in sub-Saharan Africa (+120%), and almost double in Latin America and the Caribbean (+87%). Regions with small amounts of additional cropland are western Europe, North Africa and western Asia, and southern Asia (Table 2.5). Cropland used for bioenergy purposes in the baseline scenario amounts to approximately one-third of the global cropland total in 2050 in the baseline scenario. Approximately half of the total cropland area in this scenario will be used for bioenergy crops in sub-Saharan Africa, Southeast Asia, North America, Latin America, and eastern and southeastern Europe. The share of bioenergy provision will be low in the regions with small additional potentials for cropland expansion. Figure 2.5 displays the geographic distribution and yields of energy crops in 2050 in the baseline scenario.

Table 2.5. *Land use in 2000 and in one of the scenarios for 2050*[a]

2000	Urban land (mio. km²)	Change (%)	Total cropland[b] (mio. km²)	Change (%)	Grazing land (mio. km²)	Change (%)	Forestry (mio. km²)	Unused land (mio. km²)	Total (mio. km²)
Northern Africa and western Asia	42	0	763	–	1,738	–	268	7,468	10,279
Sub-Saharan Africa	111	0	1,781	–	11,867	–	5,828	4,388	23,975
Central Asia and Russian Federation	189	0	1,572	–	6,742	–	7,155	4,774	20,432
Eastern Asia	140	0	1,604	–	5,146	–	2,121	2,522	11,533
Southern Asia	113	0	2,305	–	2,554	–	850	848	6,670
Southeast Asia	39	0	931	–	1,331	–	2,098	84	4,483
North America	337	0	2,240	–	4,473	–	4,741	6,718	18,508
Latin America and the Caribbean	64	0	1,685	–	7,932	–	8,733	1,880	20,295
Western Europe	198	0	862	–	1,130	–	1,318	147	3,655
Eastern and southeastern Europe	103	0	941	–	482	–	630	2	2,158
Oceania and Australia	23	0	540	–	3,484	–	1,216	3,121	8,385
World	**1,360**	**0**	**15,225**	**–**	**46,881**	**–**	**34,958**	**31,951**	**130,375**

(continued)

43

Table 2.5 (*continued*)

2050	Urban land (mio. km²)	Change (%)	Cropland for food (mio. km²)	Cropland for bioenergy FAO (mio. km²)	Intensified grazing land (mio. km²)	Total cropland (mio. km²)	Change (%)	Grazing land (mio. km²)	Change (%)	Forestry (mio. km²)	Unused land (mio. km²)	Total (mio. km²)
Northern Africa and western Asia	66	+59	815	4	–	819	+7	1,658	–5	268	7,468	10,279
Sub-Saharan Africa	205	+85	2,231	52	1,635	3,918	+120	9,636	–19	5,828	4,388	23,976
Central Asia and Russian Federation	197	+4	1,550	85	631	2,267	+44	6,039	–10	7,155	4,774	20,432
Eastern Asia	156	+11	1,653	41	136	1,830	+14	4,904	–5	2,121	2,522	11,533
Southern Asia	181	+60	2,359	69	–	2,428	+5	2,364	–7	850	848	6,671
Southeast Asia	51	+31	836	94	428	1,359	+46	892	–33	2,098	84	4,483
North America	400	+19	1,856	479	353	2,688	+20	3,961	–11	4,741	6,718	18,509
Latin America and the Caribbean	85	+32	1,739	298	1,117	3,153	+87	6,444	–19	8,733	1,880	20,295
Western Europe	213	+7	852	28	50	931	+8	1,047	–7	1,318	147	3,655
Eastern and southeastern Europe	103	0	744	145	278	1,168	+24	255	–47	630	2	2,158
Oceania and Australia	34	+46	669	26	100	795	+47	3,218	–8	1,216	3,121	8,385
World	**1,690**	**+24**	**15,306**	**1,322**	**4,729**	**21,356**	**+40**	**40,420**	**–14**	**34,958**	**31,951**	**130,375**

[a] For details, see text.
[b] No information on the share of cropland cultivated for food and for bioenergy purposes is available for the year 2000.

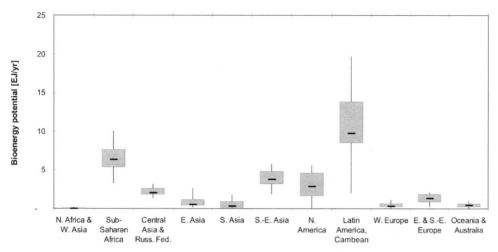

Figure 2.6. Box plots of the regional biofuel potentials from all scenarios.[3]

Figure 2.6 shows the total biofuel potential in the 11 world regions as calculated in the 43 scenarios. Taking into account the assumed losses of converting biomass into first-generation biofuels, our potentials for primary biomass translate into a global biofuel potential of 28 EJ/yr, with a statistical spread between 9 and 53 EJ/yr. Biofuel potentials are large in Latin America and the Caribbean, and sub-Saharan Africa, with 9.7 and 6.4 EJ/yr, respectively. The statistical spread, however, is much larger for Latin America, where values between 2.0 and 19.7 EJ/yr are found. North America and Southeast Asia are characterized by a smaller, but still substantial, biofuel potential. For North America, the range of results is also exceptionally large, spanning from 0 to 5.7 EJ/yr, whereas Southeast Asia has a potential between 1.9 and 5.7 EJ/yr. In most other regions, the biofuel potentials as well as the ranges of results are much smaller in absolute terms.

4. Discussion

Diets and agricultural production technologies massively influence future bioenergy potentials. The analysis presented here suggests that if such interactions are adequately considered, estimated global primary bioenergy potentials are significantly smaller than those calculated in many previous studies (Berndes et al., 2003; Hoogwijk et al., 2003, 2005; Smeets et al., 2007). According to our results, the primary bioenergy potential on farmland, obtained by applying a food-first approach, ranges from 58 to 178 EJ/yr, with over 50 percent of all scenarios falling in a range of 80–140 EJ/yr. If the conversion losses between feedstocks and first-generation biofuels are considered, this amount of primary biomass could be converted to 9–53 EJ/yr of liquid biofuels, not considering the energy used in the production process. This bioenergy potential is

---

[3] The solid line indicates the baseline scenario, the shaded box indicates the first and third quartiles (50% of all scenarios fall within that range), and the whiskers indicate minima and maxima values.

Table 2.6. *EROI in terms of biofuel output per biomass*
input for major biofuel feedstocks and biofuels

Feedstock (Biofuel)	Energy output/input ratios (Range) (J/J)	No. of case studies
Sugarcane (ethanol)	6.5 (3.1–9.3)	7
Wheat (ethanol)	2.9 (1.6–5.8)	4
Sugar beet (ethanol)	1.2	1
Corn (ethanol)	1.1 (0.8–1.7)	7
Sweet sorghum (ethanol)	0.8 (0.7–1.0)	2
Palm oil (biodiesel)	2.5 (2.4–2.6)	2
Rapeseed (biodiesel)	2.3	1
Jatropha (biodiesel)	2.2 (1.4–4.7)	6
Soybeans (biodiesel)	2.0 (1.0–3.2)	6

Source: Chapter 1.

associated with an expansion of cropland by 40 percent into the global high-quality grazing lands and a substantial increase in grazing intensity. Increasing the provision of purpose-grown bioenergy above these levels, for example, to levels up to 500 EJ/yr, as recent articles suggest (Dornburg et al., 2008), could have serious sustainability effects, including impacts on food security.

The biofuel potentials reported in Figure 2.6 do not consider the energy invested in the production process of biofuels such as the energy used for producing the fertilizers for feedstock cultivation or the energy invested in the conversion process between biomass and biofuels. A net potential, which gives the energy available for final consumption, can be derived by considering the energy return on energy invested (EROI) for the used biofuels, denoting the amount of energy that is returned in terms of biofuels per unit of energy that is used within the production process (not including the feedstock itself). Table 2.6 shows that depending on feedstock, biofuel, and conversion pathway, EROIs for biofuels vary considerably. Subtracting these inputs would considerably reduce the previously reported biofuel potentials, in particular, for crops planted in temperate regions. Some case studies have even concluded that the energy input for production is higher than the energy derived from the respective biofuel.

Our result is well in line with similar approaches that have taken constraints resulting from ecosystem degradation, carbon sequestration, or food demand into account (Haberl et al., 2010; van Vuuren et al., 2009; WBGU, 2008) but higher than the primary bioenergy potential estimates of 27–41 EJ/yr resulting from the assumption that only abandoned farmland was available for the cultivation of bioenergy crops (Campbell et al., 2008; Field et al., 2008).

In one-third out of the possible 64 scenarios, food and feed production resulting from assumptions on yield and cropland area expansion cannot sustain the global dietary requirements in 2050. Our results indicate that the quantity and quality of diets are the key factors influencing this outcome. As a general finding, it is evident that rich average diets result in low bioenergy potentials, while diets with a lower share of animal products and lower overall caloric intake might allow the production of larger amounts of bioenergy. In this scoping study, the highest bioenergy potentials were estimated in a scenario combining frugal diets with intensive land use. Such a development would require a substantive reorganization of the agricultural system and, in particular, of its societal role and self-perception. It would entail a prioritization of energy provision over the enrichment of diets, the socioeconomic and social implications of which are hardly foreseeable.[4]

Our analysis suggests that yields are the second most important factor influencing the magnitude of bioenergy potentials. Yield gains will be vital for providing enough high-quality food for a world of 9.2 billion inhabitants (Godfray et al., 2010), and even more so if substantial bioenergy potential should be realized in the mid-term future. Even our minimum yield scenario (ORG) assumes substantial yield increases in 2050 across all regions and cultivars. It should be noted that in developing countries, organic agriculture could allow for yield increases due to the currently poor nutrient status (see Diop, 1999; Halberg et al., 2006).

The baseline scenario, based on FAO forecasts (Bruinsma, 2003; FAO, 2006a), describes a development in which agricultural intensification progresses rapidly, resulting in yields in 2050 that are very high for some crops and regions. Similarly, we assumed that the yields of bioenergy crops are equal to aboveground NPP. This is in line with some studies (Campbell et al., 2008; Field et al., 2008; Johnston et al., 2009), whereas much higher yields are assumed by other authors (Smeets et al., 2007; van Vuuren et al., 2009).

It is difficult to judge whether such massive yield gains can be realized or even surpassed, as in our MAX yield assumption. Some authors argue that a continuation of past yield increases, as assumed by the FAO, seems unlikely because most of the best quality farmland is already used and rates of yield increase are already declining or stagnant as they approach biophysical limits set by soil and climate (Tilman et al., 2002). Soil degradation and depletion of soil nutrient stocks represent an additional challenge (Cassman, 1999). Past experience has shown that efforts to deploy Western agricultural technology in tropical and subtropical regions are prone to failure due to the fragile soils in these areas (Boserup, 1965; Ramankutty et al., 2002; Young, 1999).

---

[4] It is important to note at this point that the fair and frugal diet assumption of 2,800 kcal/cap/d might be considered as an extreme assumption hardly in line with global food security, in particular, in regions that already today suffer from large distribution disparities. Such a diet would require an equitable distribution of per capita food supply throughout the global population to prevent undernourishment of large fractions of the population. Both the less meat and fair and frugal scenarios would require significant cultural and attitudinal change and policy interventions. In this sense, they are included in our study to illustrate how substantial the effects of a shift away from animal products could be.

Improved management practices could help maintain yield growth, mostly because of improved stress tolerance, avoidance of nutrient and water shortages, improvement in pest control, and so on. Currently crop yields in poverty-ridden areas, such as sub-Saharan Africa, are far below world averages. This is a result of various production constraints, ranging from abiotic and biotic constraints to management and socioeconomic factors. Since, in most cases, it is a combination of constraints that hampers yield increase, especially in smallholder production systems (Devereux, 2009; Waddington et al., 2010), increasing yields by improving management practices seems promising (Connor, 2008; Edwards et al., 2007a; Tilman et al., 2002).

Large-scale irrigation may help to achieve higher yields (WBGU, 2008). However, irrigation might be limited by regional water availability, a connection that is at present also poorly understood (Gerten et al., 2008). The problem is particularly virulent in semiarid regions, where large-scale irrigation for bioenergy crops is likely to trigger social conflicts about water use (Young, 1999). Furthermore, irrigation is closely associated with sustainability issues such as soil salinization and freshwater eutrophication, as examples from the dry northeastern region of Brazil have shown (Gunkel et al., 2007) (see Chapter 7).

Another option to increase yields is the introduction of new species. For example, the introduction of genetically modified sugarcane cultures in Brazil boosted yields dramatically (Goldemberg et al., 2008). However, ecological and social consequences of such technologies can be negative, with some examples being effects on biodiversity and ecosystem health, food safety, accessibility to technology, market dependency, or commercialization. Low-income regions dominated by subsistence farming would be particularly vulnerable to such effects (Lemaux, 2008).

In any case, substantial investments and agroeconomic reorganizations will be indispensable for maintaining growth in crop yields (Kahn et al., 2009), and economic constraints may further prevent the realization of technical yield potentials (Koning and van Ittersum, 2009). This applies in particular to the bioenergy potentials of land currently used for grazing. Realizing bioenergy potentials of grazing land, as assumed in this study, might entail massive investment in agricultural technology, such as irrigation infrastructure, and will most likely be associated with sizeable social and ecological effects such as further pressure on populations practicing low-input agriculture.

The amount of land available for cropland expansion will also be critical in the interplay between bioenergy crops and food security. According to our analysis of a BAU scenario based on the FAO forecast, a global cropland expansion of 40 percent would result from strategies that aim to exploit bioenergy potentials on former grazing land of high quality. Expansions far above this average would occur in Latin America and sub-Saharan Africa but also in Southeast Asia. It is difficult to judge how realistic this assumption is. The large potentials of additional cultivable land described in many assessments (Fischer and Heilig, 1997; IIASA and FAO, 2000) have been severely criticized (Hubert et al., 2010; Waddington et al., 2010; Young, 1998, 1999). For example, in sub-Saharan Africa, several observations conflict with the assumption of abundant spare land (Young, 1999). Already today, cultivation has been extended into protected areas and nonfavorable marginal areas such as steep slopes and semiarid zones with constant drought stress (Young, 1999). Other areas, such as fallow lands or extensively used land under subsistence farming systems, are often regarded as unused or underused, without acknowledging that these areas are an integral part of the diversified and small-scale agriculture of the region. The decline in fallows due to the intensification of the cropping cycle is a further indication in this direction, and nutrient cycling studies have shown chronic negative balances, which result in a constant decline in soil fertility (IAASTD, 2009). The realization of the technical potentials will thus depend on the ability to and cost of conducting degradation-preventing measures as well as on the ability to enforce effective policies aimed at the conservation of natural reserves, aspects that are beyond the scope of this chapter.

Low-income countries might not have the economic or organizational means for such investments nor access to the necessary technologies. This might force them to sell farmland or transfer user rights to foreign investors. Over the last decade, large-scale land deals between large enterprises or even governments from the industrialized world and sub-Saharan African countries have raised awareness about this phenomenon, labeled "land grabbing," which is seen as a new form of colonialism (Cotula et al., 2009; Robertson and Pinstrup-Andersen, 2010). Even though they may not be disadvantageous to developing nations in principle, if they are not combined with sensitive and robust policy measures, these developments may be associated with multiple risks. These risks include irreversible soil degradation in sensitive ecosystems due to the introduction of industrial-scale, capital-intensive farming; shortfalls in livelihood (particularly for smallholders); a loss of indigenous knowledge; and a great deal of stress suffered by subsistence farmers, who very often do not legally own the land they cultivate. The biomass exports from these regions are also a potential threat to in-country and local food security (Robertson and Pinstrup-Andersen, 2010).

Another important aspect of land expansion, and particularly of land expansion resulting from the realization of bioenergy potentials on grazing land, is that it is likely to trigger large-scale indirect land use change such as deforestation in distant regions. Such indirect land use change effects are described for tropical regions such as Brazil, where competing land uses result in spatially disconnected but strongly interlinked effects (Mendonça, 2008; Sawyer, 2008). Furthermore, land use changes for bioenergy crops that entail deforestation are likely to accumulate carbon debts, which could cancel out or even exceed the carbon emission savings that result from substituting bioenergy for fossil fuels by several orders of magnitude (Fargione et al., 2008; Koh and Ghazoul, 2008; Searchinger et al., 2008) (see Chapter 3).

Our analysis shows that the differences in the overall feeding efficiency (referring to the input–output ratio) of the different livestock systems have a smaller influence than the factors of diet, yield, and area availability. This is because (1) our literature review suggested relatively modest differences in feeding efficiencies between intensive and organic indoor-housed livestock rearing (10%–20%; see Erb et al., 2009) and (2) there is an assumption of a significant share of subsistence livestock systems persisting until 2050. Much stronger effects would result from different assumptions on the transformation from subsistence to optimized livestock-breeding systems.

However, it should be noted at this point that extensive livestock-breeding systems with low outputs per unit of input are not necessarily inefficient or less desirable (Bradford and Baldwin, 2003; Godfray et al., 2010; Herrero et al., 2009). Besides meat and milk production, livestock fulfils a huge range of other functions, particularly in developing countries. Livestock is required to provide power for agriculture and transport, it is essential for the management of nutrients in low-input agriculture, and it provides a vital source of income in many poorer communities. Finally, the most important function of livestock in biophysical terms is the ability of ruminants to convert biomass not digestible by humans into food. Thus livestock systems that appear to be inefficient due to their input–output ratio may in fact represent well-adapted, highly efficient production systems in their respective local contexts. Commercialization and optimization of such systems might be unfavorable for food supply, food security, and overall sustainability in such regions, despite the gains in input–output efficiency.

A component of the global bioenergy potential that avoids some (but not all) of the negative effects of bioenergy, because it does not require additional area, is the *cascade utilization* of cropland residues. According to our rough approximation, 21–40 EJ/yr, or 20 to 57 percent, of the total global primary bioenergy potential of farmland might be produced from cropland residues. In particular, second-generation biofuel production could profit from this potential. However, the utilization of this potential might be associated with negative effects (e.g., for soil) if not followed cautiously (Lal, 2005). Nevertheless, the magnitude of this potential and the fact that it is associated with relatively limited additional environmental pressure (Haberl et al., 2003; Haberl and Geissler, 2000; WBGU, 2008) renders research into these issues desirable.

In the top-down analysis presented in this chapter, the decision whether a scenario is feasible is made on the basis of a comparison between global biomass demand and supply. Regional differences between production and consumption are implicitly assumed to be balanced by trade. In consequence, bioenergy potentials might be significantly smaller in cases in which these trade flows are restricted, be it due to protective tariffs or to a lack of purchasing power. Already today, 41 out of 49 sub-Saharan countries are considered low-income food-deficit countries (FAO, 2010b) that rely on imports for their domestic food consumption.

Possible effects of climate change on future yields in agriculture are not systematically considered in this scoping study. Climate change might result in increases or losses of the productivity of agroecosystems. The magnitude, and even the direction, of this effect is not well quantified because the CO_2 fertilization effect is at present only incompletely understood (Haberl et al., 2010, 2011). Preliminary calculations suggest that climate change might increase or reduce global bioenergy potentials (primary and final) by approximately 30 percent (Erb et al., 2009).

5. Conclusions

Our top-down study suggests that primarily two factors are decisive in influencing the interplay of bioenergy and food security: the development of global diets and crop yields.

Changes in diet have a strong effect on future bioenergy potential. The richer the diet is, the lower the bioenergy potential is, and vice versa. Yield changes have a strong effect on bioenergy crop potential, too, but only under ceteris paribus assumptions (i.e., constant diets). Our results indicate that if increased productivity is used to increase food consumption, the bioenergy potential may even shrink, as the influence of a richer diet (which reduces bioenergy potentials) can override the land-sparing effect of higher yields (which would increase bioenergy potentials). Viewed from another angle, if bioenergy is prioritized, then increases in food supply will be difficult to achieve. Bioenergy provision and food supply are in direct conflict, and our study indicates that developing countries will be at the very heart of this conflict.

Increasing yields will be indispensable in alleviating the direct competition between food and bioenergy provision. From this assertion a formidable challenge arises: how to increase yields while simultaneously avoiding detrimental consequences for ecosystems and social systems. Releasing the bioenergy potential in many developing regions will entail massive agricultural intensification. Intensification denotes in many cases the end of subsistence farming for pastoralists and smallholders, who are mostly responsible for providing food and maintaining food security in these regions.

Expanding agricultural land use in many developing regions will require massive investment, which might not be affordable for local governments or land users. If agricultural expansion is combined with sensitive legislation, this may enhance the risk of land grabbing, that is, the loss of food-growing farmland to bioenergy plantations, with manifold negative consequences. The latter can be a severe threat to food security, even at larger, nationwide scales.

Increasing production in these regions entails a massive expansion of agricultural land, particularly in sub-Saharan Africa and Latin America. Although this expansion can, in principle, be absorbed by existing grazing land, direct and indirect effects on the remaining pristine ecosystems are probable. This might be associated with detrimental consequences for biodiversity and, in the case of forests, massive carbon emissions.

The interrelation between food supply, food security, and bioenergy is complex, and cause–effect relationships span the entire globe. The largest energy potentials are found in regions characterized by food shortages and below world average per capita diets. Conversely, the largest bioenergy demands are expected in regions where energy potentials are low such as Europe and parts of Asia. This calls for integrated approaches that at the same time optimize land use and biomass utilization chains. Our study shows that although yield increases will be indispensable, with more moderate global average diets, there is no need to go for the highest possible yields or maximize cropland area at all costs (including the environmental, economic, social, and health impacts) to secure global food supply; rather, our calculations suggest that the world can afford to forgo some potentially possible intensification without jeopardizing food supply and still yield substantial amounts of bioenergy. Conversely, our study also shows the benefits that might result from increases in cropland yields and livestock feeding efficiency in terms of lower area demand and higher bioenergy potentials. This calls for research and technological development efforts to decouple yield gains and other efficiency improvements from adverse social, economic, and ecological effects.

Providing sufficient food, fiber, and energy for the world's growing population in a sustainable manner in the coming decades is one of the grand challenges humanity currently faces. This challenge is aggravated by the fact that not only will the quantities be decisive but so will their fair and sustainable distribution. The biophysical approach presented here allows gaining some insight on the scope of this challenge but inevitably leaves many essential aspects unaddressed. Integrating the biophysical modeling strategy with methods based on economic and ecosystem modeling would allow further insight into this emerging sustainability challenge.

Acknowledgments

The authors gratefully acknowledge funding by the European Research Council within the ERC Starting Grant 263522 LUISE, by the Austrian Science Fund (FWF) within the project P20812-G11, and by Friends of the Earth, UK, and the Compassion in World Farming Trust, UK. This study contributes to the Global Land Project (http://www.globallandproject.org) and to long-term socioecological research (LTSER) initiatives within LTER Europe (http://www.lter-europe.ceh.ac.uk/).

3

Air pollution impacts of biofuels

KRISTINA WAGSTROM
University of Minnesota, Minneapolis, USA

JASON HILL
University of Minnesota, St. Paul, USA

Abstract

This chapter reviews the impacts of increased liquid transportation biofuel production and use on air pollutant emissions and their ambient concentrations, with a focus on developing nations. Major pollutants covered include those affecting climate change (carbon dioxide, nitrous oxide, and methane) and those affecting human health (carbon monoxide, sulfur dioxide, particulate matter, ammonia, nitrogen oxides, ozone, and volatile organic compounds). Emissions and resulting ambient concentrations associated with all the stages of the full life cycle of biofuel production and use are described and, where possible, are compared against conventional fuels.

Keywords: atmospheric pollution, greenhouse gas emissions, life cycle assessment

1. Introduction

Petroleum is the most commonly used feedstock for producing transportation fuels because it is both convenient and energy dense. Through the years, petroleum has fueled global economic growth, yet it comes at the price of global conflict, exploitation of indigenous peoples, and destruction of both terrestrial and aquatic habitats. In addition, the extraction, refining, and combustion of petroleum is responsible for the emission of many air pollutants that contribute to anthropogenic climate change, harm human health, and damage ecosystems worldwide (Uherek et al., 2010).

Recently, interest has flourished in the use of biofuels as a petroleum replacement (see Chapter 1). Domestically produced biofuels have the potential to support rural communities, slow the depletion of fossil fuel resources, and avoid many of the negative environmental consequences associated with petroleum such as oil spills (Lovett et al., 2011). Similarly, biofuels may reduce greenhouse gas emissions and yield improvements to air quality if the emissions from a biofuel's production and use

53

are lower than those associated with meeting the same energy need with conventional fuels.

In reality, whether biofuels are indeed an environmentally preferable choice over conventional fuels is not well understood. What has emerged from the literature is a complex series of trade-offs related to energy security, food supply, ecosystem maintenance, water quality, and air pollution (Gasparatos et al., 2011). Overall, whether a biofuel offers a benefit over the conventional fuel it displaces depends on where and how each fuel is produced and used and what technologies are employed in mitigating their negative effects. To account fully for emissions from biofuel production and use, it is as important to consider those related to feedstock cultivation and harvesting as it is those related to conversion and fuel use. The emissions associated with feedstock cultivation and harvesting depend on both the biofuel feedstock and agricultural practices used. The net impact of increased biofuel production will depend on how and where the global industry develops. If done well, biofuels could help alleviate many of the environmental problems associated with petroleum, but if produced without regard to their potential consequences, they could both exacerbate existing problems and create new ones (Buyx and Tait, 2011; Tilman et al., 2009).

This chapter focuses on one aspect of the many potential trade-offs associated with transitioning to increased biofuel production and use, namely, biofuels' impact on atmospheric pollutant emissions and their ambient air concentrations. The potential changes in various pollutant emissions and ambient concentrations are discussed, including how they contribute to climate change and how they affect human health. Special attention is given to developing nations. Where possible, comparisons between biofuel options and current conventional fuels are presented from a life cycle perspective, as air pollutant emissions may occur throughout all stages of fuel production and use. This chapter is meant to serve as a reference list of major air pollutant species considered in the recent biofuel-focused literature, but not all air pollutants and their effects are covered.

2. Overview of atmospheric pollutants

As a starting point for understanding the impacts of biofuels on air pollution levels, it is important to understand the pertinent atmospheric pollutants that relate to gasoline and liquid biofuel production and use. Pollutants are often classified by the major risks associated with them, primarily climate change potential and negative human health impacts. In addition to these risks, there are also potential risks associated with crop damage, ecosystem destruction, and visibility deterioration.

The major greenhouse gas species of concern when considering biofuels are carbon dioxide (CO_2), methane (CH_4), and nitrous oxide (N_2O), of which the effects on human health and well-being via climate change are well described (Confalonieri et al., 2007; Tol, 2009). The pertinent species that are known to impact human respiratory and cardiopulmonary systems are carbon monoxide (CO), sulfur dioxide (SO_2), particulate matter (PM),[1] ammonia (NH_3), nitrogen oxides (NO_x), and ozone (O_3). In addition, there is also concern with changes in volatile organic compound (VOC) concentrations because they are an integral contributor to the formation of ozone and some forms of PM in the atmosphere (Seinfeld and Pandis, 2006). Other serious health impacts arise directly from emissions of two classes of VOCs, carbonyls and aromatics, of which many species are known to have carcinogenic properties.

In addition to risk classification, pollutants are also classified as either primary or secondary species. Primary pollutants are emitted directly into the atmosphere from either anthropogenic (human) or biogenic (natural) sources. The relationship between emissions changes and the resulting ambient concentration changes are often more easily understood for these pollutants as a decrease in emission rates typically translates into a decrease in ambient concentrations. Secondary pollutants are formed in the atmosphere via chemical reactions or physical processes. The concentration response to changes in emissions for these species is often more difficult to predict as the chemical and physical processes that impact their concentrations are often highly dependent on meteorology (e.g., relative humidity and temperature) and the relative concentrations of the precursor species (i.e., the species that react to form the species of interest) (Seinfeld and Pandis, 2006). Species can be both primary and secondary if they are both directly emitted and formed in the atmosphere, such as with PM.

Studies investigating the potential changes in air pollution associated with the production and use of biofuels in place of conventional fuels are becoming increasingly important for policy makers and regulators. Much of the initial work done in this area was focused on quantifying changes in emission rates from vehicles fueled by biofuel mixtures compared to gasoline, but more recent work has recognized the importance of identifying emissions associated with the whole fuel life cycle. For example, for ethanol from corn grain in the United States, the fuel-use phase, which includes tailpipe emissions and evaporative emissions from vehicles and filling stations, is responsible for over 90 percent of life cycle CO emissions but only 68 percent of VOCs, 22 percent of primary $PM_{2.5}$ (PM with diameters less than 2.5 μm), 17 percent of NO_x, 13 percent of NH_3, and <1 percent of SO_2 emissions (Figure 3.1). Table 3.1 provides a brief summary of the different pollutants covered in this chapter and their risks, main sources, and general trends associated with increased biofuel use.

2.1. Greenhouse gases

Greenhouse gases (GHG) are the atmospheric trace gases that absorb incoming infrared radiation and provide warmth to the atmospheric surface. As their concentrations change, so does the amount of radiation that is being absorbed, therefore

---

[1] PM also contributes to climate change; refer to Section 2.4.

Table 3.1. *Brief summary of the different pollutants emitted during biofuel production and use*

Pollutant species	Primary risk	Major contributing processes	Trends found in literature
Greenhouse gases (CO_2, N_2O, CH_4)	Climate change	Land use change, fossil fuel use in farming and biofuel production, fertilizer production and use	Life cycle emissions highly depend on feedstock, production technology, and any associated land use change
Carbon monoxide (CO)	Health concerns	Vehicular combustion, agricultural burning	Most studies find decreases in tailpipe emissions and increases from agricultural burning
Sulfur dioxide (SO_2)	Respiratory concerns, acid rain	Electricity-generating facilities, phosphate fertilizer production	Largest changes are expected from the change in electricity demand for fuel processing and are process dependent
Particulate matter (PM)	Respiratory and cardiopulmonary concerns, ecosystem damage, visibility deterioration, building damage	Farming, fuel processing, fertilizer production	Potential increases in rural areas; major changes expected from the change in electricity demand for fuel processing
Ammonia (NH_3)	Formation of secondary particulate matter	Fertilizer use, catalytic converters in vehicles	Likely increases with increases in fertilizer needs
Nitrogen oxides (NO_x)	Respiratory concerns, ozone formation	Vehicular combustion, fuel processing, farming and land use change, fertilizer production	Found increases in rural areas and decreases in urban areas
Ozone (O_3)	Respiratory concerns, decrease in crop yields	Chemical formation in the atmosphere	The results are varied, but most life cycle assessments indicate an increase in ozone concentrations with ethanol use
Volatile organic compounds (VOCs)	Carcinogenicity, ozone formation	Vehicular combustion, evaporative emissions, biogenic emissions	Found decreases in benzene and 1,3-butadiene emissions and increases in formaldehyde and acetaldehyde emissions

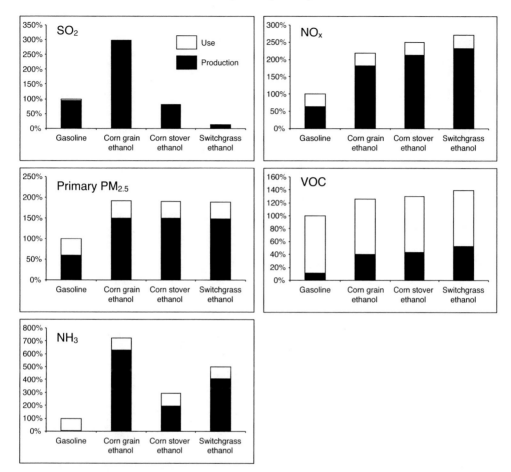

Figure 3.1. Life cycle emissions of SO_2, NO_x, primary $PM_{2.5}$, VOCs, and NH_3 for meeting a set energy need with gasoline and three biofuel options from different feedstocks. *Note*: Values are scaled to life cycle emissions of gasoline for each pollutant at 100%. Values from Hill et al. (2009) are derived from Argonne National Laboratory's GREET model (Huo et al., 2009).

leading to changes in atmospheric temperature. The main GHGs that are of importance when discussing biofuels are carbon dioxide (CO_2), nitrous oxide (N_2O), and methane (CH_4).

The carbon released as CO_2 on burning biofuels is the same carbon fixed by plants during photosynthesis. In life cycle carbon accounting of biofuels, therefore, tailpipe emissions of CO_2 are completely offset by atmospheric reductions in CO_2 from plant growth. Only trace amounts of N_2O and CH_4 are emitted on biofuel combustion. For biofuels, the focus of GHG accounting is therefore not on emissions in use but rather on both emissions in production and market-mediated effects. GHG emissions in biofuel production include those from agricultural activities, transport of feedstocks, biorefinery processes, biofuel distribution, and any supporting upstream activities. GHG emissions due to market-mediated effects include those that result from direct and indirect land use changes caused by the conversion of former food-producing croplands or native vegetation for feedstock production.

A wealth of studies exists that have estimated life cycle GHG emissions from food, animal feed, and lignocellulosic-based biofuels. These studies have been reviewed in more than a few instances (Bessou et al., 2010; Fargione et al., 2010; Hoefnagels et al., 2010; Kendall and Chang, 2009; Liska and Cassman, 2008; Sims et al., 2010; van der Voet et al., 2010). Overall trends in the climate change impacts of biofuels can be gleaned from these studies. An important caveat when interpreting biofuel GHG accounting studies is that a wide range of methodologies exist and that differences in results are, essentially, as likely to be due to the different methodologies being employed as to the different biofuel production pathways being considered (Anderson-Teixeira et al., 2011; Stratton et al., 2011; Wang et al., 2011).

The choice of feedstock crop has a large impact on the anticipated life cycle of GHG emissions as the CO_2 fluxes associated with tillage and N_2O fluxes associated with nitrogen fertilizer use are potentially large sources of GHG emissions. Biofuels produced from perennial crops (e.g., switchgrass, miscanthus), which are not tilled annually and typically use nitrogen more efficiently, tend to have lower GHG emissions than those from annual crops (e.g., corn). Biofuels produced in biorefineries that use natural gas to provide heat tend to have lower GHG emissions than biorefineries that use coal (Wang et al., 2011). The use of biomass for process heat in biorefineries tends to lead to even lower emissions. EPA (2009) found that the GHG emissions associated with fuel production varied widely and could even have cobenefits such as the displacement of current power plant emissions by excess electricity that is fed back into the grid. Methane (CH_4) is of interest because it is potentially emitted when used as a heat source for different process steps and because it serves as a material feedstock for nitrogen fertilizer production (Montzka et al., 2011).

Because of the potential for large GHG emissions associated with land use change (Fargione et al., 2008; Gibbs et al., 2008; Hertel et al., 2010; Searchinger et al., 2008), there is large uncertainty about whether food and animal feed–based biofuels such as corn grain ethanol and soybean biodiesel reduce GHG emissions relative to petroleum-derived fuels (Mullins et al., 2011; Plevin et al., 2010) (refer to Chapters 6 and 13). Many recent studies have evaluated GHG emissions associated with the life cycle of lignocellulosic biofuels and have found considerable potential for significant reductions with a variety of different feedstocks. In general, biofuels produced from wastes and residues are more likely to have GHG emissions lower than biofuels produced from dedicated bioenergy crops due to uncertainties associated with GHG emissions from land use change (Spatari and MacLean, 2010).

2.2. Carbon monoxide

Carbon monoxide (CO) is a gaseous species that is mainly present in the atmosphere as a result of incomplete combustion and, to a lesser extent, oxidation of larger organic species in the atmosphere (Finlayson-Pitts and Pitts, 2000). Vehicular combustion processes are high contributors to CO emissions because incomplete combustion is more common in vehicles than in larger stationary combustion sources such as power plants. CO is a concern to human health because it reduces the ability of hemoglobin to transport oxygen. Levels of CO found in ambient air in much of the world are of little concern to the majority of the population, but elevated ambient levels can be dangerous for individuals with preexisting heart conditions or to those in areas of the world with higher ambient CO concentrations (Jacobson, 2002).

Many studies have measured emission rates of CO from engines run on various blends of biofuels and conventional fuels. As E10 (a blend of 10% ethanol and 90% gasoline) is often employed during the winter in the northern United States to maintain lower ambient CO levels, it is not surprising that in many studies, the addition of small amounts of ethanol as an oxygenate to gasoline has been shown to decrease CO emissions (EPA, 2007). Studies aimed at determining the changes in CO emissions for higher ethanol blends from a variety of vehicles and under a variety of operating conditions have found mixed results. Kelly et al. (1996) found that E85 decreased CO emissions by 12 to 24 percent, while Jacobson (2007) found a slight increase (5%) in CO emissions from E85 combustion. In addition, Graham et al. (2008) conducted a statistical analysis of the relevant results available in literature and found a 16 percent reduction in CO emissions for E10 blends but no significant change in CO emissions for E85 blends. For biodiesel blends, researchers have seen significant reductions in CO emissions during vehicular operation (McCormick, 2007), even as high as a 50 percent reduction for B100 (100% biodiesel) (EPA, 2002).

Studies that look beyond the tailpipe to include full life cycle emissions for CO have also found mixed results. The majority of these studies show an increase in the prepump CO emissions for ethanol, whether from corn grain or cellulosic feedstocks, and, as described earlier, a decrease in fuel usage emissions compared to gasoline (Hess et al., 2009; Huo et al., 2009). Hess et al. (2009) found that when agricultural burning was accounted for in the life cycle emissions for sugarcane feedstocks, ethanol use led to a several-fold increase in CO emissions compared to emissions from the gasoline life cycle. This is an important consideration in countries where burning is common agricultural practice (e.g., Brazil; refer to Chapter 6) or is associated with land clearing for increased demand for cropland (e.g., Indonesia; refer to Chapter 9). Sheehan et al. (2004) found increases in CO when considering the full life cycle of using ethanol from corn stover in the midwestern United States. Similar studies for biodiesel have largely shown an increase in CO emissions for a full life cycle analysis of biodiesel over traditional diesel (Delucchi, 2006).

2.3. Sulfur dioxide

Sulfur dioxide (SO_2) is a primary gaseous pollutant – over 70 percent of its emissions in the United States come from fossil fuel–powered electricity-generating facilities.

SO_2 can cause respiratory irritation when inhaled as it can dissolve in the moist air of the respiratory system to form an acidic solution. In addition, SO_2, or its oxidation product, sulfuric acid, can dissolve in cloud droplets to form an acidic solution resulting in acid rain. Acid rain can have devastating impacts on sensitive ecosystems both through immediate impacts of the dilute acid solution coming into contact with surfaces of leaves and through the accumulation of acid in soils and bodies of water over time (Jacobson, 2002). Oxidation of SO_2 and subsequent condensation of the oxidation product is also the major source of the sulfate that is found in particulate matter.

Both ethanol and biodiesel contain little to no sulfur, so it is expected that tailpipe emissions of SO_2 will be lower for vehicles fueled with biofuels than for those fueled with gasoline and conventional diesel. As vehicular SO_2 emissions are quite small compared to those from electricity-generating facilities, this will likely not have a large impact on SO_2 concentrations in the atmosphere. Hess et al. (2009) hypothesized that benefits in SO_2 reduction from displacing conventional fuels with biofuels might be more pronounced in countries with higher sulfur contents in liquid fuels. More likely, the most significant changes in SO_2 emissions due to biofuel use will occur as part of the fuel production stage, considering that the energy requirements to process each different fuel and biofuel feedstock are different. Hess et al. (2009) found that life cycle emissions of E85 produced from cellulosic feedstock, specifically corn stover and sugarcane, resulted in a decrease in SO_2 emissions compared to those from gasoline, while E85 produced from corn grain resulted in an increase in SO_2 emissions. In some studies, cellulosic E85 even resulted in a decrease in overall SO_2 emissions from the baseline value as excess electricity, produced as a coproduct from the ethanol plant, would be fed back into the grid and displace SO_2 from coal-fired power plants (Spatari et al., 2010). These impacts will likely be strongest in areas with high fractions of electricity production supplied by coal, including China, India, and South Africa. Carrying out a full life cycle assessment, Sheehan et al. (2004) found increases in SO_2 emissions in the midwestern United States for corn stover ethanol use, and Rettenmaier et al. (2010) found increases in SO_2 emissions for sugar beet ethanol in Europe.

2.4. Particulate matter

Atmospheric PM is a collection of fine particles suspended in air ranging in size from 1 nm to as large as 100 μm. They are emitted from natural sources, such as sea salt and dust, or from anthropogenic sources, such as diesel trucks and power plants. PM has a complicated composition as it is comprised of a variety of different chemical species. Primary PM is emitted directly into the atmosphere from a variety of sources, with combustion being a large contributor in many areas, particularly in developing nations. Secondary PM can be formed in the atmosphere when gaseous species with low volatilities are able to condense onto preexisting PM. These gaseous species can either be those that are directly emitted (e.g., ammonia condenses to form ammonium salts) or oxidized versions of gaseous primary species (e.g., SO_2 is oxidized to form sulfuric acid and condenses to form sulfate salts). Many organic species also undergo oxidation reactions that lower their volatilities and may condense to form additional PM.

Dozens of studies have found positive correlations between PM concentrations and mortality and hospital admission rates for respiratory illness and heart disease (Dockery et al., 1993; Klemm et al., 2000). Other significant impacts of increased PM concentrations include decreases in visibility (Pilinis, 1989), damage to buildings (Esbert et al., 2001; Marinoni et al., 2003), and ecosystem damage (Seinfeld and Pandis, 2006).

In addition to its health impacts, PM also affects climate change. PM can directly impact the radiative balance of Earth by absorbing incoming and outgoing radiation (particularly black carbon) or reflecting incoming radiation back out of the atmosphere (the *aerosol direct effect*). PM can also impact the radiative balance by changing the lifetime and radiative balance of clouds (the *aerosol indirect effect*). Both impacts are complex and can potentially have a warming impact in some locations and a cooling impact in others (Forster et al., 2007).

Biodiesel has been found to have lower PM tailpipe emissions than conventional diesel (EPA, 2002). For ethanol fuel blends, there has been little research on changes in tailpipe PM emissions as the PM emissions from both gasoline and ethanol are low (EPA, 2007). Lara et al. (2005) utilized principal component analysis to study PM concentrations near the Brazilian city of Piracicaba to show that 60 percent of the PM found during the study was attributable to burning of sugarcane fields and that an additional 14 percent was related to soil dust, primarily from agricultural practices (refer to Chapter 6). They also found significant increases in black carbon concentrations during the burning season, which can have implications for climate change. Allen et al. (2004a), also studying high sugarcane-producing areas of Brazil, observed peak PM concentrations up to 240 $\mu g/m^3$ during the burning season. The implications of these abnormally high concentrations can be seen in studies that have shown increases in hospital admissions during the agricultural burning season in these regions (Arbex et al., 2010), particularly among the young and elderly (Cançado et al., 2006). These findings have significant ramifications for PM concentrations in areas of South America, Africa, and Asia that still practice agricultural burning. Studying the life cycle emissions from ethanol production in India, Kadam (2002) found that using bagasse-derived ethanol as a fuel additive may result in a decrease in PM emissions.

Hill et al. (2009) investigated the life cycle PM concentrations expected from ethanol use in the United States with a variety of ethanol feedstocks. They found that when compared to life cycle emissions from gasoline, corn grain ethanol resulted in a significant increase in PM emissions, while cellulosic feedstocks (corn stover, miscanthus, mixed prairie plant species, and switchgrass) led to PM emission decreases. Kusiima and Powers (2010) calculated the anticipated societal costs associated with a variety of fuel sources and found high PM costs associated with ethanol, with the majority of the PM emissions associated with fuel processing, farming, and fertilizer production (these results were not compared to those for gasoline). Finally, Huo et al. (2009) also found increases in PM emissions for corn grain ethanol in place of gasoline but noted that the main PM increases were in rural areas and anticipated decreases were in many urban areas.

2.5. Ammonia

Ammonia (NH_3) is a basic gaseous pollutant primarily emitted from animal waste, the breakdown of organic materials, fertilizer production and use, and industrial processes (Seinfeld and Pandis, 2006). The main risk associated with ammonia in the atmosphere is the role it plays in the formation of secondary PM. Ammonia neutralizes sulfuric and nitric acids to form ammonium sulfate and ammonium nitrate salts that condense to form PM.

The main ammonia source associated with biofuels is ammonia volatilization from nitrogen fertilizers applied to croplands to grow biofuel feedstocks. For this reason, the emission of ammonia associated with biofuels is highly variable by season (Goebes et al., 2003). Emissions of ammonia also depend greatly on the amount and type of fertilizer applied and how efficiently the crop utilizes the nitrogen in the fertilizer (Bouwman et al., 2010). Bouwman et al. (2002) found that globally, 14 percent of the nitrogen applied as synthetic fertilizers and 23 percent of that applied as animal manure is lost due to volatilization of ammonia. In addition, they found that emissions in developing nations tend to be higher than in industrialized countries due to higher temperatures and the higher use of urea and ammonium salts. They found 18 percent and 7 percent volatilization of ammonia in developing and industrialized countries, respectively, for synthetic fertilizer and 26 percent and 21 percent, respectively, for animal manure. Because some crops use nitrogen in fertilizers more efficiently and require less fertilizer application, the type of crop that is being grown can have a large impact on the ammonia emissions (Miller et al., 2006). Another important factor in ammonia volatilization is the technique that is used to apply the fertilizer to the soil, though different application methods can result in more ammonia runoff into groundwater (Aneja et al., 2009).

Another source of ammonia emissions is vehicle tailpipes (Durbin et al., 2004; Kean et al., 2000), although this is a minor source compared to nitrogen fertilizer production and use. These emissions are the result of a reaction on the surface of the three-way catalytic converter. Unlike most pollutants, ammonia emissions increased during the initial implementation of three-way catalytic converters but have started to decline again in recent vehicle models (Burgard et al., 2006).

2.6. Nitrogen oxide

The term NO_x refers to a variety of nitrogen oxide species that exist in the atmosphere. The two primary contributing species to NO_x concentrations are nitrogen monoxide (NO) and nitrogen dioxide (NO_2). NO_2 has been shown to cause airway inflammation in individuals exposed to high concentrations (Jacobson, 2002). High NO_x concentrations tend to be most prevalent near roadways as vehicular sources are the main contributors. In addition to its direct impacts, NO_x plays an important role in ozone formation. This role is highly nonlinear, meaning that lowering NO_x concentrations may increase, decrease, or have no effect whatsoever on ozone levels in certain areas. The relative composition of the atmosphere, particularly the ratio between NO_x and VOC concentrations, plays an important role in determining which of the end results mentioned earlier will actually occur (Finlayson-Pitts and Pitts, 2000). Finally, NO_x can also be oxidized in the atmosphere to form nitric acid, which will potentially condense on PM to form nitrate salts and can also contribute to acid rain (Seinfeld and Pandis, 2006).

Many studies have found increases in NO_x emissions from increased proportions of biodiesel in blended fuels (EPA, 2002; McCormick, 2007; Morris et al., 2003). The results from studies investigating NO_x emissions from different biofuel blends have been varied. In a survey of the literature, Yanowitz and McCormick (2009) found that the use of E85 reduced emissions of NO_x by 54 percent on average, but individual studies, such as that of the National Renewable Energy Laboratory (NREL, 1999), have found that emissions of NO_x increased by 33 percent. Many additional studies have also found values within this range. Much of this variability is likely related to differences in testing procedures, including engine load, vehicle type, and operating temperature, all of which potentially have a large impact on the levels of emitted pollutants, particularly of NO_x and CO. In a similar literature review, Graham et al. (2008) found that on average, NO_x emissions decreased by 45 percent with E85 but remained unchanged with E10 blends.

Studies that have considered NO_x emissions throughout the biofuel life cycle have found that NO_x emissions associated with farming, fuel processing, and fertilizer production could potentially outweigh any potential NO_x emissions decrease associated with biofuel use (Hess et al., 2009; Hu et al., 2004). Martinelli and Filoso (2008) estimate high NO_x emissions from the approximately 4.9 million ha of sugarcane fields in Brazil that still utilize burning. These emissions could contribute to the high nitric acid content of rainwater in the area (Lara et al., 2001). There is also concern that next-generation biorefineries may emit high levels of NO_x. On a per volume basis, NO_x emissions from lignocellulosic ethanol demonstration facilities are approximately an order of magnitude higher than currently observed emissions from commercial corn grain ethanol facilities and petroleum refineries (Jones, 2010). Kusiima and Powers (2010) found that NO_x emissions from fertilizer production, farming, and processing were the main contributors to the societal costs associated with ethanol. As much of the fertilizer used globally is produced in developing nations, these emissions have the potential to affect air pollutant concentrations in those areas. Huo et al. (2009) found increases in NO_x concentrations in rural areas but decreases in urban areas. As many rural areas are NO_x limited, these increases could very well lead to increases in localized ozone concentrations, but as the ozone concentrations in these areas are often already relatively low and there are fewer people to be exposed, the net impact on human health may be less severe.

A large set of unknowns related to the air pollution impacts of biofuel production comes from the potential changes in NO_x emissions due to land use change. It is likely that land use changes will have particularly large impacts in regions where rainforests and peat swamps could be potentially converted to croplands to meet additional demand. Hewitt et al. (2009) investigated the potential for increases in NO_x emissions in Thailand where native rainforests are replaced with oil palm plantations, and found potential increases in NO_x emissions that would likely lead to increased O_3 concentrations as these palm plantations are also large sources of VOCs.

2.7. Ozone

Reducing ambient ground-level ozone (O_3) concentrations is a complicated process for several reasons. Ground-level ozone is not emitted directly but is instead formed via a series of chemical reactions involving two chemical precursors, NO_x and VOCs, in the presence of sunlight. The chemical reactions that generate ozone are complex and interdependent. Lowering emissions of a precursor will not necessarily yield an improvement in ozone concentrations and depends strongly on the relative abundance of the two precursors (NO_x; VOCs) as well as on local conditions such as temperature (NRC, 1999). In some cases, emission reductions can actually increase ozone concentrations. Existing research shows that in areas with an abundance of VOCs (*NO_x-limited regimes*), reducing NO_x improves ozone but reducing VOCs has little impact. In areas with an abundance of NO_x (*VOC-limited regimes*), reducing VOCs improves ozone but reducing NO_x often actually increases ozone concentrations, including the peak ozone concentrations regulated by many governments. This non-linearity is the result of competing reactions at higher NO_x concentrations (Seinfeld and Pandis, 2006). VOC-limited regimes typically exist in urban and suburban areas, whereas NO_x-limited regimes typically exist in more rural areas.

Ozone exposure increases susceptibility to respiratory infections, medication use by asthmatics, and hospital admissions for individuals with respiratory disease (Halonen et al., 2010). Ozone may also contribute to premature death (Bell et al., 2006; Jerrett et al., 2009), reduced crop yields, and harm to sensitive ecosystems (Van Dingenen et al., 2009).

Of the pollutants that cause the greatest concern, should an increase in biofuel use occur, ozone is the most complex. For the reasons explained earlier, it is difficult to determine potential changes in ozone concentrations from changes in pollutant emissions. For this reason, it is necessary to incorporate some form of air quality modeling into any study focused on potential ozone concentration changes. Like many other pollutants, studies related to changes in ozone concentrations associated with increased biofuel production and use show mixed results. Luo et al. (2009) found no significant increase in photochemical oxidation potential – a metric related to ozone production – expected from a life cycle analysis of sugarcane-derived ethanol, while the U.S. Environmental Protection Agency (EPA, 2007) found maximum ozone concentration increases (up to 0.33 ppb) with an average increase of 0.153 ppb over the entire United States in high-ethanol-use areas for corn grain ethanol. Jacobson (2007) found increases of over 2 ppb in the Los Angeles area, accounting for changes in tailpipe emissions that could lead to a 9 percent increase in ozone-related mortality, hospitalization visits, and asthma in the region. This reiterates that even though a 2 ppb increase seems relatively small compared to the 75 ppb U.S. National Ambient Air Quality Standard, it may still have a significant impact on human health. In addition, Cook et al. (2010) found increases of up to 1 ppb when investigating increases in biofuel use in the United States. The findings were highly spatially dependent and found ozone decreases in a few highly populated areas. Alhajeri et al. (2011) found modeled changes in maximum ozone concentrations ranging from −2.1 to 2.8 ppb when modeling the impacts associated with replacing gasoline with E85 in Austin, Texas.

In studies that have modeled potential changes of ozone in urban areas, including the study by Jacobson (2007), very large increases in ozone concentrations have been predicted. For instance, Milt et al. (2009) predicted maximum increases of up to 16 ppb in peak ozone concentrations with the replacement of gasoline and diesel with E10 and B10, respectively, in Bangkok, Thailand. Ginnebaugh et al. (2010) found increases of 7 ppb in ozone in the summer and 39 ppb in the winter in southern California. These high predicted increases are likely the result of the low mixing rates between the modeled domain and the surrounding atmosphere coupled with the high acetaldehyde and formaldehyde emissions associated with ethanol fuel usage.

Most studies on biofuels and ozone have focused on locations within the United States. Future research in other areas of the world would shed light on the complexity of ozone formation and its impacts as a result of biofuel production and use. For example, based on studies of Thailand oil palm plantations, land use change may well lead to increases in ozone in some areas (Hewitt et al., 2009).

2.8. Volatile organic compounds

Hundreds of species fall into the category of VOCs. These species have a variety of properties, and many of them play an integral role in ozone formation. Sources of VOC species are as varied as the species themselves. However, the most important biofuel-related sources are evaporative emissions from vehicles, incomplete combustion emissions from vehicles, and biogenic emissions from plants.

The impacts of VOCs are threefold. First, many are harmful to human health as they are carcinogenic. Second, they contribute to the formation of ground-level ozone. Third, many can be oxidized and condense on preexisting PM to create secondary organic aerosols that are also associated with serious health impacts. This section discusses the changes in the total emissions of these species as they relate to increased ozone formation potential. As measuring and reporting these species individually can often be unrealistic, there are three common ways VOCs are grouped and measured for emissions purposes: nonmethane hydrocarbons (NMHC), nonmethane organic gases (NMOG), and total organic gases (TOG). NMHCs include all organic gases comprised of only carbon and hydrogen, whereas NMOGs contain all organic gases except methane. TOGs comprise NMOGs and methane.

Results from studies investigating differences in VOC emissions from biofuels and conventional fuels are mixed. Many studies indicate that higher biofuel blends, including both ethanol and biodiesel, lead to decreases in NMHC emissions from vehicular use (Graham et al., 2008; Kelly et al., 1996; Yanowitz and McCormick, 2009) but also indicate increases of NMOG emissions as these include ethanol and carbonyl emissions from incomplete combustion (Jacobson, 2007). Increases of NMOG emissions of up to 63 percent have been found in tailpipe emission studies in which E85 was used in place of gasoline (Winebrake et al., 2001). NMOG emissions have been found to increase with increasing fuel ethanol content (Durbin et al., 2007). These species can have important impacts on O_3 concentrations and are of great importance when considering changes in air quality from the use of biofuels. It is also expected that VOC emissions will change significantly in response to land use change, particularly in areas where rainforests or peat swamps could potentially be converted to croplands. For instance, Hewitt et al. (2009) found that palm trees grown on plantations in Thailand have higher VOC emissions than the native rainforest they replaced. This increase in VOCs along with the potential increase in soil NO_x emissions from increased fertilizer use could lead to substantial increases in ozone concentrations in these areas. Conversely, Eller et al. (2011) demonstrated that growing switchgrass may have lower VOC emissions than those expected from growing other feedstocks. More research is needed to understand VOC emissions from the cultivation of different biofuel feedstocks.

Four chemical species are of particular interest when discussing VOC emissions from biofuels. These species have both carcinogenic properties and strong ozone formation potential. They are, in order of decreasing Cancer Unit Risk Estimate, as defined by the EPA, 1,3-butadiene, formaldehyde, benzene, and acetaldehyde (Jacobson, 2007). Most studies related to the changes in these emissions focus on laboratory tests of vehicles running on different fuels and fuel blends. These tests indicate that high ethanol fuel blends, such as E85, are expected to produce benzene emissions that are approximately 60 to 85 percent lower than those from gasoline (Black et al., 1998; Graham et al., 2008; Kelly et al., 1996; Winebrake et al., 2001). Emissions of 1,3-butadiene have also been shown to decrease by 70 to 80 percent with higher ethanol blends (Graham et al., 2008; Kelly et al., 1996; Winebrake et al., 2001).

Unlike benzene and 1,3-butadiene emissions, a shift from gasoline to E85 blends could increase formaldehyde emissions by 20 percent (Kelly et al., 1996) to 315 percent (Winebrake et al., 2001). Acetaldehyde emissions are predicted to increase by over 10 times the current vehicular emission levels (Graham et al., 2008; Jacobson, 2007; Kelly et al., 1996; Winebrake et al., 2001). Some studies have found increases in acetaldehyde emissions in excess of 3,000 percent (Black et al., 1998; Magnusson et al., 2002). These results have been found to align with measurements of formaldehyde and acetaldehyde concentrations in Brazil, where ethanol use is high, as the ambient concentrations of acetaldehyde and formaldehyde are higher than in the rest of the world and the concentrations of aromatics such as benzene are lower than in much of the world (Anderson, 2009; Nguyen et al., 2001). Correa et al. (2003) found particularly high concentrations of formaldehyde (1.5–54.3 ppb) and acetaldehyde (2.4–45.6 ppb) near roadways in Rio de Janeiro, Brazil. Using an air quality model, Jacobson (2007) found increases of over 0.7 ppb and 0.07 ppb for acetaldehyde and formaldehyde, respectively, in a modeling study of E85 use. These increases are particularly significant for acetaldehyde, which had ambient concentrations around 2.5 ppb in this study. These increases are most problematic concerning their ozone formation potential, and Jacobson (2007) did find significant increases in ozone in the same study.

In addition to the studies mentioned earlier, Correa and Arbilla (2008) found biodiesel blends to have increased acetaldehyde, formaldehyde, acrolein, acetone, propionaldehydes, and butylaldehyde emissions and decreased benzaldehyde emissions when compared to emissions from conventional diesel combustion. These species have varied levels of ozone formation potential and associated cancer risks.

Another group of organic species of interest when considering biofuels are polycyclic aromatic hydrocarbons (PAHs). These compounds, like those mentioned previously, are also carcinogenic. Many studies have indicated that PAH concentrations are expected to decrease with increasing biofuel use (de Abrantes et al., 2009; McCormick, 2007; Turrio-Baldassarri et al., 2004). Conversely, Martinelli and Filoso (2008) theorize that high PAHs in certain areas of Brazil are likely attributable to sugarcane burning. Some rural areas have concentrations that are higher than those typically found in large cities.

3. Conclusions

Around 90 percent of global biofuel production occurs in the United States and Brazil (Timilsina and Shrestha, 2010; Wiens et al., 2011), and so it is not at all surprising that few studies on the air pollution impacts of biofuels have focused on developing nations. Much of what has been learned from the United States and Brazil can be used to inform biofuel policies in developing nations, but because of the location- and timing-specific factors described earlier, further insight into the issues associated with growing biofuel production in developing nations will need to be nation specific.

Studies that have investigated the potential air quality impacts of the production and use of biofuels have shown that displacing conventional fuels will likely lead to some localized improvements in air quality. Varied findings are as much the result of the differing experimental setups and modeling approaches used in the studies reviewed in this chapter as they are the result of the differences among the effects of the fuels themselves as they are deployed for vehicular transportation. It is also important to note that studies often do not account for emerging engine and emission control technologies. Many findings related to vehicular emissions for the different pollutants discussed in this chapter have the potential to change dramatically as engines are redesigned to better accommodate new fuels and meet new emission standards (Twigg, 2005). These changes in emissions will likely lead to changes in ambient concentrations. For instance, Correa et al. (2010) found decreases in ambient formaldehyde and acetaldehyde concentrations in Brazil from 2004 to 2009 that were associated with the implementation of new engine technologies.

Any air pollution benefits associated with transitioning from conventional fuels to biofuels may also be augmented through the use of improved agricultural practices, better conversion technologies, and the production of advanced liquid biofuels beyond ethanol and biodiesel. So-called green hydrocarbons derived from lignocellulosic biomass have the potential to be produced with much greater efficiency than the current generation of biofuels (Regalbuto, 2009), which may translate into a reduction of air pollutants across their entire life cycle. Future studies will help increase our understanding of what air quality changes to anticipate as the use of biofuels becomes more prevalent.

Acknowledgments

We would like to thank Julian Marshall and Christopher Tessum for their comments.

4

Water for bioenergy: A global analysis

P. WINNIE GERBENS-LEENES, ARJEN Y. HOEKSTRA, AND
THEO H. VAN DER MEER
University of Twente, Twente, Netherlands

Abstract

Agriculture is by far the largest water user. This chapter reviews studies on the water footprints (WFs) of bioenergy (in the form of bioethanol, biodiesel, and heat and electricity produced from biomass) and compares their results with the WFs of fossil energy and other types of renewables (wind power, solar thermal energy, and hydropower). WFs for bioenergy vary, depending on crop type applied, production location, and agricultural practice. The most water-efficient way to generate bioenergy is to use biomass for heat generation, with electricity generation being the second best option. Biofuel production requires roughly twice as much water as bioelectricity. Regarding biofuels, bioethanol has smaller WFs than biodiesel. For example, the WF of rapeseed biodiesel is four times larger than the WF of sugarcane ethanol and seven times larger than the WF of sugar beet ethanol. Global weighted ethanol WFs increase in the order of sugar beet, potato, sugarcane, maize, cassava, barley, rye, paddy rice, wheat, and sorghum and range between 60 and 400 m^3/GJ. For sugar beet, maize, and sugarcane, differences between regions are large. The European Union, northern Africa, and the United States have relatively small WFs for ethanol from sugar beet and maize, while eastern Europe has large WFs. Global weighted average biodiesel WFs increase in the following order: palm oil (95 m^3/GJ), soybean and rapeseed (400 m^3/GJ), and jatropha (570 m^3/GJ). Conversely, the WFs of fossil fuels are relatively small. Finally, the WF of hydropower varies widely between 0.5 and 850 m^3/GJ. Our results provide new insight into the impacts of bioenergy on the use and pollution of freshwater. This knowledge is a valuable contribution to future research and for policies concerning energy needs, freshwater availability, and the choice whether to allocate water to food or energy production.

Keywords: renewable, water footprint

1. Introduction

Freshwater is essential for the functioning of society. Water use in agriculture, industry, and households has increased sharply in the twentieth century (Shiklomanov, 1997). Human activity consumes and pollutes great amounts of water, particularly through agricultural production (Hoekstra and Chapagain, 2007). Freshwater is a scarce natural resource. Most water on Earth is saline and cannot be used for societal needs. The oceans contain about 97.5 percent of available water in the form of saltwater. Of the remaining 2.5 percent of freshwater, most is not accessible because it forms part of ice or snow covers (Shiklomanov, 1997). Although the amount of water on the planet is constant, the annual freshwater supply in the form of precipitation is limited.

Natural precipitation is the main provider of agricultural water. This is the so-called *green water* (Hoekstra et al., 2011). When precipitation is insufficient, farmers can apply irrigation, or *blue water*.[1] Today, about 80 percent of the agricultural water requirements are met by precipitation, with the rest withdrawn from other sources such as rivers and lakes (de Fraiture and Berndes, 2009). These withdrawals account for 70 percent of all human water use (UNEP, 2009).

Freshwater is becoming, more and more, a global resource because water-intensive products are traded on global markets. International trade results in a spatial disconnection between consumers and the water resources used for making consumer products. Water footprint (WF) research shows the relationship between consumer goods and water consumption along supply chains, thereby addressing the link between consumption and production. By doing this, WF research offers a new perspective on how a consumer or producer relates to the use of freshwater systems (Hoekstra et al., 2011). The WF concept provides a tool to calculate water needs for consumer products and provides an indication of the total amount of freshwater used, directly and indirectly, along product supply chains (Hoekstra et al., 2011). The WF of a product, for example, bioethanol, is the volume of freshwater used to produce the ethanol, measured over the complete supply chain. Important water-intensive products are crop and livestock commodities, natural fibers, and bioenergy.

The next decades will see an increased demand for food (Tilman et al., 2002; Bruinsma, 2003) as well as an increased demand for biofuels (Stromberg et al., 2010). The corresponding necessary growth of agricultural output can be achieved in three ways: (1) an increase of agricultural land areas, (2) an increase of yield levels per unit of land (increase of land productivity), or (3) an increase of cropping intensities (e.g., by increasing multiple cropping and shortening fallow periods) (see also Chapter 2). If agricultural land areas are increased, water use will probably increase by the same

---

[1] The irrigation sector has increased enormously in past decades and is currently the largest water user, accounting for 61% of total water withdrawal globally. Between 1900 and 1995, the irrigated area expanded fivefold, from 50 million to 250 million ha. Half of these irrigated areas are located in just four countries: China, India, the United States, and Pakistan (Shiklomanov, 1997).

factor, given that water input per unit of land usually remains the same. The increase of yield levels or cropping intensities might also increase water use in those cases in which water is the limiting factor for crop growth.

Bioenergy production may divert land, water, and other resources away from the production of food and feed (Fischer et al., 2009). In many countries, agricultural water use competes with other uses such as urban supply and industrial activities (Falkenmark, 1989), causing the aquatic environment to show signs of degradation and decline (Postel et al., 1996). Crop growth (for biomass production) requires freshwater, and agricultural activity associated with feedstock production is by far the largest user of water, followed by industrial activities (WWAP, 2009). In general, increased biofuel production will probably require more water (Berndes, 2002; de Fraiture et al., 2008), and a shift from fossil energy toward bioenergy might put additional pressure on freshwater resources.

Today, some of the world's most important agricultural areas show signs of water scarcity[2] (de Fraiture and Berndes, 2009), such as northern India, Pakistan, and northern China (Shah et al., 2007). China and India will account for one-third of the world population and will demand one-third of the world's energy supply by 2030 (de Fraiture and Berndes, 2009), so they aim to partly replace transport fuels from fossil sources by biofuels such as bioethanol and biodiesel (MNRE, 2008; Yang et al., 2009) (see Chapters 1 and 10). This is expected to increase water scarcity because China and India have already overexploited their natural water resources (de Fraiture et al., 2008; Muller et al., 2008). Sufficient water for agriculture is available in Latin America and sub-Saharan Africa (Muller et al., 2008), excluding South Africa (Jewitt et al., 2009). All the preceding suggests that biofuel-related water consumption might aggravate water scarcity in many countries. In all, about 30 developing countries face water scarcity, and it is expected that by 2050, over 50 developing countries will suffer from water shortages (Fischer et al., 2009). It is therefore important to have insight into the relationship between agricultural output, water consumption, and water availability to properly allocate the water to food or biofuels.

Biofuel production does not only affect the quantity of water resources but can also affect the quality of such resources (Stromberg et al., 2010). Apart from water, other important agricultural inputs for feedstock production include nutrients (such as nitrogen and phosphorus) and agrochemicals for controlling pests, diseases, and weeds. When agricultural yields increase, the demand for nutrients expressed per unit area also increases (de Wit, 1992). Part of these inputs leach to water bodies and cause water pollution (Simpson et al., 2009; Stromberg et al., 2010; UNEP, 2009). Ethanol production, for example, has serious implications for coastal water quality and will almost certainly worsen already serious hypoxic conditions in many locations around

---

[2] Water shortages are the result of a mismatch between demand for freshwater and its availability over space and time.

the world (Simpson et al., 2009). Sugarcane expansion is one of the main drivers of increased fertilizer and agrochemical use in Brazil, which has been linked to water pollution and ecosystem deterioration (Martinelli and Filoso, 2008) (see Chapter 6).

In this chapter, we employ the WF concept to assess the water requirements of different biofuel production practices and compare them with the WFs of fossil energy carriers (natural gas, coal, and crude oil). Initially, we summarize the WF methodology (Section 2), and we review the main bioenergy WF studies that have estimated WFs per unit of bioenergy (m^3/GJ) (Section 3). We cover biofuels (bioethanol and biodiesel) as well as heat and electricity generated from biomass. Subsequently, we compare the results generated using the WF methodology with the results of studies that have used other methodologies. Next, we compare WFs of bioenergy with WFs of fossil energy carriers, nuclear energy, and the WFs of renewables (wind power, solar thermal power, and hydropower). The chapter gives WFs of various types of bioenergy in meters cubed per unit of energy (m^3/GJ) and covers the main producing countries, including developing countries, transition countries, and industrialized countries.

2. Methodology

The WF is a multidimensional indicator, giving water consumption volumes by source and polluted volumes by type of pollution. The WF of a product is defined as the volume of freshwater used for its production at the place where it was actually produced (Hoekstra et al., 2011). In general, a product's actual water contents are negligible compared with its WF. For many products, such as bioenergy, the water used (or consumed) during the agricultural production stage makes up the bulk of the product's total life cycle water use.

The WF concept includes three components – green, blue, and gray water – and distinguishes between direct and indirect water use, taking into account the water use along supply chains. The components of WFs are specified geographically and temporally. The green WF refers to consumption of green water resources, that is, rainwater stored in the soil as soil moisture and water that stays on top of the soil and on the vegetation. The blue WF refers to consumption of blue water resources, that is, fresh surface water and groundwater. The gray WF refers to pollution and is defined as the volume of freshwater required to assimilate the load of pollutants based on existing ambient water quality standards (Hoekstra et al., 2011). The gray component of the WF is shown in equation (4.1):

$$\mathrm{GrayWF} = \frac{\alpha * \mathrm{AR}}{\dfrac{C_{\max} - C_{\mathrm{nat}}}{Y}}, \qquad (4.1)$$

where AR is the chemical application rate to the field per hectare (kg/ha), α is the leaching–runoff fraction, $C_{\max}$ is the maximum acceptable concentration of the

---

Box 4.1. Sample WF analysis for sugarcane ethanol in Brazil and India

Brazil is the largest sugarcane producer, with 29 percent of global sugarcane production, followed by India, which produces 21 percent (FAO, 2008a). Often the sugarcane crop is harvested manually. At the plant, the stalks are chopped into pieces and washed to remove trash. Next the cane is crushed, producing a juice, the ingredient for ethanol and sugar, and a residue, the bagasse (Cheesman, 2004).

Process water use is small compared to agricultural water use. Traditional water use in a sugarcane mill is about 21 m^3/ton of processed cane (Macedo, 2005). New techniques have decreased water use to 0.92 m^3/ton of cane. The São Paulo State Plan on water resources estimated the water use in 1990 at 1.8 m^3/ton of cane (Macedo, 2005). For Brazil and India, we have calculated a total WF, green, blue, and gray, of 209 and 256 m^3/ton. This means that in Brazil, the process water is about 1 percent of the total water use in the whole supply chain. For India, we also estimate that process water use is almost negligible when compared to water use over the entire supply chain.

---

pollutant (kg/m^3), C_{nat} is the natural concentration for the pollutant considered (kg/m^3), and Y is the crop yield (ton/ha).

The pollutants generally consist of fertilizers (nitrogen, phosphorus, etc.) and agrochemicals. One has to consider only the waste flow to freshwater bodies, which is generally a fraction of the total agricultural application to the field. One needs to account for only the most critical pollutant, that is, the pollutant for which the preceding calculation yields the highest water volume (Hoekstra et al., 2011).

The WF of biofuels is dominated by the agricultural phase, that is, feedstock production. Other processes during the biofuel's life cycle, such as feedstock transportation and processing, are much less water intensive. Box 4.1 provides a simplified WF analysis for sugarcane ethanol in Brazil and India.

It should be noted that all WF studies presented in this chapter have been based on the assumption that crop water use is equal to crop water requirements. However, when actual water availability is lower and water stress occurs, the WF analysis overestimates crop water use. This assumption differs from the method applied by Chiu et al. (2009), who estimated the water requirements of U.S. ethanol based on measured irrigation and arrived at lower values. In their study, Mekonnen and Hoekstra (2010) find lower WFs because they also account for water shortages, thus estimating the actual water consumption. With respect to agricultural yields, the studies presented and reviewed in this chapter have taken into account actual yields, which, in many cases, can be increased in the future without increasing water use per unit of product. This means that future WFs per unit of energy can actually be significantly smaller than the ones currently calculated. Conversely, regarding the efficiency of obtaining electricity or biofuels from biomass, the studies have made optimistic assumptions by taking theoretical maximum values, or values that refer to the best available technology. This means that the resulting WFs are conservative.

3. Overview of bioenergy WF studies

3.1. Bioethanol

Gerbens-Leenes and Hoekstra (2012) performed a detailed study of bioethanol WF (green, blue, and gray) for the main producing countries as well as the main producing U.S. states. Countries and feedstocks included in this analysis are as follows:

- *Sugarcane*. Argentina, Australia, Belize, Brazil, China, Colombia, Cuba, Egypt, Ethiopia, Guatemala, India, Indonesia, Morocco, Pakistan, Peru, the Philippines, South Africa, Thailand, the United States, Venezuela, and Vietnam
- *Sugar beet*. Belgium, China, the Czech Republic, Denmark, Egypt, France, Germany, Iran, Italy, Japan, Morocco, the Netherlands, Poland, the Russian Federation, Serbia, Spain, Turkey, Ukraine, the United Kingdom, and the United States
- *Maize*. Argentina, Brazil, Canada, China, Egypt, France, Germany, India, Indonesia, Italy, Mexico, Nigeria, the Philippines, Romania, South Africa, Spain, Thailand, Ukraine, and the United States

An earlier study assessed green and blue WFs for all bioenergy types, including bioethanol (Gerbens-Leenes et al., 2009a). That study also calculated WFs for carbohydrate-rich crops that do not yet have a large-scale application for ethanol production. These crops were barley, cassava, potato, rice, rye, sorghum, and wheat. For the purpose of this chapter, we use data from Gerbens-Leenes et al. (2009a) and adopt the gray WFs of Gerbens-Leenes and Hoekstra (2012). For the United States, we present data from Gerbens-Leenes and Hoekstra (2012).

Mellko (2008) calculated the WFs of biofuels, including those of bioethanol and biodiesel, for the European Union and other large producing countries. Mellko applied the WF methodology and derived WF crop data from Chapagain and Hoekstra (2004) that were updated later (Gerbens-Leenes et al., 2009a, 2009b). Yang et al. (2009) have calculated the total WF of bioethanol from sugar beet grown in China. Chiu et al. (2009) estimated the amount of embodied water for ethanol in the United States and calculated the blue WF using the measured irrigation water. A fourth study included for comparison is the study of Berndes (2002), who has also given values for crop evapotranspiration per unit of bioenergy. The most recent study considered in this chapter is by Mekonnen and Hoekstra (2010); it provides a comprehensive global database of green, blue, and gray WFs of crops and derived crop products, including bioethanol and biodiesel, at a spatial resolution of 5 × 5 arcminute.

3.2. Biodiesel

Gerbens-Leenes et al. (2009a) have calculated the WFs of biodiesel from rapeseed and soybean for the main producing countries:

- *Rapeseed*. India, Bangladesh, Pakistan, Australia, the United States, Canada, China, Poland, the Czech Republic, France, Denmark, the United Kingdom, and Germany
- *Soybeans*. India, Indonesia, China, Bolivia, the United States, Canada, Brazil, Argentina, Paraguay, and Italy

For jatropha, Gerbens-Leenes et al. (2009c) and Hoekstra et al. (2009) have calculated green and blue WFs for locations distributed over the *Jatropha curcas* belt (between 30°N and 35°S), including Brazil, Indonesia, Nicaragua, Guatemala, and India. For palm oil, van Lienden et al. (2010) calculated the WF for the Philippines, Thailand, Indonesia, Malaysia, and Honduras. For the purposes of this chapter, we adopt the WF estimates of biodiesel from Gerbens-Leenes et al. (2009a, 2009c) and van Lienden et al. (2010).

3.3. Heat and electricity generation from biomass

Crop burning can provide heat for different societal uses. Gerbens-Leenes et al. (2009b) have calculated the WFs of heat generation from the combustion of biomass (m^3/GJ) from different sources, including poplar (a tree); miscanthus (a nonfood crop); and cassava, coconut, cotton, groundnuts, maize, palm oil, potato, wheat, rapeseed, sugar beet, sugarcane, sunflower, and soybean (crops that can be used as food or feed). That study included four countries: the Netherlands, Brazil, the United States, and Zimbabwe. Gerbens-Leenes et al. (2009a) calculated the WF of heat generation from four biomass categories, including the following:

- *Starch crops*. Barley, maize, rice, rye, sorghum, wheat, cassava, potato
- *Sugar crops*. Sugar beet and sugarcane
- *Oil crops*. Rapeseed and soybean
- *Trees*. Jatropha

Significant amounts of heat are released during biomass combustion. This heat can be converted into electricity, with the amount of generated electricity depending on the efficiency of the power plant. Theoretically, there is a maximum amount of electricity that can be generated in a power plant (Blok, 2006). A new technology is the gasification of biomass for electricity production, termed *biomass fired integrated gasifier/combined cycle* (BIG/CC). It can be applied for biomass with a moisture content below 70 percent and an ash content below 10 to 20 percent (DM) (Faay, 1997). Temperatures are relatively low, however, at 720 K, reaching an efficiency of

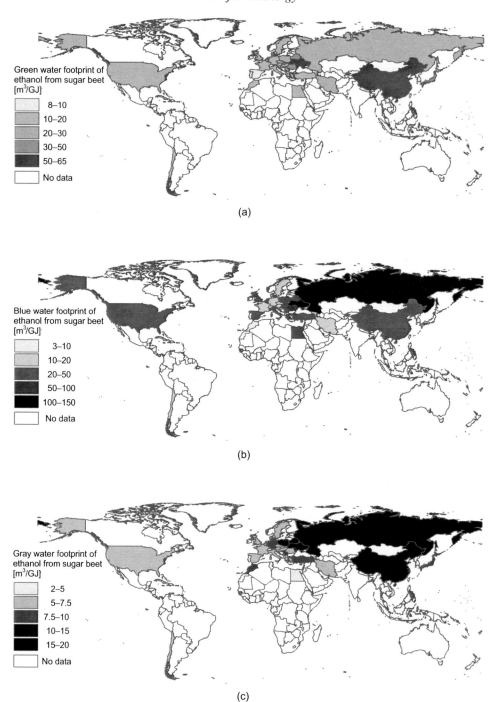

Figure 4.1. Weighted average (a) green, (b) blue, (c) gray, and (d) total WFs of ethanol from sugar beet (m³/GJ) for five regions. (See color plate.)

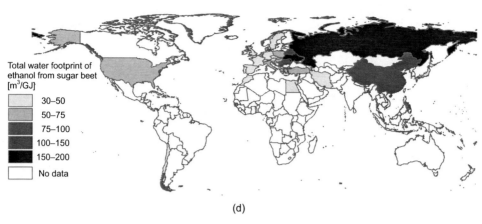

(d)

Figure 4.1 (*continued*)

59 percent (Blok, 2006). The WF results presented in this chapter are based on an electricity generation efficiency of 59 percent.

3.4. Other energy carriers

At present, important energy carriers include fossil energy carriers (crude oil, coal, and natural gas), uranium, and electricity from hydropower (IEA, 2006). Promising renewables are solar and wind energy. We compared the WFs of biofuels with the WFs of these important energy carriers. For crude oil, coal, natural gas, and uranium, we derived data from literature (Gleick, 1994). For electricity from hydropower, we estimated the global WF by dividing the global evaporation of reservoirs (Shiklo-manov, 2000) by the hydroelectric generation (Gleick, 1993) for the year 1990. Next, we compared these results with information on WFs of hydropower from Mekonnen and Hoekstra (2011).

4. Results

4.1. Bioethanol

Figure 4.1 shows the weighted average green, blue, gray, and total WFs (m³/GJ) of bioethanol from sugar beet for five regions (northern Africa, the European Union, the United States, Asia, and eastern Europe). The global average is 70 m³/GJ (green WF, 36%; blue WF, 51%; and gray WF, 13%). Morocco and Egypt have relatively small green WFs. Green WFs are large for Ukraine and China. Figure 4.1 shows that blue and gray WFs in particular are much greater in eastern Europe than in the other regions. The contribution of eastern Europe to global production, however, is only 11 percent so that these large WFs have a limited impact on the global WF. The largest contribution to global production is from the European Union. On the basis of

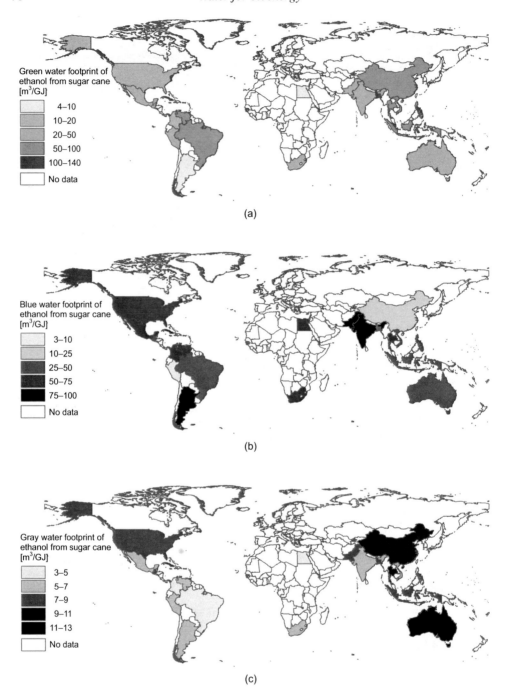

Figure 4.2. Weighted average (a) green, (b) blue, (c) gray, and (d) total WFs of ethanol from sugarcane for four regions. (See color plate.)

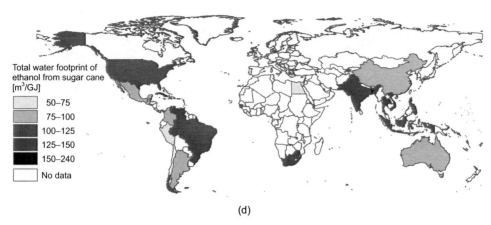

(d)

Figure 4.2 (*continued*)

FAO (2008a) production data, 54 percent of global sugar beet production is generated in the European Union.

Figure 4.2 shows weighted average green, blue, gray, and total WFs of bioethanol from sugarcane for four regions (northern Africa, represented by Egypt; Australia; the United States; and Asia). The global weighted average value is 115 m³/GJ (green WF, 43%; blue WF, 50%; and gray WF, 7%). The WFs are below average in Egypt and Australia, about average in the United States, and above average in Asia. Argentina, Pakistan, and India have large blue WFs compared to the other countries because very small quantities of green water are available. The gray WF is large in Australia, at 13 m³/GJ, whereas values for Egypt and the United States are only 3 m³/GJ and 5 m³/GJ, respectively. The contribution of Australia to global production, however, is small, given that 43 percent of global production takes place in the Americas (mainly in Brazil) and an additional 45 percent in Asia. It is worth noting that the WFs of sugarcane ethanol of the major bioethanol producers in the developing world are relatively higher than global weighted averages. In particular, Brazil shows a WF of 210 m³/GJ (green WF, 55%; blue WF, 41%; and gray WF, 4%), South Africa shows a WF of 306 m³/GJ (green WF, 33%; blue WF, 61%; and gray WF, 6%), and India shows a WF of 256 m³/GJ (green WF, 33%; blue WF, 61%; and gray WF, 6%).

Figure 4.3 shows the global weighted average green, blue, gray, and total WFs of bioethanol from maize for five regions (the European Union, the United States, Africa, Asia, and eastern Europe). The total global weighted average WF is 120 m³/GJ (green WF, 56%; blue WF, 36%; and gray WF, 9%). The WFs are below average in the western European Union and the United States and above average in Mexico, Nigeria, Ukraine, Romania, India, Indonesia, and the Philippines. A comparison of Figures 4.1 and 4.3 shows the relatively large WFs for bioethanol from sugar beet and maize grown in eastern Europe. For sugar beet and maize, the green WF is twice as large as the green WF in the European Union. For sugar beet, the gray WF is two and a half times the European Union value. For maize, the gray WF is 13 percent larger than

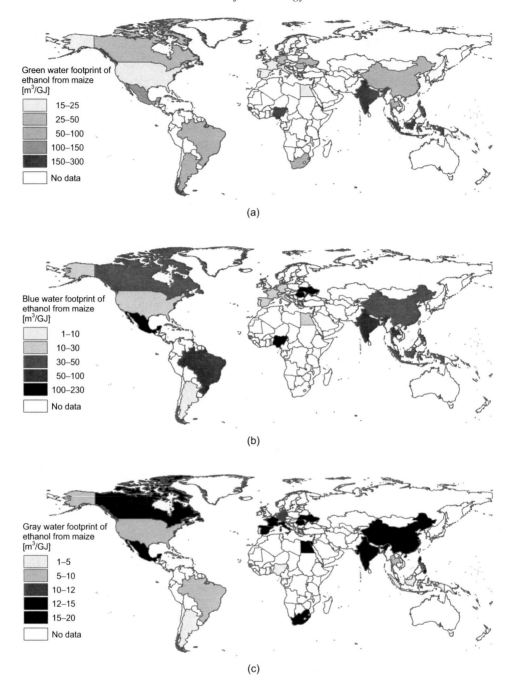

Figure 4.3. Global weighted average (a) green, (b) blue, (c) gray, and (d) total WFs of ethanol from maize for five regions. (See color plate.)

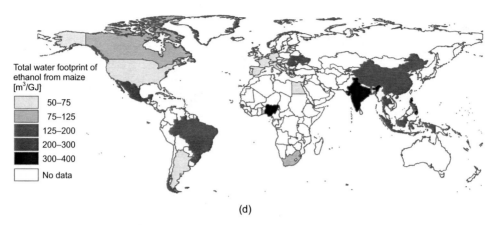

(d)

Figure 4.3 (*continued*)

in the European Union. The contribution to global production, however, is small and mainly comes from Asia (25%) and the United States, which contributes 56 percent.

Figure 4.4 shows the green, blue, gray, and total WFs of bioethanol from maize in the United States for 10 states. The weighted U.S. average WF is 54 m^3/GJ (green WF, 44%; blue WF, 37%; and gray WF, 19%). The results show some variation but are smaller than the global weighted average, with a green WF of 67, a blue WF of 43, and a gray WF of 11 m^3/GJ.

Figure 4.5 shows the global average green and blue WFs of bioethanol from maize, sugar beet, sugarcane, potato, cassava, barley, rye, paddy rice, wheat, and sorghum. The WF of potato bioethanol is 100 m^3/GJ, while the WF of sorghum bioethanol is more than four times the WF of potato bioethanol. The remaining crops have WFs between these two extremes. Figure 4.5 also shows the large variation among green and blue WFs.

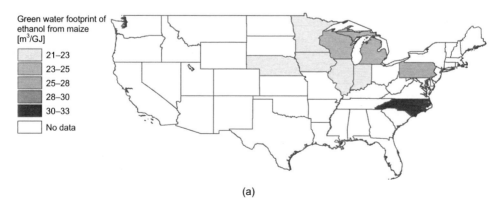

(a)

Figure 4.4. (a) Green, (b) blue, (c) gray, and (d) total WFs of ethanol from maize in the United States for 10 states. (See color plate.)

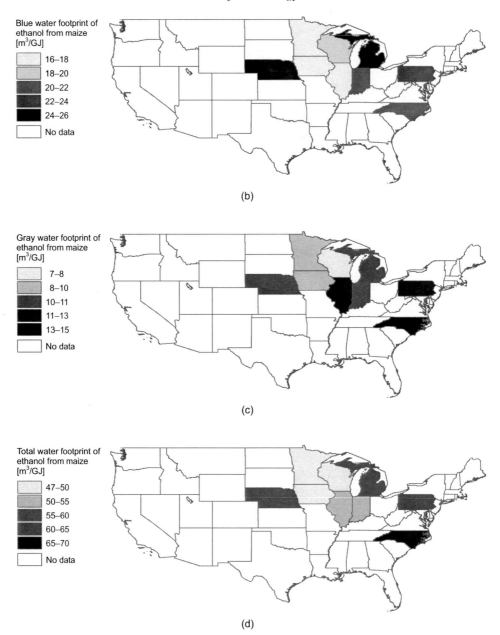

Figure 4.4 (*continued*)

4.2. Biodiesel

Table 4.1 gives the green, blue, and total WFs of jatropha, palm oil, rapeseed, soybean oil, and sunflower oil biodiesel in different areas of the world. The WFs of biodiesel from jatropha oil are greatest, followed by WFs of soybean and rapeseed oil. The WFs of biodiesel from palm oil and sunflower oil are the lowest. Table 4.1 also

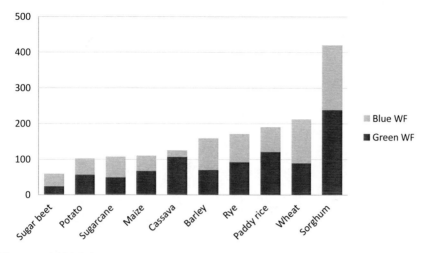

Figure 4.5. Global average green and blue WFs of ethanol from sugar beet, potato, sugarcane, maize, cassava, barley, rye, paddy rice, wheat, and sorghum.

Table 4.1. *Green, blue, and total WFs of jatropha, palm oil, rapeseed, soybean oil, and sunflower oil biodiesel for different locations*

		Green WF	Blue WF	Total WF
Jatropha oil[a]	India	575	1,116	1,691
	Guatemala	156	174	329
	Nicaragua	120	187	307
	Indonesia	184	109	293
	Brazil	160	91	251
	Average	**239**	**335**	**574**
Rapeseed oil[a]	India	256	591	847
	China	185	238	422
	Germany	78	49	127
	Global weighted average	**165**	**245**	**409**
Soybean oil[a]	United States	227	216	443
	Argentina	128	115	244
	Italy	108	117	225
	Global weighted average	**177**	**217**	**394**
Palm oil[b]	Philippines	70	54	124
	Thailand	53	44	97
	Indonesia	64	26	90
	Malaysia	67	1	68
	Honduras	87	9	96
	Average	**68**	**27**	**95**
Sunflower oil[c]	**Average European Union**			**93**

Note: Units are m^3/GJ oil.
[a] *Source:* Gerbens-Leenes et al. (2009a).
[b] *Source:* van Lienden et al. (2010).
[c] *Source:* Mellko (2008).

Table 4.2. *Global weighted average water footprint of heat from total biomass for 12 crops*

Crop	Global weighted average WF (m³/GJ)		
	Total WF	Green WF	Blue WF
Sugar beet	27	11	16
Maize	30	18	12
Sugarcane	32	15	17
Barley	42	20	22
Paddy rice	48	28	21
Wheat	56	23	32
Potatoes	60	33	27
Cassava	87	75	13
Soybeans	102	46	56
Sorghum	107	61	47
Rapeseed	226	91	135
Jatropha	234	97	136

shows significant differences among locations and averages for countries. In India, for example, the WF of biodiesel from jatropha oil is twice the WF of biodiesel from rapeseed oil. In Indonesia, the WF of biodiesel from jatropha oil is three times the WF of biodiesel from palm oil.

4.3. Heat and electricity generation from biomass

Table 4.2 includes the global weighted average green, blue, and total WFs of heat generation from biomass combustion for 12 crops. There is a large variation among the different crops. The total WF of heat generation from total sugar beet biomass is almost 10 times smaller than values for rapeseed and jatropha (whole fruit yield).

Table 4.3 provides the WFs of heat generation from total biomass combustion for a tree (poplar) and five crops (miscanthus, sunflower, sugar beet, sugarcane, and maize) for three countries. Table 4.3 shows that poplar and miscanthus (which are also feedstocks for second-generation biofuels) do not have a more favorable WF than food crops. In the Netherlands, the WF of heat generated from maize combustion is half the WF of heat generated from poplar or miscanthus combustion. In the United States, maize is the most favorable crop in terms of WF, whereas in Brazil, the most favorable crop is sugarcane.

When WFs for bioethanol are compared to WFs for heat, results show that it is much more favorable to generate heat than to produce biofuels. Depending on crop type, WFs for heat generation are much lower than bioethanol WFs. For example, the WFs of bioethanol from cassava are 50 percent higher than the WFs of heat generated

Table 4.3. *Water footprint of heat for one tree type, poplar, and five crop types, miscanthus, sunflower, sugar beet, sugarcane, and maize, grown in three countries*

Trees and crops	WF of heat generation from total biomass combustion (m^3/GJ)		
	The Netherlands	United States	Brazil
Maize	9	21	38
Sugar beet	16	25	–
Sugarcane	–	31	29
Miscanthus	20	37	49
Poplar	22	42	55
Sunflower	27	61	54

from cassava, while the WFs of ethanol from barley are four times the WF of heat from barley. Also for soybean, rapeseed, and jatropha, biodiesel WFs are two to four times greater than the WFs for heat generation.

The WF of bioelectricity was estimated from the WF of total crop biomass, including stems and leaves, assuming an efficiency of 59 percent for the conversion of heat into bioelectricity. The WF of electricity can be derived from the data given in Table 4.2 and the conversion efficiency of 59 percent.

4.4. Comparison with other studies

Table 4.4 provides a comparison of the findings from Mellko (2008) with the results presented in this chapter. Although the studies do not always cover the same countries, the results for total WFs are similar. Additionally, Mellko gives a value for sunflower oil. The data presented in this chapter show the same pattern as in the recent study by Mekonnen and Hoekstra (2010) but tend to be a bit higher, which can be explained by the fact that the latter study included the effect of water shortages in agriculture, an effect not included in the data presented here.

Yang et al. (2009) have calculated the total WF of ethanol from sugar beet grown in China to be 96 m^3/GJ, while we arrived at 97 m^3/GJ. In the U.S. context, Chiu et al. (2009) measured actual irrigation and arrived at values between 0.3 m^3/GJ (Iowa and Kentucky) and 91.4 m^3/GJ (California), with a weighted average of 6 m^3/GJ. This value is small compared to the value of 20 m^3/GJ for the U.S. bioethanol blue WF calculated by Gerbens-Leenes et al. (2009a). The main reason behind this discrepancy is that the study by Chiu et al. (2009) considered more states and included states, such as California, that have a dry season. Conversely, the study by Gerbens-Leenes et al. (2009a) estimated the WF of only the 10 main producing states.

Table 4.4. *Comparison of WFs estimated in different studies*

	Mellko (2008)	m³/GJ	Gerbens-Leenes et al. (2009a, 2009b)[1] and van Lienden et al. (2010)[2]	m³/GJ
Ethanol				
Barley	EU 27	126	14 largest European producers[1]	178
Sugar beet	EU 27	51	18 largest European producers[1]	49
Wheat	EU 27	108	9 largest European producers[1]	196
Maize	China, India, United States	101	China, India, United States[1]	170
	United States	51	United States[1]	44
Sugarcane	India	78	India[1]	118
	South Africa	96	South Africa[1]	98
	Brazil	76	Brazil[1]	99
	Average India, South Africa, Brazil	83	Global weighted average[1]	108
Biodiesel				
Sunflower	13 EU countries	93	NA	–
Rapeseed	EU 27	133	6 European countries[1]	139
Palm oil	Indonesia, Malaysia	86	Philippines, Thailand, Indonesia, Malaysia, Honduras[2]	95

Berndes (2002) has also given values for crop evapotranspiration per unit of bioenergy and has arrived at a large variation in values. Table 4.5 compares the results of Berndes (2002) with the results of Gerbens-Leenes et al. (2009a).

4.5. Comparison with the WF of other primary energy carriers

Table 4.6 gives an overview of the energy carriers' WFs derived from the literature (Gleick, 1994). Although the data are rather old, they give an indication of the WFs

Table 4.5. *Comparison of crop evapotranspiration per unit of bioenergy from different studies*

Crop	Berndes (2002)	Gerbens-Leenes et al. (2009a)
Sugar beet	71–188	22–174
Maize	73–346	41–377
Sugarcane	37–155	47–225
Wheat	40–351	50–1,001
Rapeseed	100–175	127–847

Note: Units are in m³/GJ.

Table 4.6. *Water footprint of different energy carriers*

Energy carrier	Average WF (m³/GJ)
Wind energy[a]	0.0
Nuclear energy[a]	0.1
Natural gas[a]	0.1
Coal[a]	0.2
Solar thermal energy[a]	0.3
Crude oil[a]	1.1
Hydropower[b,c]	0.5–850
Bioethanol[d]	165
Biodiesel[d]	313

Note: Units are in m³/GJ.
[a] *Source:* Gleick (1994).
[b] *Sources:* Gleick (1993) and Shiklomanov (2000).
[c] *Source:* Mekonnen and Hoekstra (2011).
[d] Global weighted average.

of conventional energy carriers. WFs of crude oil, coal, natural gas, uranium, solar thermal energy, and wind electricity generation are all smaller than the WFs of biomass. For fossil fuels, it should be noted, however, that the water required over time to grow the vegetation that has accumulated and turned into fossil fuel is excluded from the figures presented. For a fair comparison between the water footprint of bioenergy and fossil fuels, this historically accumulated water consumption should be accounted for.

For hydropower, Gerbens-Leenes et al. (2009b) have found an average global WF of 22 m³/GJ, while Mekonnen and Hoekstra (2011) have arrived at an average value of 68 m³/GJ. WF values per unit of generated electricity, however, show enormous variation. For example, the Lubuge power plant in China uses only 0.5 m³/GJ of generated electricity. Conversely, the Akosombo dam and Kpong power plant in Ghana use 850 m³/GJ of generated electricity, which is the highest WF for any energy source discussed in this chapter so far. WFs of hydropower and bioenergy are much higher than WFs of other energy sources. The global weighted average WF for bioethanol is 165 m³/GJ, and for biodiesel, it is 313 m³/GJ. When these data are also compared to the results presented elsewhere in the chapter, the large WF of bioenergy, especially for the first generation biofuels, becomes clear.

5. Discussion

In assessing the WFs of heat, electricity, and biofuels, all studies took into account the WF of the *gross* energy output from crops. Energy inputs in the production chain, such as energy requirements in the agricultural system (e.g., energy use for the production of fertilizers and pesticides) or the energy use during the industrial biofuel production process, were excluded. For high-input agricultural systems, energy input is substantial (Giampietro and Ulgiati, 2005; Pimentel and Patzek, 2005) so that net energy yields are smaller than presented here. This means that this overview underestimates the WF of bioenergy from agricultural systems with relatively large energy inputs.

The WFs presented in this chapter are based on rough estimates of freshwater requirements in crop production, in combination with theoretical maximum conversion efficiencies in heat, bioelectricity, and biofuel production. The studies have integrated data from several sources, each adding a degree of uncertainty. Meteorological data, for example, are averages over several years rather than representing data for a specific year and do not reflect annual variations. Calculations of crop water requirements are sensitive to input of climatic data and assumptions concerning the start of the growing season. The data on energy carriers from the literature (Gleick, 1994) give an indication of water requirements but are probably outdated. Therefore results are indicative. However, the differences among the WFs of different energy carriers are so great that they support general conclusions with respect to relative WFs of different types of bioenergy, crops, and countries.

It is worth mentioning that the WF of second-generation biofuels will be higher than the WFs of heat generation because the biomass needs to be converted into biofuel, which will probably have a conversion efficiency of less than 100 percent. How much of the WF of a crop that delivers both food and second-generation energy will be allocated to the energy component depends on the value of the energy derived from 1 kg of harvested crop relative to the value of food coming from the same kilogram of crop.

Our results show that the WFs of bioenergy and hydropower are large. A policy-relevant question is whether (and to what extent) water should be used for food, fibers, or fuel. This is especially relevant in developing countries with increasing populations, such as China and India, where the demand for food will increase. Large biofuel and hydropower programs may need large amounts of water, making the water unavailable for food production. Another issue is the sustainability of energy with large water requirements. Whether the WF related to the production of bioenergy and hydropower is sustainable depends on two criteria: the geographic context and the characteristics of the production process itself (Hoekstra et al., 2011). A WF is unsustainable when the process is located in a so-called hotspot, a catchment where during a certain period of the year environmental water needs are violated or pollution exceeds waste assimilation capacity. For example, when ethanol from sugarcane is produced in northern India, an area where water stress occurs, this is unsustainable. A WF is also considered unsustainable when the WF of the process can be reduced or avoided altogether. One could argue that allocating water to bioenergy or hydropower with large WFs is unsustainable because other renewables (e.g., sun and wind) have much smaller WFs. If the choice is made to produce bioenergy, however, the agricultural practices chosen should produce the feedstock in the most water-efficient way. The reduction of green WFs can be achieved by increasing land productivity. Blue WFs can be reduced with more efficient irrigation or by selecting alternative crops. Gray WFs can be reduced by applying fewer chemicals (e.g., thanks to the use of precision agriculture). We have shown the large differences in gray WFs among countries. An example of this is provided by eastern European countries, which have relatively large gray WFs, indicating an inefficient use of chemicals. Large improvements are possible under such scenarios.

6. Conclusions

The most water-efficient way to generate bioenergy is to use total biomass, including parts without a large economic value, and generate heat. The generation of electricity is the second best option. The production of first-generation biofuels requires about twice as much water per unit of energy as bioelectricity. When comparing different biofuel practices, it is more water efficient to produce ethanol than it is to produce biodiesel. The WF of a typical biodiesel energy crop, rapeseed, is four times larger than the WF of ethanol from sugarcane and seven times larger than the WF of ethanol from sugar beet.

WFs for similar energy carriers can show large variation, depending on the crop type applied, production location, and agricultural practice. The global weighted WF for ethanol increases in the following order: sugar beet, potato, sugarcane, maize, cassava, barley, rye, paddy rice, wheat, and sorghum; it ranges between 60 and 400 m^3/GJ. For sugar beet, maize, and sugarcane, the differences among regions are large. The European Union, northern Africa, and the United States have relatively small WFs for ethanol, whereas eastern Europe has large WFs for ethanol from sugar beet and maize. The global weighted average WF of biodiesel increases in the following order: palm oil (95 m^3/GJ), soybean and rapeseed (400 m^3/GJ), and jatropha (570 m^3/GJ).

When the WFs of bioenergy are compared to the WFs of hydropower or conventional fossil fuels, there are large differences. Hydropower in particular shows a wide range of values for WFs per unit of electricity generated, ranging from 0.5 to 850 m^3/GJ. The WF of heat generated from sugar beet is 25 times larger than the WF of crude oil, while the WF of jatropha biodiesel is 6,000 times larger than the WF of natural gas. A shift toward bioenergy brings a need for more water. A shift toward hydropower can also bring a need for more water, but this depends on the climate and the way in which the power is generated. The results presented in this chapter have led to new insight with respect to the large impact of bioenergy on the use of freshwater resources. This knowledge can make a valuable contribution to research concerning energy needs and freshwater availability in the near future.

5

The challenges of estimating tropical deforestation due to biofuel expansion

YAN GAO
Hokkaido University, Sapporo, Japan

MARGARET SKUTSCH AND OMAR MASERA
Universidad Nacional Autonoma de Mexico, Morelia, Mexico

Abstract

In this chapter, we attempt to assess the impacts of biofuel development on tropical deforestation by analyzing available tropical deforestation and biofuel production data. We identify the main biofuel production hotspots in Africa, South America, and Southeast Asia. We find that this task is extremely difficult because of the following methodological challenges. First, biofuel production may have either direct or indirect impacts on deforestation, and indirect deforestation can take place at both national and international levels. Second, measuring both deforestation and biofuel production accurately is difficult due to the lack of standard definitions and the lack of updated data sets with sufficient spatial resolution and global coverage. Third, many feedstocks used for biofuel production are multipurpose since they are produced for both food and fodder and for fuels; thus decisions on how much feedstock is devoted to any of the uses vary considerably over time, depending on market prices. Assessing the share of biofuel in the deforestation caused by such multi–end use crops is difficult; there are several different ways to allocate this burden. Fourth, deforestation is often caused by multiple drivers, of which in any given area biofuels (or biofuel production) may be just one. This chapter reviews the methodological difficulties in estimating the relationship between biofuel and deforestation in detail and considers both the well-established biofuel feedstocks, such as sugarcane for ethanol (in Brazil and Argentina) and palm oil for diesel (in Malaysia and Indonesia), and the emergent feedstocks, such as jatropha, which is expanding in sub-Saharan Africa, India, and Latin America.

Keywords: biofuel hotspot, deforestation, land use change, remote sensing, tropics

1. Introduction

Plans to expand biofuel production have sparked debates concerning whether biofuel feedstock production threatens food security and reduces land-based income generation and whether it is resulting, or will result, in a growth in deforestation rates (Ravindranath et al., 2009; Schubert et al., 2008). The latter concern is, to a large extent, related to the additional carbon emissions that result from forest clearing but also has to do with broader concerns linked to sustainability in the sense of loss of natural heritage and biodiversity and decreases in the environmental services and goods that forests provide to local populations. This chapter shares these latter concerns and seeks to explore the spatial interactions between biofuel feedstock production and deforestation.

There are several underlying problems when it comes to assessing the implications of biofuel development on land use change and, specifically, on deforestation. The first relates to the availability and quality of recent data on deforestation at the global level and on biofuel production. Biofuel data are problematic as regards both geographical location of feedstock plantations and the level of production. The second problem has to do with the multipurpose nature of feedstocks since most of them are used both for food and fuel consumption (e.g., soya, which is used for food and cattle feed as well as for biodiesel production). The third is linked to the fact that biofuel production may have either a direct or indirect effect on deforestation and that indirect deforestation can take place at the national but also at the international scale. The fourth challenge is that deforestation is often caused by multiple drivers, of which, in any given area, biofuel may be just one. These challenges suggest that making simple spatial correlations between biofuel production and deforestation is likely to be a difficult task. To assess the extent to which it is possible, we conducted a comprehensive review of both global deforestation data and biofuel production areas in Latin America, Asia, and Africa. The study focuses on developing countries and does not analyze or include data from North America or Europe.

The aim of this chapter is to unravel the challenges that arise when estimating tropical deforestation due to biofuel expansion. Section 2 provides a short literature review of the current evidence linking biofuel expansion and deforestation. Section 3 presents the global tropical deforestation and biofuel expansion hotspots that are identified in the literature. Finally, Section 4 discusses in depth the key methodological challenges that arise when attempting to estimate tropical deforestation due to biofuel expansion.

2. Review of the biofuel-induced deforestation literature

There is an important debate going on in the popular media on the links between biofuel development and deforestation. Two contradicting perspectives dominate the discussion. On one side, environmental perspectives, including the Global Forest Coalition, FERN, Greenpeace, and conservation scientists, argue that biofuels will increase greenhouse gas (GHG) emissions (Rosenthal, 2008), destroy tropical forests (GFC, 2010), cause conflicts with local communities (Lacey, 2009), and undermine food security (Demirbas, 2009; Fearnside, 2001; Ribeiro and Matavel, 2009). On the other side, proponents of the biofuels industry argue that in addition to reducing the use of fossil fuels and emissions and providing jobs and income opportunities, biofuels are grown almost entirely on agricultural or pastoral land and thus do not involve deforestation (Goldemberg, 2007). The divide between these two points of view is large, but it is important that the two extreme positions are juxtaposed and discussed.

Brazil, being the largest producer of biofuels among developing countries (see Chapter 6), has been at the center of the biofuel–deforestation debate. In a simplified perspective, some argue that sugarcane expansion in the south of the country is leading to growing expansion of soybean in the center west, which in turn is displacing cattle herds farther into the Amazon region, thus inducing growing deforestation (Nepstad et al., 2008). In contrast, others argue that there is a lack of evidence for this argument and that bioethanol production does not lead to deforestation since more than 85 percent of the planted sugarcane in Brazil is located more than 2,000 km from the Amazon forest (Sawaya and Nappo, 2009). Contradictory arguments also prevail for soybean expansion in Mato Grosso, Brazil. Branford and Freris (2000) conclude that the expansion of soya plantations is a cause of deforestation resulting in various social problems. In contrast, others argue that the Brazilian soya industry has little to do with the clearing of forests and has an important role in promoting regional economic development (Brown et al., 2005; Goldemberg, 2007; Goldemberg and Guardabassi, 2009).

The debate between the pro-biofuel and anti-biofuel camps exists also in the palm oil sector, as evidenced by the report "Palm Oil, the Sustainable Oil" circulated by a support group for the industry (World Growth, 2009). This report categorically denies that oil production causes deforestation or emission of GHG due to land use change. However, the integrity of the report has been heavily criticized as it is based on highly selective or simply biased use of data and facts (Laurance et al., 2010a). Other reports argue that the expansion of oil palm plantations has indeed caused deforestation in tropical countries, especially Malaysia and Indonesia (Friends of the Earth, 2008; Koh and Wilcove, 2008) (see Chapter 10).

It is noteworthy that more balanced views have also emerged in the academic literature regarding the relationships between biofuel development, deforestation, and forest degradation. These nuanced views analyze both the pros and cons of biofuel development, suggesting that – within reasonable limits – expansion of biofuel feedstocks might be possible while protecting forest resources (Demirbas, 2009; Gibbs et al., 2008). In addition, the Roundtable on Sustainable Biofuels, the Roundtable on Sustainable Palm Oil, and the Roundtable on Sustainable Soy have emerged as formal initiatives involving producers, industry, government officials, and experts to actively seek ways in which responsible and sustainable production of biofuels can be promoted. The truth, however, is that to date, little in-depth research has been carried

Table 5.1. *Identified established biofuel hotspots*

Country	Province	Biofuel	Feedstock	Plantation areas[a]	Source
Argentina	Santa Fe	Biodiesel	Soya[b]	3.6 million ha (2006)	Lamers (2006)
Brazil	São Paulo	Bioethanol	Sugarcane	4.2 million ha (2006)	Meloni et al. (2008)
Brazil	Mato Grosso	Biodiesel	Soya[b]	5.7 million ha (2008)	Meloni et al. (2008)
Colombia	Valle de Cauca	Bioethanol	Sugarcane	41,000 ha	Toasa (2009)
Colombia	Zona Oriental	Biodiesel	Palm oil[b]	121,135 ha	FEDEPALMA (2009)
Indonesia	W. Kalimantan	Biodiesel	Palm oil[b]	367,619 ha	Potter (2008)
Malaysia	Sabah	Biodiesel	Palm oil[b]	1,165,412 ha	MPOB (2010)

[a] The area shown for these feedstocks is used for both biofuel production and for food/fodder.
[b] In all these cases, only a small fraction of the total oil produced is actually converted into biodiesel.

out on the spatial links between biofuel development and deforestation at a global scale.

3. Current biofuel–deforestation hotspots

3.1. Main biofuel hotspots at the global level

The identification of biofuel hotspots is based on available secondary information. Well-established biofuel hotspots are presented in Table 5.1. For each selected hotspot, there is information such as the country and province where the hotspot is located, the type of biofuel and feedstock, and information on the size of the plantation areas. The authors found that the information on plantation areas varies in different sources, and we reference here the seemingly most reliable data source. Hotspots include Santa Fe State in Argentina (Lamers, 2006); Mato Grosso State in Brazil (APROSOJA, 2010); São Paulo State in Brazil (Meloni et al., 2008); and Valle de Cauca in Colombia, where sugarcane is produced for bioethanol (Toasa, 2009). Zona Oriental in Colombia is another important hotspot for biodiesel production from palm oil (FEDEPALMA, 2009). In Asia, West Kalimantan, Indonesia, and Sabah, Malaysia, are two important hotspots where palm oil is the main feedstock (MPOB, 2010; Potter, 2008).

Meanwhile, there are also many emerging biofuel hotspots, some of which are presented in Table 5.2. All of the newly emerging biofuel hotspots are associated with jatropha as a biodiesel feedstock. Since jatropha is utilized almost exclusively for biofuel production, the area used for growing biodiesel feedstock corresponds with the entire plantation area. In Latin America, we noted the cases of Minas Gerais, Brazil, and Mexico, where biodiesel production from jatropha is starting on a very

Table 5.2. *Identified emerging biofuel hotspots*

Country	Province	Biofuel	Feedstock	Plantation Areas	Source
Brazil	Minas Gerais	Biodiesel	Jatropha	13,500 ha established, 1.3 million planned (2010)	GEXSI (2008a)
Mexico	Yucatán, Chiapas, Michoacán	Biodiesel	Jatropha	13,000 ha established, 100,000 ha planned (2010)	Field survey
India	Tamil Nadu	Biodiesel	Jatropha	100,000 ha	GEXSI (2008b)
Ghana	Brong Ahafo, Northern Ashanti	Biodiesel	Jatropha	7,000–8,000 ha established, 100,000 ha planned	GEXSI (2008c)
Madagascar	South West	Biodiesel	Jatropha	30,000 ha	Loos (2009)
Mozambique	Gaza, Inhambane Zambezi Delta	Biodiesel	Jatropha	10,000 ha	Ribeiro and Matavel (2009)
Tanzania	Kisarawe	Biodiesel	Jatropha	13,100 ha	Sulle and Nelson (2009)
Zambia	Southern Province	Biodiesel	Jatropha	35,222 ha	Schoneveld et al. (in press)

small scale in the states of Yucatán, Michoacán, and Chiapas (GEXSI, 2008a; field survey, UNAM). In Asia, India is expanding its production of jatropha for biodiesel production (GEXSI, 2008b). In Africa, jatropha production has been identified in Ghana, Madagascar, Mozambique, Tanzania, and Zambia (GEXSI, 2008c; Loos, 2009; Ribeiro and Matavel, 2009; Sulle and Nelson, 2009; Schoneveld et al., in press) (see Chapter 13).

The identified established and emerging global biofuel hotspots are presented in a global map (Figure 5.1). Because no information is available that indicates the exact geographic locations of the hotspots, they are represented by the states where they are located. The biofuel hotspot data are very preliminary. For the cases that use multipurpose feedstocks, the plantation area data do not represent the feedstock used for biofuel but rather the total plantation area of the feedstock.

3.2. Main deforestation hotspots at the global level

This section illustrates the global deforestation hotspot data using MODIS Vegetation Cover Conversion deforestation data produced using MOD44A data, designed and generated at the University of Maryland's Department of Geography (Carroll et al., 2006). These data constitute the best available global map of deforestation fronts currently available. It is an alarm product, to be used as an indicator of changes and not as a means to measure change. As a consequence, the deforestation data used consist of information on "change" and "no change," covering an area somewhat larger than

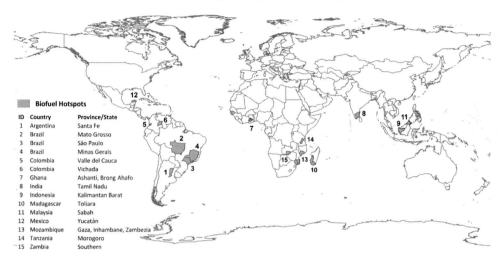

Figure 5.1. A preliminary map of global biofuel hotspots in tropical developing countries at subnational level.[1] (See color plate.)

the tropics (between 30°S and 30°N). The mapping method for deforestation is derived from the original space partitioning method and relies on decision tree classification to determine antecedent vegetation conditions, which are then compared to current vegetation conditions (Zhan et al., 2002).

We built a mosaic based on 68 MODIS images in GeoTIFF format, from which a global tropical deforestation map was derived, as shown in Figure 5.1. This map only provides referential information about the location of deforestation because the resolution of the images is not sufficient to calculate the actual quantity of deforestation taking place at each identified site. A red area indicates that deforestation has been identified at the location but does not represent the size of the area deforested. To facilitate the display, the deforested areas are actually exaggerated, though the true information can be accessed and managed in any GIS software.

Estimates from this map almost certainly underestimate the true situation, giving estimates of deforestation for Brazil as 3,790,000 ha/yr, for Indonesia as 52,000 ha/yr, and for Mexico as 76,000 ha/yr for the period 2000–2005. Compared with data from the Food and Agriculture Organization (FAO) (RFA, 2008; country statistics), these figures are low for Indonesia and Mexico but about 20 percent higher for Brazil. This may reflect the weakness of MODIS data for detecting deforestation, but as noted earlier, the FAO statistics are also not entirely accurate. Figure 5.2 shows that the typical mosaic pattern of deforestation in Indonesia is not well picked up by this methodology, and very importantly, the conversion of natural forest to oil palm plantation in Indonesia and Malaysia is not recognized. This is further illustrated by Figure 5.3.

---

[1] Only the main state or region is shown per hotspot for illustration purposes.

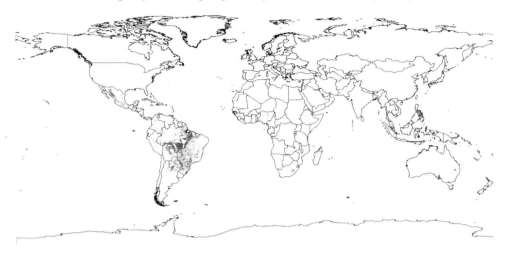

Figure 5.2. Global deforestation map (2001–2005). (See color plate.)

Figure 5.3 shows the deforestation in Sabah, distinguished by satellite images of distinct spatial resolutions. Figure 5.3(a) is the digital global image with high spatial resolution (<5 m) from which the pattern of deforestation was clearly distinguished; the oil palm plantations at different grown stages and the intact forests can also be well distinguished. Figure 5.3(b) is the same area one year later, represented by a Landsat ETM+ image with 30 m spatial resolution. Even with the missing data, shown as the black strips in the image, the deforestation pattern can still be distinguished. Figure 5.3(c) is the same area represented by MODIS MOD09 data with 500 m spatial resolution. The white frame in the three images indicates the area that has been deforested and planted with oil palm at different stages. Outside of this frame, the forests are mainly intact. The detailed land cover type in the red frame is shown in Figure 5.3(c). In this figure, the pattern of the deforestation detected in Figures 5.3(a) and 5.3(b) was nowhere to be seen due to the very coarse spatial resolution, and the detected deforestation area is seriously underrepresented. This helps us understand why the deforestation in Indonesia was not picked up in Figure 5.1.

3.3. Spatial analysis of tropical deforestation and biofuel development

To assess the effects of biofuel development on deforestation, we would need to compare the loss of forest area with the increase in feedstock area devoted to biofuels within the same period. While global deforestation data are available for the period 2001–2005, data on global biofuel hotspots are comparatively much poorer. First, the data on the hotspots are from different dates. Second, the data represent a single point in time, not the change between two different time periods. Third, the location of the biofuel feedstock plantations in those hotspots is unknown. In fact, the jatropha- based biofuel hotspots, such as those detected in Africa, Mexico, and Brazil, are more recent

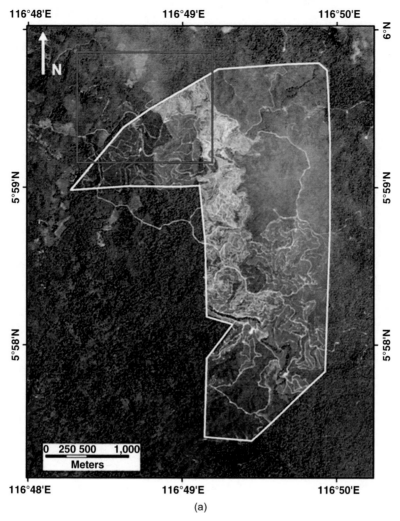

Figure 5.3(a). Deforestation in Sabah, Malaysia, represented by satellite images from three different sensors with different spatial resolutions: image of DigitalGlobe from Google Earth, August 2007. (See color plate.)

than the deforestation map. For all these reasons, an analysis of spatial coincidence at the global level cannot be conducted, though it brings out the possibility to conduct analyses at finer scales.

Looking at the scope of this study from a methodological perspective and considering the difficulties regarding the data, cause–effect mechanisms, product end use, and so on, it appears that the most suitable approach would be to observe land use changes at the required spatial and thematic resolution over a representative sample rather than attempting wall-to-wall global map coverage. This way, the land use changes at local scales can be captured, and if crop plantation area, crop type, and product destiny (i.e., energy or food) during the same period were also observed, in-depth information

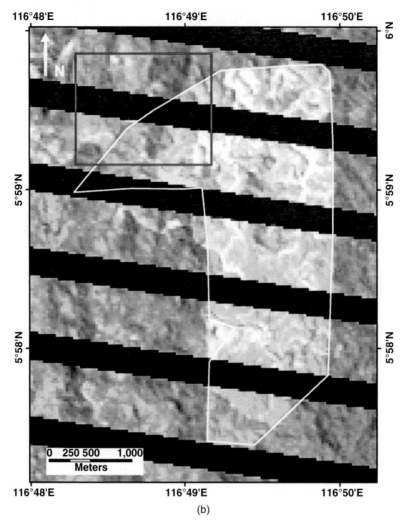

Figure 5.3(b) Deforestation in Sabah, Malaysia, represented by satellite images from three different sensors with different spatial resolutions: color composite image of Landsat-7 ETM+, July 2008. (See color plate.)

can be provided about the developments that are occurring in particular regions where there is a known risk of deforestation.

The impacts of biofuels on deforestation depend greatly on the particular feedstock that is used for biofuel production. For example, in Latin America, sugarcane is generally expanding on lands that were cleared for agriculture a long time ago (see Chapter 6). Thus expanded production of ethanol from sugarcane is unlikely to cause direct deforestation. On the other hand, soya is in general a pioneer crop that is frequently produced on the agricultural frontier either on forestlands cleared for this purpose or in areas cleared for pasture and beef production (see Chapter 6). In Asia, oil palm plantations (Malaysia, Indonesia) are often found in rainforest areas specifically cleared for this purpose (see Chapter 10). Hence, to find the spatial relation between

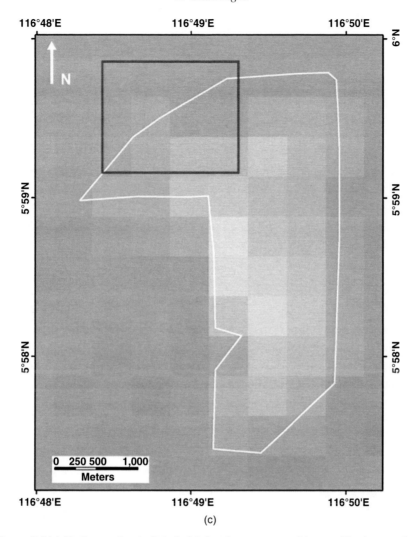

Figure 5.3(c) Deforestation in Sabah, Malaysia, represented by satellite images from three different sensors with different spatial resolutions: color composite image of MODIS MOD09, August 2007. (See color plate.)

deforestation and biofuel development, sample density and strata should be adequate to represent the expanded soya or oil palm plantations.

4. Challenges

4.1. Methodological challenges related to biofuels

Two issues that present serious challenges in estimating the role of biofuel production in deforestation are (1) the lack of information about the location of biofuel production and (2) the fact that many feedstocks used for biofuels have other uses as food or fodder and, in most cases, these other uses tend to dominate.

4.1.1. Limited availability of production data

At present, no good, universal, and easily accessible global databases exist, either on production of feedstock or on production of ethanol or biodiesel at a subnational level. Biofuel is a relatively new commodity, and because of the commercial interests involved, whatever data are available are often not in the public domain.

To correlate biofuel production with deforestation, it would be necessary to work at the subnational level and in spatial terms, and ideally, data would be needed on

1. Different types of feedstock production at a relatively detailed level of disaggregation (e.g., at the municipal level) both in terms of area and crop yield
2. Clear indications of how much of each feedstock in each location is processed into biofuel and how much is used for other purposes (food, fodder, soap, cosmetics, etc.); if such long-term data can be retrieved, then they could be compared to time series data on deforestation

Unfortunately, these data are simply not available. Databases that provide information on the volumes of biofuels processed only have data at a high level of aggregation (usually national totals per year). For example, F. O. Licht (http://www.agra-net.com/portal2/), which is probably the most comprehensive source, provides information on the national production of the major biofuels, and only in a few countries (e.g., Brazil) is it broken down by state. Data at lower levels of aggregation type (district, municipal) are simply not available for most countries.

4.1.2. Multipurpose nature of biofuel feedstocks

Many biofuel feedstocks are multipurpose. For example, palm oil is used to produce both biodiesel and food or cosmetic products; sugarcane is used for both bioethanol and food. Data on feedstock plantation area and yield do not provide an indication of the proportion of the produced feedstock that is used for biofuel purposes. The fuel and nonfuel distinction is crucial, but there are hardly any data available that would enable the spatial identification of the dedicated plantations for the most prominent feedstocks (sugar, soya, palm oil, maize). Crops that are usually intended only for biofuel production, such as jatropha, are an exception.

In addition, data relating to individual biofuel processing plants tend to be limited because of commercial secrecy. It is possible to get data on plant capacity, usually in broad ranges, that is, large, medium, and small, but data on actual production levels are not available. However, a simple count of the number of processing plants listed by country gives some notion of the level of biofuel activity. The location of the biofuel plants is only a proxy indicator of biofuel production, and thus it is difficult to relate to deforestation. For example, in Brazil and Argentina, there is a concentration of biofuel production plants near the coastal cities. This has probably more to do with the centers of demand for biofuels and opportunities for their export, as feedstock can be transported overland to the processing centers from various parts of the country. The location of biofuel processing plants is in no way a reliable indicator of the location of feedstock production. Moreover, as the quantity of biofuel produced at individual plants is rarely available, it is not possible to calculate backward the quantity of the feedstock in the supply area.

The uncertain link between the location of feedstock cultivation and the location of processing plants is a result of the variable procedures adopted during the processing of different biofuel feedstocks. Biofuel feedstocks such as sugarcane, soya, and palm oil follow different paths in their processing. Sugarcane, if used to produce fuel, is directly pressed to produce a solution of sugar, which is then converted to ethanol, usually at the same plant. Because of the weight of the cane, it is usual that sugar ethanol is processed close to the feedstock production areas. Soya, conversely, first needs to be pressed and separated into soy meal and soy oil (roughly 80%–20%). The oil is then processed to produce biodiesel. Because soy oil is relatively compact in terms of commercial value per metric ton and can be transported relatively easily, the processing plants for biodiesel may not be located in the same area as the crushing plants.

A further complication relating to the multipurpose use of biofuel feedstocks is how to calculate the burden of the biofuel in the total deforestation caused by the feedstock. This can be illustrated by the case of biodiesel from soy in Mato Grosso, see Box 5.1.

4.2. Challenges related to measuring tropical deforestation at the global scale

Deforestation is a complex process, and getting reliable deforestation estimates at the global level remains a challenge. The main issues to deal with are related to (1) the different definitions of deforestation, (2) the poor reliability of the available global deforestation data, (3) the limited time series data for recent years, and (4) the restrictions in the spatial resolution of the satellite images for mapping deforestation at the global level. Finally, there is difficulty involved in attributing deforestation dynamics to particular drivers.

4.2.1. Data in global databases and lack of standard definitions about deforestation

There is as yet no universally accepted definition of deforestation, which makes it difficult to make comparative analyses across countries. The United National Framework Convention on Climate Change defines deforestation as occurring when the canopy cover of a forested area falls below a minimum threshold already selected by each country, in the range between 10 and 30 percent, with some attendant height and area thresholds (Achard et al., 2007). The U.S. Department of Agriculture Forest Service considers deforestation a nontemporary change of land use from forest to other land use or depletion of forest crown cover to less than 10 percent. Clear cuts (even with stump removal) if shortly followed by reforestation for forestry purposes are not considered deforestation. The difficulty is therefore to distinguish those losses that are temporary and part of a sustainable cycle from those that are permanent and contributing to long-run increased atmospheric carbon dioxide.

Box 5.1 Allocating the deforestation burden to biofuels – A case study of biodiesel from soya in Mato Grosso, Brazil

In cases where feedstock is not used solely for biofuel production but for several activities, allocating the share of the different end products to overall deforestation poses some methodological problems. This is illustrated here for the case of soya in Mato Grosso, Brazil.

It has been estimated that between 16 and 20 percent of the deforestation in Mato Grosso is attributable to clearance for agricultural purposes; the rest is due to grazing (Morton et al., 2006; Nepstad et al., 2009). We make the conservative estimate that 100 percent of the cleared land in the region will be planted with soy, as it is by far the most important crop in the region.

Approximately 18 percent of the total primary product (i.e., of the soy bean) by weight is oil; the rest is cake or meal, primarily used for cattle feed. However, not all of the oil is used for biodiesel. Much is processed for cooking oil or cosmetics. In the case of Brazil, we estimate the maximum share of soy oil that is processed into biodiesel to be about 36 percent (on the basis of the quantity of soy oil that is needed to meet Brazil's biodiesel 5% mix requirement; Hall et al., 2009). Roughly one-quarter of all Brazil's soy is produced in Mato Grosso, so it is reasonable to estimate that 8 percent of Mato Grosso's soy oil is destined to be processed into biodiesel.

There are at least three possible ways to allocate the burden of deforestation to this biodiesel:

1. *Deforestation allocated on the basis of the share of primary product weight.* In this case, the share of deforestation caused by biodiesel is 8 percent of 18 percent, or about 1.5 percent of the deforestation caused by soy; this would represent less than 0.5 percent of all deforestation in the state.
2. *Deforestation allocated on the basis of economic value.* It can be argued that deforestation is more likely to be affected by the relative economic value of the different products rather than by their relative weight and that therefore the shares of deforestation should be weighted by market prices. Per metric ton, soy oil trades at approximately three times the value of soy meal (the ratio was USD 599/ton to USD 209/ton in 2006 and USD 1,423/ton to USD 466/ton in 2008, according to World Bank commodity trading records). Using this as the basis for the calculation, the proportion of the soy deforestation that should be allocated to biodiesel would be 8 percent of 75 percent, or 6 percent of the deforestation attributable to soy; this is equivalent to not more than 1.25 percent of all deforestation in Mato Grosso.
3. *Deforestation allocated in terms of actual area sown with soy.* A more conservative estimate would suggest that soy oil and soy meal are in fact inseparable; without the one, there is none of the other. As a result, the total area sown for soy should be considered deforested for soy oil. On this basis, the deforestation burden attributable to biodiesel would be 8 percent of the total soy area, or 1.6 percent of all the deforestation taking place in the state of Mato Grosso.

There is no accepted methodology for making this calculation. The three alternatives are presented here simply to illustrate the difficulties and the fact that the result – the area said to be directly deforested as a result of biofuel feedstock cultivation – can vary by a factor of 3, depending on which method is selected. However, whichever method is used, it appears that biodiesel, while certainly contributing to deforestation, plays a rather small role when compared to other uses of soy and to nonsoy causes of deforestation in Mato Grosso.

During the last 50 years, FAO reports have been the main, and often the only, reference for discussion and analysis of forest and deforestation data at the regional and global levels. However, the FAO data have uneven quality and inconsistent definitions across nations, which makes the data difficult to compare and verify (Drigo et al., 2009; Jepma, 1995; Rudel et al., 2005; Stokstad, 2001; Zahabu, 2008).

Moreover, FAO's data on forest cover and deforestation are reported as aggregate figures at the national level. On the basis of data from FAO (2006b), it is estimated that about 11.8 million ha per year were lost worldwide during the period 2000–2005; 80 percent of total deforestation took place in tropical Africa and tropical America, and the global deforestation rates have remained almost constant between 1990 and 2000 and over the period 2000–2005 (Drigo et al., 2009). In fact, deforestation rates have increased within tropical Asia and Latin America, while they have decreased for Africa. However, the estimation of deforestation rates at country level offers little insight into the causes and mechanisms behind this phenomenon. In particular, it is not possible from the FAO database to ascertain where in the country deforestation is occurring, which is critical when attempting to relate deforestation to biofuel production.

The data are, moreover, out of date, and it is to be expected that changes have occurred in the last five years. In particular, there was a reduction in deforestation rates in Brazil in this period. The availability of new FAO pantropical data based on classified Landsat scenes provides a spatial view of where deforestation has occurred in the past and has recently been utilized (Gibbs et al., 2010) to show that between 1980 and 2000, 55 percent of new agricultural land came from intact forests and 28 percent came from disturbed forests. Unfortunately, these data do not throw light on the changes that have occurred in the last decade. Hansen et al. (2008) used a probability-based sampling method to estimate gross forest clearance (i.e., not taking into account any regrowth or new plantation) and found, like Gibbs et al. (2010), that deforestation tends to be highly concentrated. The main deforestation hotspots identified include the Amazon basin and Southeast Asia (Indonesia and Malaysia).[2]

---

[2] Also included are parts of the boreal forests of the Northern Hemisphere. These, however, are not included in the scope of our analysis.

4.2.2. Limited availability of global maps and images

Before remote sensing techniques became widely available, field surveys were the only way to obtain accurate spatially explicit deforestation data. Forest inventories of this kind are expensive, laborious, and time consuming. Remote sensing has been commonly used in most countries for the last 20 years, and the increasing availability of remote sensing images and techniques facilitates the production of global deforestation maps, independent of national assessments. This requires, first, that satellite images with appropriate spatial, spectral, and temporal resolutions be selected. Higher spatial resolution images, on one hand, cover smaller areas with lower spectral and temporal resolutions and are expensive to obtain. On the other hand, lower spatial resolution images cover larger areas at lower cost and have higher spectral and temporal resolutions. So, for wall-to-wall mapping of deforestation at the global level, the use of lower spatial resolution data seems to be the only viable alternative (notwithstanding the limitation posed on the observable land cover changes). The main implication of image selection with a given resolution relates to the minimum area that can be identified as deforested. For example, images with a spatial resolution of 250 m cannot detect Earth objects that are smaller than 30 ha. For reasons of cost, mapping deforestation at the global level is generally carried out using relatively coarse spatial resolution images, which can indicate where large-scale clearance has taken place. It never will pick up smaller or scattered clearances. For example, the diffuse increase of small-scale farming and short-fallow shifting cultivation goes largely undetected in coarse-resolution data, although in many tropical countries of Africa and Asia, this is the main process of deforestation (Drigo et al., 2009; FAO, 1996). Even more serious are the constraints of coarse-resolution data on the mapping of forest degradation. For example, reduction in the canopy density, even if significant, is extremely difficult to map with Moderate Resolution Imaging Spectral Radiometer (MODIS) imagery, which has spectral bands with spatial resolutions of 250 m and 500 m. Other forms of degradation, in which there is a biomass loss under the canopy, are virtually impossible to detect.

Nevertheless, there are some interesting products in this respect, such as the global tree cover percentage data and the change maps based on MODIS satellite data (Hansen et al., 2005). The Global Land Cover Facility at the University of Maryland mapped global deforestation from 2001 to 2005 (GLCF, 2010). The World Resources Institute (WRI) mapped tree cover change in four tropical locations, Brazil, Cambodia, central Africa, and Indonesia, from 2000 to 2006 (Adam et al., 2007). The particular advantage of the WRI approach is that they have used MODIS images (500 m) to identify the locations of deforested areas and then zoomed in on these locations using higher spatial resolution images, to allow the calculation of the deforestation areas. But while these data are very useful for the four locations mentioned, they do not constitute a global data set.

Besides the data selection, the study of land cover changes over large and diverse landscapes is not a simple task. Except for particularly simple conditions such as, for instance, the large and squared clearings that are common in the Brazilian Amazon, land cover changes are small, elusive events whose reliable detection requires evaluation processes far more rigorous than normally accepted for conventional mapping purposes. The comparison of two land cover maps independently produced over the same areas will inevitably give differences that include true changes as well as other differences resulting from the different interpretation procedures adopted in each mapping process (Drigo et al., 2009).

4.2.3. *Difficulties in numerically (and spatially explicitly) ascribing deforestation to particular drivers*

The drivers of deforestation are very diverse and may vary by (and within) countries. A number of important studies have attempted to generalize and pull together large numbers of local studies (Angelsen and Kaimowitz, 1999; Geist and Lambin, 2002). This work is hampered, however, by the fact that most of the information on drivers is not quantitative. As a result, few direct quantitative correlations can be made linking certain quantities of deforestation to particular activities. Angelsen and Kaimowitz (1999) showed that, when looking at proximate causes, deforestation is often associated with the presence of more roads, higher agricultural prices, lower wages, and a shortage of off-farm employment. Also, they considered it likely that policy reforms included in the current economic liberalization and adjustment efforts would increase the pressure on forests. They pointed out, however, that many research studies have adopted poor methodologies and low-quality data. This makes the drawing of clear conclusions about the role of macroeconomic factors, and that of other underlying factors inducing deforestation, difficult. Geist and Lambin (2002) identified four broad clusters of direct causes: agricultural expansion, wood extraction, infrastructure extension, and other factors. Besides the direct causes, they found that underlying economic factors are prominent driving forces for tropical deforestation (81%); institutional factors are involved in 78 percent of the cases of deforestation, and technological factors are involved in 70 percent. In addition, cultural, sociopolitical, and demographic factors are relatively less important drivers of deforestation and have different effects in different regions. They concluded that there is no universal link between cause and effect in analyzing the drivers of deforestation.

The causes and driving forces are often region specific, which means that deforestation dynamics are shaped by geographical and historical contexts. As Hansen et al. (2008), in their update on global deforestation, make clear, understanding these proximate drivers is crucial not only to understanding how to deal with deforestation but also to understanding where it is likely to occur in the future. These reviews, however, do not examine the role of biofuels in deforestation. The principal reason for this is that biofuel development (with the exception of Brazil) only started in most places in the last five years, while the studies are based on data from the 1990s.

4.3. Methodological challenges related to the direct and indirect land use change effects of biofuel production

Biofuel production could result in both direct and indirect land use change. Given that biofuel feedstocks are essentially crops, direct land use change occurs when biofuel feedstock production is directly established on forest land, clearing existing forests. To analyze the direct spatial relation between deforestation and biofuel development, both the deforestation data and the expansion of the biofuel feedstock data need to be at the same scale and for the same period of observation.

Indirect land use change (iLUC) occurs, for example, when the establishment of biofuel displaces food or pasturelands for livestock production, which then moves to other regions and is responsible there for deforestation (see Chapter 6 for the case of Brazil). These indirect impacts may take place even on different continents (Kim et al., 2009).

There have been several recent attempts to model and quantify the iLUC effects, including GHG emissions from the Agriculture Simulation (GreenAgSiM) model (Dumortier et al., 2009; Searchinger et al., 2008), the Food and Agricultural Policy Research Institute (FAPRI) model (Fabiosa et al., 2009), and the Global Biomass Optimization Model (GLOBIOM) (Havlík et al., 2011; Schneider and McCarl, 2003), among others.

Calculating the scale of iLUC is problematic given the complexities of the economic and social systems that connect biofuel production with land conversion throughout the world. Searchinger et al. (2008) estimated that over a 10-year period, the allocation of 12.8 million ha of U.S.-grown maize to produce ethanol in the United States would result in a need for 10.8 million ha of new cropland around the world. This is expected to involve mostly forest clearance. The iLUC emissions alone are higher than estimated emissions from the gasoline the biofuel would replace. This study has been widely contested on grounds of the many uncertainties in the modeling (e.g., Dumortier et al., 2009; Liska and Perrin, 2009). Dumortier et al. (2009) predicted that the payback period of corn ethanol's carbon debt was sensitive to assumptions concerning land conversion and yield growth and can range from 31 to 180 years. However, although the exact size of the iLUC impacts may be in doubt, other studies back up the general trend and direction of the findings. Delucchi (2003) estimated that adding iLUC emissions originating from within the United States to the cumulative emissions intensity of corn ethanol would significantly increase their GHG intensity. Hertel (2008) suggested that 57 billion L/yr of additional corn ethanol produced in the United States would result in a global increase in land clearance of 4 million ha.

Most of the models so far developed are econometric and do not predict the *location* of the iLUC with any specificity. One exception is a Brazilian study by Lapola et al. (2010), which used a spatially explicit modeling approach and found that while direct LUC from the expansion of biofuel production in Brazil to meet its 2020 targets is very small, the iLUC may cause a forest loss of 12.9 million ha in the period 2003–2020, mostly due to the expansion of pasture lands into forested area, and have mapped these areas. Most of the direct LUC occurs from rangeland to soybean in the state of Mato Grosso and from rangeland to sugarcane in the state of São Paulo, moving livestock production farther toward the Amazon, where it causes deforestation (iLUC).

The importance of iLUC effects has been acknowledged by policy makers. So far, estimates depend on complex modeling approaches, and the models have not been empirically tested. In particular, estimating the spatial extent of iLUC effects of biofuel production is still very challenging (Brinkmann Consultancy, 2009; Kammen et al., 2007; Mathews and Tan, 2009). However, most simulations suggest that iLUC is not negligible and will need to be included in future analysis.

5. Conclusions

Biofuels are being produced on a large scale in only a limited number of locations at the moment, but if production were to increase, the impact on deforestation would have to be seriously taken into consideration. The global analysis presented in this chapter indicates that the relationship between biofuel development and tropical deforestation is complex and difficult to pin down in spatial terms. Limited data availability, lack of time series with sufficient resolution at the global scale, the multipurpose nature of many feedstocks, and the very recent boost in biofuel production in most countries and regions preclude a quantification of the phenomenon. In addition, the fact that biofuels have both direct and indirect impacts on land use change makes this task even more complex. The primary problems lie with the available data, particularly on biofuels (data on biofuel have such low geographical precision that it is not possible to link them to maps of deforestation). However, the tools (interpretation of MODIS images) are also problematic, and as noted, global models of indirect land use change due to biofuels have not managed to capture the spatial dimension of this process.

Looking at the scope of this study from a methodological perspective and considering the difficulties as regards data, cause–effect mechanisms, product end use, and so on, it appears that the most suitable approach would be to observe land use change at the required spatial and thematic resolutions using representative samples. We have shown that wall-to-wall global map coverage is virtually impossible to obtain. However, using samples, land use changes at different scales can be captured. If increments of crop area, crop type, and product density are also observed in these sample areas, the plausible relation between deforestation and feedstock production could be monitored quantitatively. This local-leveled analysis can provide in-depth information about biofuel developments that are occurring in particular regions with high deforestation risk.

Acknowledgments

This chapter was connected with an EC-funded project titled "Bioenergy, Sustainability, and Trade-offs: Can We Avoid Deforestation While Promoting Bioenergy?" which is managed by the Centre for International Forestry Research (CIFOR). The views expressed are those of the authors and do not reflect the official opinion of the European Union. The contributions from Dr. Rudi Drigo, FAO Italy, and from Dr. Pablo Pacheco, CIFOR, are highly appreciated. Thanks also go to José Antonio Navarrete Pacheco MSc, for elaborating Figures 5.1 and 5.2.

Part Two

The case of Brazil

6

The Brazilian bioethanol and biodiesel programs: Drivers, policies, and impacts

ALEXANDROS GASPARATOS
Oxford University, Oxford, UK

MATTEO BORZONI
Scuola Superiore Sant'Anna, Pisa, Italy

RICARDO ABRAMOVAY
University of São Paulo, São Paulo, Brazil

Abstract

This chapter introduces the origins, evolution, and impacts of the Brazilian bioethanol and biodiesel programs. The economic, environmental, and social impacts of both programs are critically discussed alongside their initial targets. The discussion of the bioethanol program aims to capture the main lessons learned from the Brazilian experience in order to inform the development of bioethanol schemes around the world, particularly considering the gradual recognition of ethanol as an international market commodity. Additionally, special attention is paid to the reasons behind the apparent failure to promote biodiesel as an effective strategy toward the diversification of biofuel feedstocks and the benefit and empowerment of poor family farmers.

Keywords: Brazil, soybean biodiesel, sugarcane bioethanol

1. Introduction

Brazil is a global agricultural powerhouse with abundant fertile land, an already established agricultural infrastructure, and favorable environmental conditions that allow the cultivation of a large diversity of crops. It is of no surprise that Brazil has chosen to aggressively pursue the production of first-generation biofuels. In certain respects, Brazil has been the pioneer of large-scale liquid biofuel production and use as a substitute for conventional fossil fuels. In 2008, Brazil was the second largest producer of bioethanol (behind the United States) and the fourth largest producer of biodiesel (behind Germany, France, and the United States) (IEA, 2011a). Considering the impressive progress of its biodiesel program, Brazil will most likely overtake these countries in the near future (see Chapter 1).

Although Brazil started experimenting with ethanol fuel in the 1920s, its ambitious bioethanol program has been, for the most part, a response to the mounting energy security concerns in the aftermath of the first oil crisis. Several sources suggest that the first energy crisis motivated the Brazilian government to assess the potential of local renewable resources (e.g., sugarcane) as an alternative to conventional transport fuel, subsequently promoting policies to spur their development (Puppim de Oliveira, 2002).

The first major policy initiative for the promotion of bioethanol was the 1975 Proálcool program. Since then, the Brazilian bioethanol sector has radically transformed on both the technical and the institutional levels. Major landmarks included the abolition of government subsidies in the early 1990s, the introduction of the flex-fuel vehicle (FFV) in 2003, and the designation of Brazilian sugarcane as an advanced biofuel by the U.S. Environmental Protection Agency (EPA) in 2010.

Recently, the Brazilian government has enacted a 20 percent minimum to 25 percent maximum blending mandate (E20–E25)[1] (Ministry of Agriculture Portaria no. 143 of 2007). This makes Brazil the only country in the world that has succeeded in substituting significant quantities of conventional fossil fuel with biofuels.

In fact, for several interconnected reasons, the Brazilian bioethanol experiment can be seen as a success (e.g., Fischer et al., 2009) that several other countries have attempted to replicate (see Chapter 12). Most recently, the Brazilian government's strategic plan is to transform ethanol into an internationally traded commodity that can enter, under favorable terms, into protectionist markets such as those of the United States, the European Union, and Japan (Franco et al., 2010).

Brazil's biodiesel program, conversely, has a much shorter history. Even though biodiesel research began as early as the 1970s, it was not until late 2004 that the National Program for the Production and Use of Biodiesel (Programa Nacional de Produção e Uso de Biodiesel; PNPB) was established. The PNPB seeks to achieve three main objectives:

1. To set up a biodiesel production system that will guarantee competitive supply, quality, and price
2. To further promote energy security through feedstock diversification
3. To strengthen family farming

This suggests that PNPB's motivation revolves, more than anything else, around energy security and rural development. The initial target set by the Brazilian government was a 5 percent substitution of conventional diesel by biodiesel (B5) by 2013. However, the PNPB was taken up so quickly and effectively that this initial target was

---

[1] The blending mandate is linked to sugarcane production and can change according to sugarcane yield. For example, in 2010, the blending mandate decreased from 25% to 20% for three months as a result of low sugarcane production and high ethanol prices.

attained by January 2010 (2.4 billion L of biodiesel), and Brazil is now moving fast toward becoming one of the world's preeminent biodiesel producers.

The impressive penetration of bioethanol and biodiesel into Brazil's national energy system and the even more ambitious plans for biofuel expansion present both opportunities and potential pitfalls. Despite certain proven environmental and socioeconomic benefits, both the bioethanol and biodiesel programs have been linked to negative impacts on the environment and society.

From this starting point, the aim of this chapter is to provide an overview and a critical discussion of the main drivers, policies, and impacts of the Brazilian bioethanol and biodiesel programs. Considering the already significant literature on bioethanol production and use in Brazil, this chapter does not aim to provide a comprehensive review of the literature. Instead, this chapter seeks to identify bioethanol's main impacts and discuss them alongside key emerging issues, including Brazil's attempt to create an international ethanol market. The impacts discussed are mainly those observed in the state of São Paulo, which is the hub of Brazilian ethanol (see Chapter 7 for an in-depth discussion of bioethanol's impacts in Brazil's northeastern region). Conversely, owing to its relatively brief existence, there is little literature relating specifically to PNPB's impacts, so this chapter provides a rather comprehensive review of these impacts.

2. The bioethanol program

2.1. History and main policies

The 1975 Proálcool program was the first major policy initiative to promote ethanol use in the transport sector. Between 1976 and 1992, the Brazilian government made mandatory the blending of ethanol with gasoline in different ratios ranging from 10 to 22 percent (E10–E22). To meet these early blending mandates, the first series of policies aimed at increasing distillery capacity and fuel conversion rates by providing massive subsidies to facilitate investment in sugarcane plantations, mills, and distilleries (Puppim de Oliveira, 2002). Between 1975 and the mid-1980s, ethanol was delivered to Petrobras[2] and, through the provision of subsidies, was added to gasoline and sold as a fuel for the new fleet of vehicles that was being developed by Brazil's automotive industry.[3]

This intervention was decisive in that it enabled the sector to achieve a clearly descending learning curve (Goldemberg et al., 2004a) and obtain the technical competence that characterizes it today. Bermann et al. (2008) estimate that in the first two decades of the Proálcool program, the sector's overall investment amounted to

---

[2] Petrobras (Petroleo Brasileiro S/A) is a public corporation under federal government control and the largest energy company in Brazil. Petrobras held the oil sector monopoly until 1997.

[3] During the 1980s, Brazil was the only country in which a large part of the car fleet ran exclusively on ethanol.

USD 11.7 billion, of which USD 7.4 billion came from the Brazilian government, largely in the form of subsidized credits. Plummeting fossil fuel prices in the 1980s made the program extremely costly as the Brazilian state had to cover for the difference between a declining gasoline price and the price of alcohol. This caused the production of Brazilian alcohol-powered cars to drop to a minimal level (UNICA, 2011a). Gradually, the role of the Brazilian government shrank. The closing of the Sugar and Alcohol Institute, in 1990, marked the end of the government's direct monitoring of mill–supplier relations. As a consequence, at that point the government essentially ceased to directly determine the rules that govern the sugar and ethanol market.

Today Brazilian bioethanol does not receive any public subsidies, with rural production credit coming mainly from the private sector. Even though there is still strong integration with Petrobras, the high fossil fuel prices and the ethanol sector's significant productivity gains no longer require Petrobras to subsidize the product to allow its financial viability. The role of the government is now limited only to financing programs offered by the National Bank for Economic and Social Development (Banco Nacional do Desenvolvimento).

The bioethanol sector is considered to be among the most attractive sectors for new investments in the Brazilian economy, despite the 2008 economic crisis. According to Wilkinson and Herrera (2008a), foreign capital investments in the bioethanol sector come not only from companies that are already involved in the sugar and alcohol business (e.g., Tereos and Louis Dreyfus, Cargill) but also from consortiums and funds with no previous links to the industry (e.g., George Soros, Vinod Khosla). In the last decade, the involvement of foreign capital in the sector has increased from almost zero to 25 percent of total cane crushed (Pellegrini and Oliveira, 2011).

2.2. Production trends

Sugarcane and ethanol production have increased by 155.8 percent and 77.6 percent, respectively, since 1990 (UNICA, 2011b, 2011c). Even though ethanol production remained constant for the greater part of the 1990s, it experienced a drop between 1998 and 2001 before starting to increase, particularly since 2003, when the FFV was introduced (Figure 6.1).

In 2008, the country's southeastern region accounted for 69.8 percent of sugarcane and 69.0 percent of ethanol production. Other important producing areas are the Northeast (11.1% of sugarcane; 7.0% of ethanol) and the Center West (11.0% of sugarcane; 14.8% of ethanol) (UNICA, 2011b, 2011c). On the state level, the state of São Paulo is by far the most important sugarcane and ethanol production center, representing 60.9 percent and 59.0 percent of national production, respectively (UNICA, 2011b, 2011c). According to the Agricultural Census of Brazil, in 2006, nine of the top 10 sugarcane-producing municipalities were located in the state of São Paulo (IBGE, 2009a).

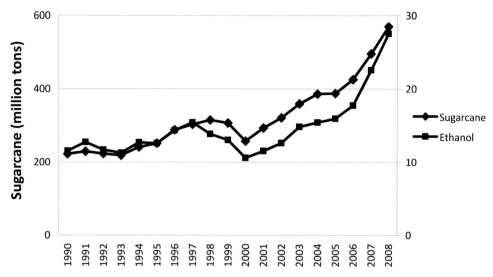

Figure 6.1. Sugarcane and ethanol production since 1990. *Source*: UNICA (2011b, 2011c).

The already impressive sugarcane and ethanol expansion is expected to accelerate in the next decade. Data from the Agricultural Census of Brazil show that in 2006, sugarcane roughly occupied 6.5 million ha (IBGE, 2009a). In 2020, sugarcane production is expected to occupy more than 14 million ha (Sachs, 2007).

It is interesting to note that contrary to cereals and oleaginous crops (refer to Section 3), sugarcane is a highly perishable product that needs to be processed immediately after harvest. This technical limitation, compounded by Brazil's prevailing land tenure structure, explains to a large degree why two-thirds of sugarcane is produced in integrated crop–cattle farming units owned by the sugarcane mills themselves, while only one-third is produced by independent suppliers.

Actually, these producers' independence remains, in most circumstances, theoretical. In fact, the mills lease the independent producers' lands, incorporate them into their existing land, and undertake all productive operations, from sowing to harvesting. In sugarcane production, the space occupied by family farming is practically zero, contrary to what is observed in the production of grains and even in parts of the cattle ranching industry (e.g., dairy). This comes in contrast to other developing nations, such as India, where sugarcane is grown on small establishments. In Brazil, the predominance of the *latifundium* (large plantation) in the sugarcane sector is almost absolute. An indication of this is the fact that the cattle and crop areas in which the mills operate are constantly consolidating. A mill's average area in São Paulo increased from 8,000 ha (in 1970) to more than 12,000 ha presently, with several plantations occupying areas that range between 40,000 and 50,000 ha (Veiga Filho and Ramos, 2006). The average size of sugarcane plantations larger than 20,000 ha was 31,000 hectares in 2003, up by 9 percent since 1995. Meanwhile, during the same period, the average size of plantations smaller than 1,000 ha decreased from 476 ha to 376 ha (Veiga Filho and Ramos, 2006). This consolidation occurs not only through land acquisitions but also through the leasing of land formerly worked by independent farmers.

It should be noted that the largest Brazilian sugarcane group owns 23 mills and roughly represents 10 percent of the total national production, with the largest 30 groups owning 91 mills and representing 50 percent of national sugar production (Pellegrini and de Oliveira, 2011). This suggests that despite a certain consolidation, there are still some small groups in the sector.

Apart from sugarcane bioethanol, there is also the potential for producing second-generation bioethanol in the country. Even though there is currently no commercial production of second-generation bioethanol, and given its relative low economic competitiveness when compared to first-generation practices, there are numerous R&D efforts from different Brazilian institutions (Furtado et al., 2011; IEA, 2010a; Soccol et al., 2010). The most promising feedstocks include sugarcane residues (bagasse, tops, and leaves), other agricultural residues (from corn, soybeans, rice, etc.), and residues from the forestry sector (IEA, 2010a). Of these, sugarcane residues provide perhaps the largest readily available feedstock for second-generation biofuel production in the country. However, the growing importance of bagasse for electricity cogeneration (see Section 2.3.1 and Chapter 12) will play a major limiting factor to second-generation bioethanol production from this resource. Agricultural residues (particularly from soybeans) can also play an important role, but their transportation from rural areas will pose a great challenge (IEA, 2010a).

2.3. Energy and environmental impacts

2.3.1. Energy and climate

Several life cycle analyses (LCAs) have shown that Brazilian bioethanol exhibits some of the highest energy returns on investment (EROIs) when compared to other biofuel production practices around the world (see Chapter 1). In some cases, the EROI can be as high as 9.3 (Boddey et al., 2008), 8.5 (Macedo et al., 2004; Smeets et al., 2008), 8.3 (de Castro Santos, 2007), and 8.2 (Pereira and Ortega, 2010) (see Chapter 1). Even when taking into account some of Brazil's most energy-intensive biofuel production practices, Brazilian bioethanol exhibits higher EROIs than most other first-generation biofuel production practices, for example, 3.9 (Oliveira et al., 2005; Smeets et al., 2008) and 3.1 (Oliveira et al., 2005; Smeets et al., 2008) (see Chapter 1). A meta-analysis of different biofuel LCAs has shown that Brazilian sugarcane ethanol also exhibits very high fossil energy improvements, sometimes well over 90 percent, when compared to other biofuel practices around the world (Menichetti and Otto, 2009).

Electricity cogeneration from bagasse burning in sugar mills is expected to further boost energy gains from Brazilian bioethanol in the future. Bagasse is the remains of sugarcane stalk after it is crushed to extract sugarcane juice. As a by-product of sugar and ethanol production, bagasse is traditionally burned in high-pressure boilers to provide electricity, which is primarily used in sugar mills (Pellegrini and de Oliveira, 2011) (see also Chapter 12). However, bagasse burning is also becoming a popular source of electricity generation supplied in the national grid.[4] In 2010, 6.3 GW of electricity were produced through bagasse burning, of which 75 percent was used by the sugar mills themselves and the rest sold to the national grid (Pellegrini and de Oliveira, 2011). As a direct result, Brazilian sugarcane bioethanol exhibits some of the highest greenhouse gas (GHG) savings when compared to other first-generation practices (Menichetti and Otto, 2009; Zah et al., 2007) (see also Chapter 3). As mentioned (Section 2.2), bagasse is targeted as a potential feedstock for second-generation ethanol production (Botha and von Blottnitz, 2006; Cardona et al., 2010). However, any bagasse directed toward second-generation biofuel production would no longer be available for electricity generation, thereby diminishing bagasse's contribution as a source of electricity.

If direct and indirect land use and cover change (LUCC) is factored into the LCAs, Brazilian bioethanol can incur carbon debts. Fargione et al. (2008) determined that, for sugarcane ethanol produced in the Cerrado woodland, the time required to repay the carbon debt would be 17 years. RFA (2008) reports a 3- to 10-year payback time for sugarcane bioethanol from agricultural lands and 15–39 years from previously forested lands. Lapola et al. (2010) have calculated that by 2020, sugarcane expansion in Brazil might create a carbon debt of up to 44 years, mainly due to indirect LUCC (direct LUCC will also contribute but not significantly). In their model, Lapola et al. (2010) suggest that replacing rangeland in the south of the country with sugarcane cultivation might push the rangeland frontier into the Amazon and cause significant deforestation, subsequently increasing GHG emissions. Conversely, ICONE (2009) conducted a large-scale modeling exercise using the Brazilian Land Use Model and predicted that there would be no significant future deforestation due to sugarcane expansion in the Southeast.[5]

As a result of these relatively low carbon debts and high GHG savings (consistently over 50%, including direct and indirect land use change emissions), the EPA recently designated Brazilian bioethanol an "advanced biofuel" (EPA, 2010).

2.3.2. Air quality

Bioethanol use can have positive impacts on ambient air quality. Bioethanol's wide diffusion in the transport system has been partly credited for air quality improvements in the city of São Paulo, the country's largest metropolitan area (Goldemberg, 2008).

---

[4] Bagasse use in the Brazilian energy system has increased by almost 140 times since 1970 (MME, 2011).

[5] Deforestation rates have been falling since 2004. In fact, to a certain extent, the focus on deforestation prevention efforts seems to be found in the discussion surrounding forest law rather than the bioethanol debate.

The rapid introduction of FFVs since 2003 has led to the gradual dephasing of older, more polluting and less energy efficient vehicles.

However, certain production practices can be significant emitters of atmospheric pollutants. For example, Martinelli et al. (2002) have identified sugarcane burning[6] as a major source of atmospheric pollution in certain parts of São Paulo State, and particularly of particulate matter with aerodynamic diameters of < 2.5 μm ($PM_{2.5}$) and <10 μm (PM_{10}) (Cançado et al., 2006; Castanho and Artaxo, 2001; Lara et al., 2005; Martinelli et al., 2002), polycyclic aromatic hydrocarbons (PAHs) (Martinelli and Filoso, 2008), and NO_x (Oppenheimer et al., 2004). Respiratory illness–related hospital admissions are increasing in parts of the state where sugarcane burning is still widespread with such admissions being more prevalent for children. Two to three times more hospital admissions occur during the burning season (e.g., Cançado et al., 2006; Uriarte et al., 2009).

Such problems are expected to disappear in the near future, following the recent agreement between the Brazilian government and large ethanol producers to ban burning as a harvesting practice in the state of São Paulo between 2013 (in the nonmountain areas) and 2017 (in the mountain areas). Sugarcane production practices in newly converted areas of Cerrado are expected to follow similar patterns in the future. In addition, to air quality benefits, this ban could have a significantly favorable effect on the sector's GHG emission reduction efforts (de Figueiredo et al., 2010).

2.3.3. Water quality

Sugarcane cultivation and ethanol production have been blamed for reducing water quality in several parts of São Paulo State in the past. Gunkel et al. (2007) have linked water heating, acidification, increased turbidity, oxygen imbalance, and high coliform bacteria levels in the Ipojuca River to the increased production and treatment of sugarcane in the catchment area. Filoso et al. (2003) have reported high nitrogen loading across rivers that contain sugarcane plantations in their catchment areas. Such impacts on freshwater ecosystems result largely from the fertilizer-intensive nature of sugarcane cultivation (FAO, 2004).[7] Fertilizer use is expected to be eliminated as sugarcane production is modernized across the country.

Sugarcane plantations also use copious amounts of agrochemicals that endanger ecosystems and human health (Lara et al., 2001; Lehtonen, 2010). Lehtonen (2010) identifies a number of acute and chronic health symptoms that are associated with exposure to pesticides that are routinely used in sugarcane agriculture. Smeets et al. (2008) discuss several cases in which bad pesticide practices in sugarcane farming have resulted in the poisoning or even death of agricultural workers. Lehtonen (2010)

---

[6] Burning is widely used to assist sugarcane harvesting. An estimated 80% of Brazilian sugarcane is burned for this purpose each year.

[7] Sugarcane expansion has been identified as an important driver of fertilizer use increase in the country (Martinelli and Filoso, 2008).

tracks several studies of bad practices in sugarcane cultivation and the ways in which they have resulted in water and soil contamination. Martinelli and Filoso (2008) have identified banned agrochemicals attributable to sugarcane agriculture in sediments and fish in the Piracibaba River.

Finally, Martinelli and Filoso (2008) cite several case studies that link sugarcane burning with the acidification of streams and rivers as well as with the detection of PAHs in lake sediments.

2.3.4. Biodiversity

Water pollution caused by sugarcane plantations can result in significant biodiversity loss in the riparian ecosystems that are in the vicinity of sugarcane plantations. However, in the state of São Paulo and beyond, direct and indirect LUCC induced by sugarcane expansion can be a much more significant driver of biodiversity loss.

The Brazilian Forest Code stipulates that every farming establishment must pre-serve a portion of forested area within its borders. In the Southeast, this set-aside forested area should be 20 percent of the establishment's land. However, there are strong indications that most sugarcane producers do not comply with this obligation. In this regard, Gonçalves and Castanho Filho (2006) argue that compliance with this regulation would entail renouncing close to 3.7 million ha of highly productive land, with a loss of crop and cattle gross earnings estimated at BRL 5.6 billion. Neverthe-less, failure to comply with this policy, and particularly with the conversion of riparian ecosystems can decrease biodiversity in the state of São Paulo (Martinelli and Filoso, 2008). It is feared that the degradation of highly biodiverse riparian ecosystems can further reduce water quality, which, in turn, will further threaten biodiversity and human well-being (Martinelli and Filoso, 2008).

However, the future expansion of sugarcane cultivation can pose an even more significant threat to biodiversity. This is because much of Cerrado's land, labeled as "degraded" land, is actually characterized by rich biodiversity that could be negatively affected by sugarcane expansion. Predictions reveal that sugarcane expansion in the Brazilian Southeast might induce direct and indirect LUCC in the Cerrado region (Lapola et al., 2010; Smeets et al., 2008; Sparovek et al., 2007) as well as in the Amazon (Lapola et al., 2010), leading to biodiversity loss in these two highly biodiverse areas.[8]

2.4. Socioeconomic issues

Bioethanol production in Brazil is historically based on three main factors: (1) large plantations (*latifundia*); (2) extensive monocultures; and (3) highly intensive, low-wage labor (formerly slave labor, with current practices that remain very far from what the International Labor Organization considers decent labor standards). Some

---

[8] Conservation International identifies the Cerrado as one of the world's 25 biodiversity hotspots.

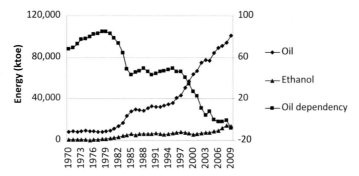

Figure 6.2. Trends of domestic oil production, bioethanol production, and oil dependency. The right-hand *y* axis refers to oil dependency (in %). *Source*: IEA (2011a).

of the social impacts discussed in this section started unfolding before the Proálcool program. However, understanding the evolution of the sugarcane sector can shed light on the present impacts of Brazil's bioethanol program. An understanding of these impacts can guide policies in other countries that pursue sugarcane ethanol production (see Chapter 12).

2.4.1. Macroeconomic effects

The penetration of bioethanol and, more recently, of biodiesel into the Brazilian transport system has undoubtedly contributed to the decreasing dependence of the Brazilian economy on foreign oil. According to data from the International Energy Agency (IEA, 2011a), in 2008, biofuels constituted about 21.0 percent of all road transport fuel consumed within the country. Actually, biofuel production, combined with the discovery of oil reserves in the country, has significantly decreased the need for imported fuel (Figure 6.2). This has had a positive effect on the trade balance and foreign exchange reserves. According to de Castro Santos (2007), the value of avoided gasoline imports is about 200,000 barrels a day, which translates into USD 2 billion per year.

In addition to these positive impacts on domestic energy security, the bioethanol program had several ripple effects in the Brazilian economy. First of all, biofuel-increased demand for sugarcane has been an agent of innovation in the sugarcane and ethanol sectors. When the Proálcool began, each hectare of sugarcane yielded approximately 2,000 L of ethanol, whereas currently the figure is closer to 6,000 L (Goldemberg, 2008). This increase in productivity was possible because of the development of new sugarcane varieties that are more suited to Brazil's weather conditions, the improvement of sugar extraction–vinasse recovery–fermentation, and the cogeneration of power using bagasse (Furtado et al., 2011). This was, to a large extent, a direct result of a sectoral innovation system set up in the state of São Paulo that has been largely private in nature since the 1990s (Furtado et al., 2011).

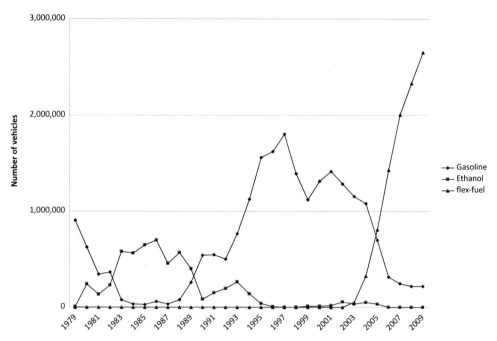

Figure 6.3. Number of vehicles in Brazil by engine type. *Source*: UNICA (2011a).

The bioethanol boom has also been an agent of innovation in the Brazilian automotive industry. The industry benefited tremendously from this fuel shift, initially producing neat ethanol cars (running on E100) and subsequently flexible-fuel engines, fit to run on any blend of gasoline ethanol. Since their introduction in 2003, FFVs have captured the market and, in 2009, accounted for 92.3 percent of all new vehicles sold in the country (Figure 6.3).

Finally, it should be mentioned that the sugarcane industry is a major employer within Brazil. Trade unions and business associations estimate that approximately 1 million jobs have been created by the sugarcane industry (*Repórter Brasil*, 2009). Even though most of these jobs are low-skill jobs (e.g., cane cutting), the technological innovation in the sector has boosted high-skill job opportunities to an appreciable level having positive multiplying effects in many regions.

2.4.2. Income and labor standards

The sugarcane sector has a notorious reputation for the labor practices it employs. Even though some bad practices, including a lack of formal contracts, slave labor, and child labor, are still prevalent, incidences of these practices are decreasing, and there is a clear improvement, particularly in the state of São Paulo (Balsadi, 2007). Key to this improvement has been the organized struggle of the sector's salaried workers, most of them represented by the Rural Employees' Federation of the State of São Paulo. In particular, child labor fell from 14.7 percent in 1992 to 0.8 percent in 2004. In 2004, 93 percent of the workers residing in municipal centers had work contracts and enjoyed social security benefits.[9] Additionally, workers' revenues from salaried work in the sugarcane sector increased. The percentage of workers who earned above the minimum wage increased from 57.6 percent in 1992 to 69.6 percent in 2004. In fact, the average salary in São Paulo's sugarcane plantations is higher than the average salary paid in other large-scale crop production sectors (Smeets et al., 2008).[10] However, even these higher-than-average wages are not high enough to allow workers to escape poverty (e.g., Martinelli and Filoso, 2008). Most important, these increased wages do not necessarily come with a decrease in workload or an improvement in working conditions. Cane cutters are on several occasions still paid using the same antiquated method – by meter harvested (Wilkinson and Herrera, 2010). It is estimated that, while in 1969 a worker harvested an average of 3 metric tons of sugarcane per day, currently a harvest of less than 10 or 12 metric tons a day is deemed inadequate and can put the job security of the cutter at risk (Ramos, 2006). Correcting for inflation, this represents a decrease in the harvester's pay from BRL 2.73 per metric ton in 1969 to BRL 0.86 per metric ton in 2005 (Ramos, 2006). The structure of this payment scheme combined with the loss of purchasing power in 2008 resulted in several cane cutter strikes in the state of São Paulo that year (*Repórter Brasil*, 2009).[11]

Evidently, such productivity gains entail long hours of physically arduous work. Cane cutters make an average of 30 scythe blows per minute over a working day that will extend for 10 to 12 hours. Usually the sugarcane harvest is performed by young individuals who come from very poor and distant regions such as the Jequitinhonha Valley in Minas Gerais and the arid and semiarid zones of the Northeast. Before being hired, they are submitted to a productivity test that screens only the strongest and fittest. Martinelli and Filoso (2008) collected numerous studies that report the negative health effects (sometimes leading to death) of this highly intensive manual job. Furthermore, an average of almost 70 working accidents are reported every day in the whole country, most of them reported in areas with limited mechanization such as the Northeast (*Repórter Brasil*, 2009). Slave labor also remains a very troubling issue for the industry in these areas, with labor inspectors freeing 1,911 workers in 2009 and 2,553 workers in 2008 (*Repórter Brasil*, 2009, 2010a).

The mechanization of agricultural production is expected to eliminate such bad working practices. Presently almost 70 percent of São Paulo's harvest is mechanized. Increased mechanization is also foreseen in the Center West region of the country

---

[9] Workers living and working outside of the municipalities' government center exhibit far worse social conditions. For example, in 2004, only 40% of them had work contracts.

[10] The same applies when comparing the ethanol industry to other industries (e.g., sugar, food, and beverages) in the state of São Paulo.

[11] On several occasions, these strikes were ended violently by the companies, intimidating the workers (*Repórter Brasil*, 2009).

with most of the new mills expected to be 100 percent mechanized. Yet, despite increased mechanization, inhumane working conditions are still prevalent in areas of new investment, such as Mato Grosso do Sul, particularly for migrant and indigenous workers (Wilkinson and Herrera, 2010).

However it is also expected that increased mechanization will affect employment opportunities in the sector as a whole. On the one hand studies suggest that 171,000 high-skilled jobs will be generated during this transition in the sugarcane industry. On the other hand, it is expected that once slash-and-burn techniques are entirely banned, mechanization will eliminate 420,000 low-skill jobs by 2014 in São Paulo alone. Indeed, low-skilled migrants are expected to disappear in São Paulo's plantations in 10 years' time, signs of which are already visible. For example, in 2008, the sugarcane complex laid off more people than it will be able to recruit for new tasks arising from the mechanization. More worryingly it is expected that only a small proportion of the low-skilled cane cutters who will lose their jobs during this transition will be covered by the Brazilian government's planned retraining schemes (Wilkinson and Herrera, 2010).

2.4.3. Land tenure

The history of the sugarcane sector is marked by disputes between landowners and workers over the workers' access to land. Until the late 1960s, the sugarcane mills catered to the workforce's needs based to a great extent on the model of the dweller (*morador*). In addition to working on the plantation, the workers had their dwellings (*morada*) inside the mill or on the farm premises, where they also grew their own food. However, since the 1950s (and especially since 1964, a result of the dictatorial regime that had weakened trade unions), there was a massive expulsion of workers from their homes. This led to the construction of self-contained neighborhoods in the suburbs of medium-sized cities situated in the sugarcane-producing regions, which gave rise to the concept *trabalhadores de ponta de rua* (end-of-street workers).

Essentially the access of poor plantation workers to housing and to areas apt for food production has, to a large extent, been eliminated. This, in turn, seriously aggravated the living conditions of those who relied on salaried work in the sugarcane sector. Additionally, it restricted land ownership to a few large landowners, further increasing the high land concentration that was a key characteristic of the sector to begin with. As a result, the apparent lack of recent land tenure conflicts in São Paulo State observed by different scholars can be attributed to this prior consolidation of land into the hands of a few large landowners (Smeets et al., 2008).

If rights to housing represented a constraint to sugarcane expansion, the ever more important presence of sugarcane in agrarian reform settlement areas in the state of São Paulo today can be seen as land reform in reverse. Many of those who were granted land rights during the agrarian reforms are expected to begin profiting by leasing their land to mills. Ramos (2006) shows that a settler leasing a 15-ha plot in the municipality of Promissão, in the state of São Paulo, earns an income of nearly BRL 500 per ha, for a six-year contract, during which the farmer ceases any activity whatsoever on his own plot. Hardly any other farming activity would be as profitable for the farmer. This essentially transforms families settled under the land reform scheme into land leasers. In this type of contract, the farmers cede part of their land for sugarcane cultivation, while the companies take charge of all the productive tasks, from preparing the soil to harvesting. Even though such lease contracts are frequent in the pulp and paper industry, in the case of sugarcane, the crop occupies the entire plot, and as a result, the farmer has to lease the entire plot rather than only part of it.

As a consequence, Novo et al. (2010) report a fourfold increase in the price of agricultural land in São Paulo State since 1999. They attribute this, to a large extent, to sugarcane expansion.

2.4.4. Food security

Sugarcane-cultivated land in the state of São Paulo has been increasing over the past decade at an average of 590,000 ha/yr (IEA, 2011b). Considering that in that period, the state's land under agricultural activities was already almost entirely used, it can be inferred that sugarcane expansion came at the expense of other agricultural land uses. Gonçalves et al. (2007) show that the main land use replaced by sugarcane was low-productivity pasture and, to a lesser extent, peanut and rice cultivation. Data suggest that since 2006, approximately 90 percent of new sugarcane area has come at the expense of pastureland (Novo et al., 2010). Furthermore, combined with a crisis in the beef chain, sugarcane expansion might have contributed to the reduction of milk farming in the state (Novo et al., 2010). Other affected crops included tomatoes and oranges in São Paulo and coffee in São Paulo, Espírito Santo, and Minas Gerais (Smeets et al., 2008). On the other hand bean, corn, poultry, and egg production was not affected by sugarcane expansion in the state of São Paulo.

Nonetheless, there is no consistent indication that increasing the production of bioethanol represents a serious and direct threat to national food security. Instead, it has been suggested that biofuel expansion in the state might have had an indirect positive effect on domestic food security due to higher incomes (Smeets et al., 2008). Conversely, there are some concerns that bioethanol expansion in Brazil in response to growing domestic and international demand might (Chapter 8), in the future, affect food security outside the country owing to possible increases in sugar prices (Koizumi, 2009; Koizumi and Ohga, 2009; Mitchell, 2005).

3. The biodiesel program

3.1. Main drivers and policies

The PNPB was launched in 2004. The Brazilian government has enacted a 5 percent blending mandate (B5), which essentially guarantees biodiesel demand for private companies. The legal basis on which the PNPB was formed allows the National Council for Energy Policy (Conselho Nacional de Politica Energética; CNPE) to change this blending mandate depending on feedstock supply, technology, and the involvement of small farmers. This allows for rather flexible changes in the blending mandate because blending changes are not submitted to congress for voting. Conversely, this same flexibility may make the legal provision unstable and malleable to the interests and influence of the government (Garcez and Vianna, 2009).

The PNPB's stated objectives revolve primarily around social inclusion, competitive price, and feedstock diversification. Considering the preceding, energy security and rural development are the two main drivers of the Brazilian biodiesel program. Environmental considerations, conversely, seem to be given much less attention.

Currently, Petrobras is in charge of the mixture under the supervision of the CNPE. Petrobras can acquire the biodiesel either through the market or through auctions. The National Agency for Petroleum, Natural Gas, and Biocombustibles (Agência Nacional do Petróleo, Gás Natural e Biocombustíveis; ANP) is in charge of administering the auctions, which are organized into two main rounds. In the first round, the bulk of the biodiesel is auctioned, and participation is limited to the companies that have been awarded a social seal. In the second round, a much smaller quantity of biodiesel is auctioned, with participation open to companies without the social seal. This seal is awarded by the Ministry for Agrarian Development and attests that the company supplies its feedstock from family farmers. For biodiesel companies in the Northeast, at least 50 percent of the feedstock processed by the industry has to come from family farming to be awarded the social seal. In the South and Southeast regions, the required amount of family farmer–supplied feedstock is 30 percent, and in the Center West, it is 15 percent. The social seal requires companies to provide farmers with technical assistance as well as the guarantee of a preestablished price. In return for these socially minded practices, the company is endowed with fiscal exemptions. It is important to remark that the social seal does not contain any environmental criteria. The seal merely guarantees that the biodiesel feedstock comes from family farms but it does not indicate any conditions under which the raw material must be produced.

Enforcing the provision of technical assistance to family farmers is rather difficult. Despite the numerous incentives that have been developed to try to enhance the participation of poor family farmers, the fact remains that consolidated soybean farms account for more than 90 percent of current feedstock production.

3.2. Production trends

In 2010, 42 percent of Brazilian biodiesel was produced in the Center West region, also the main soybean production center, while only 9 percent was produced in the Northeast. Figure 6.4 shows each region's contribution to biodiesel production.

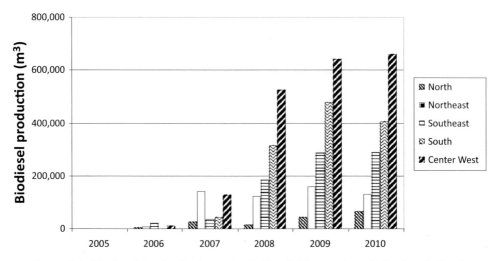

Figure 6.4. Biodiesel production by region in Brazil. The graph excludes the biodiesel production of the last four months of 2010. *Source*: ANP (2010a).

The currently authorized production capacity can serve as an indicator of the different regions' future contributions to biodiesel production (Garcez and Vianna, 2009). Figure 6.5 shows how the contribution of the North and Northeast is expected to be minimal compared to other regions since in 2010, their output amounted to just 17 percent of the authorized production capacity. It should be stressed that most of the feedstock produced in the Northeast comes from much capitalized soybean areas in the Cerrado region (located in the western part of Bahia State). The participation of poor family farmers in the program is practically nonexistent.

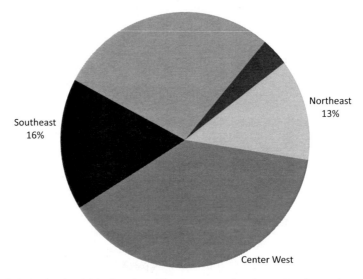

Figure 6.5. Authorized biodiesel production capacity in each of Brazil's regions. *Source*: ANP (2010b).

The private sector became immediately interested in the opportunities created by the PNPB. The production capacity of the transesterification plants soon exceeded the demand for Brazilian biodiesel, and by the end of 2010, the production capacity reached 5.5 million m^3/yr (ANP, 2010a). This quantity was enough to substitute 10 percent of the diesel consumed within the country. As a result, of the 62 plants authorized to operate in the country, only 35 were operating during the whole of 2010 (ANP, 2010a, 2010b).

Soybean oil is the main feedstock used for biodiesel production in Brazil. Since the beginning of the PNPB, the contribution of soybean oil has remained a consistent 80 percent approximately, while beef tallow amounted to roughly 15 percent (ANP, 2010c). The remaining feedstock needs are covered by cotton seed oil, canola oil, and other vegetable oils. Brazil's experience with palm cultivation is quite limited, but owing to oil palm's very high oil productivity, palm oil could become an important feedstock in the future. Kaltner (2005) reports that there are 3 million ha of easily accessible deforested land in the Amazon region that could be cultivated with oil palm.

3.3. Energy and environmental impacts

3.3.1. Energy and climate

As discussed in the previous section, soybean has emerged as the most dominant biodiesel feedstock in the country. There are few LCA studies that have calculated the energy performance of Brazilian soybean biodiesel (Cavalett and Ortega, 2010). Unfortunately, given the very distinct production conditions found in Brazil,[12] the findings of LCAs conducted in other countries are not transferable to the Brazilian context. Table 6.1 presents an LCA for soybean biodiesel and reports the total energy requirement of soybean production–transport–crushing, soybean oil transesterification, and biodiesel transport.[13]

According to Table 6.1, the net energy ratio of soybean biodiesel is 1.1. This value is higher than 1 (i.e., the amount of energy delivered is higher than the energy invested) but is much lower than competing energy sources such as sugarcane (3.1–9.3) (see Section 2.3.1) and oil and gas, with a value of around 20 (Cleveland, 2005). In fact, for any given amount of energy delivered, only 9 percent is the result of the soybean biodiesel system, with the remaining 91 percent consumed by other energy sources.

---

[12] Brazil exhibits one of the highest soybean productivities in the world as well as a very low use of fossil fuel for electricity production due to most of the electricity being produced by hydroelectric dams (MME, 2010).

[13] In our calculations, no energy input was allocated for glycerine production given that in 2006 (the beginning of PNPB), the total amount of glycerine produced within the country amounted to just 14,000 tons (ABIQUIM, 2007). The current level of biodiesel production entails the production of about 265,000 tons of glycerine per year, 7.5 tons more than the domestic consumption (Fairbanks, 2009). This significant coproduction of glycerine provoked a substantial decline in glycerine price, from 4 R$ in 2007 to 1.8 BRL in June 2009.

Table 6.1. *Soybean biodiesel energy balance*

Inputs	Unit	Q/ha/yr	MJ/unit	Total energy	Source
Agricultural production[a]					
Machinery	Kg	25	92.5	2,312.5	Scholz et al. (1998)
Diesel[b]	Lt	65	45.1	2,933.7	Boustead and Hancock (1979)
Phosphorus	Kg	78.8	4.6	361.3	West and Marland (2002)[c]
Potassium	Kg	78.8	5.9	463.8	West and Marland (2002)[c]
Lime	Kg	375	1.3	489.3	Shapouri et al. (2004)
Seeds	Kg	69	4.8	331.0	Following a common procedure[d]
Herbicides	Kg	4.8	233.8	1,122.2	West and Marland (2002)[c]
Insecticides	Kg	3.2	242.7	776.7	West and Marland (2002)[c]
Electricity	KWh	34	5.7	193.5	Coltro et al. (2003)
Farm buildings	m^2	0.09	1,800	162	Macedo et al. (2008)
Soybean transport					
Machinery	Kg	2	127.3	259.4	Scholz et al. (1998)
Diesel	Lt	5.2	45.1	234.6	Boustead and Hancock (1979)
Crushing					
Buildings	m^2	0.0005	1,800	0.9	Macedo et al. (2008)
Machinery	Kg	0.3	70.5	20.5	Scholz et al. (1998)
Diesel	Lt	62	45.1	2,799.7	Boustead and Hancock (1979)
Electricity	KWh	202.8	5.7	1,153.7	Coltro et al. (2003)
Water	Kg	234.5	0.0026	0.6	Haguiuda and Veneziani (2006)
Hexane	Kg	7	45.4	317.2	Ahmed et al. (1994)

128

Biodiesel production					
Buildings	m²	0.01	1,800	23.9	Macedo et al. (2008)
Machinery	Kg	1.2	70.5	89.7	Scholz et al. (1998)
Methanol	Kg	74.6	36.3	2,709.4	Kamahara et al. (2010)
Catalyst (CH_3ONa)	Kg	8.8	39.1	345.5	Sheehan (1998)
Electricity	KWh	24.1	5.7	137.4	Coltro et al. (2003)
Water	Kg	261.2	0.003	0.7	Haguiuda and Veneziani (2006)
Wood	Kg	229.3	13.2	3,035.3	MME (2010)[e]; Oliveira and Seixas (2006)[f]
Fuel oil	Kg	0.8	51.7	41.4	Boustead and Hancock (1979)
Biodiesel transport					
Machinery	Kg	0.4	127.3	47.5	Scholz et al. (1998)
Diesel	Lt	1.0	45.1	43.0	Boustead and Hancock (1979)
Total inputs[g]	MJ			17,836.0	
Output[h]	Kg	533.1	36.95	19,699.5	Sheehan (1998)
EROI				1.1	

[a] Background data were obtained from Cavalett and Ortega (2010).

[b] Diesel density was assumed 0.84 g/mL (Cavalett and Ortega, 2010).

[c] Adjusted for the Brazilian electricity efficiency reported by Coltro et al. (2003).

[d] The energy cost of seed production was estimated to be 150% of the agricultural phase.

[e] For direct energy and wood density.

[f] For indirect energy.

[g] The energy cost of soybean crushing was proportionally attributed to the energy content of soybean oil (the main products of soybean crushing are soybean oil and soybean cake). Using data from Domalsky et al. (1986) and assuming a cake yield of 77% (from Abiove, 2010), it was assumed that the energy content of soybean oil was 40.2% of the combustion value of soybean oil and soybean cake. The inputs of the crushing phase are consequently multiplied by 40.2%.

[h] Soybean productivity was assumed to be 2,911 kg/ha based on estimates by CONAB for the 2009–2010 cropping season. Oil yield was assumed to be 19%, as calculated by Abiove (2010).

This low net energy ratio implies that the GHG savings from Brazilian soybean biodiesel are small. LCA meta-analyses have shown that Brazilian soybean biodiesel emits larger amounts of GHGs than other first-generation biofuel practices (Panichelli et al., 2009). In some cases, Brazilian soybean biodiesel is a net GHG emitter (Menichetti and Otto, 2009; Zah et al., 2007). This is because most Brazilian GHG emissions come from deforestation and land use change, which accounted for 58 percent of national GHG emissions in 2005 (MCT, 2009). As a result, comparative LCAs such as those conducted by Smeets et al. (2009) reveal highly variable GHG emission savings from soybean in South America (ranging from −87% to 44%) due to the sensitivity of his LCA on the reference-land system used.

It should be noted that the main area subject to soybean cultivation expansion is the Cerrado, a region characterized by savannah vegetation. If soybean expansion provokes deforestation in the Cerrado region, then the carbon debt that is incurred will nullify biodiesel's modest GHG displacement effect. It is calculated that biodiesel produced from soybean grown on converted Cerrado savannah and tropical forest would require 37 and 319 years, respectively, to repay its carbon debt (Fargione et al., 2008). Even though there is currently no indication of increased deforestation caused by biodiesel expansion at the country scale, there are fears that soybean, oil palm, and jatropha expansion for biodiesel production could result in direct and indirect LUCC and, consequently, in significant carbon debts (Lapola et al., 2010). For example Finco and Doppler (2010), in their study area (in the state of Tocantins), found that family farmers growing castor bean deforested, on average, 0.5 ha (releasing 22.5 metric tons of CO_2) and that those cultivating jatropha deforested 0.72 ha (emitting 80 metric tons of CO_2).

3.3.2. Air quality

Ambient air pollutant emissions due to diesel combustion affects significantly air quality in Brazil's urban centers. Biodiesel is essentially sulfur free and its combustion releases lower quantities of particulate matter (PM) and volatile organic compounds (VOCs) compared to conventional diesel (EPA, 2002). However, if the objective is to reduce exhaust emissions, other, more effective alternatives should be evaluated. For example, a comparison between the exhaust emissions of biodiesel and of compressed natural gas reveals that the previously mentioned polluting elements are much lower for natural gas vehicles (DOE, 2010). Sulfur abatement technologies could therefore be introduced to mitigate biodiesel's impact on air quality.

On the other hand, biodiesel combustion emits slightly more nitrogen oxides than standard diesel. This aspect should be also considered properly because nitrogen oxides are a main precursor of tropospheric ozone, an atmospheric pollutant that constitutes one of the most serious air quality problems faced by the city of São Paulo (CETESP, 2009).

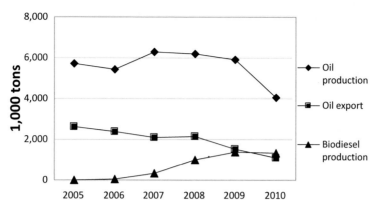

Figure 6.6. Soybean oil and biodiesel production trends in Brazil. *Sources*: ANP (2010a), Abiove (2010).

3.4. Economic issues

According to the initial plans of the PNPB, the contribution of nonsoybean feedstocks should have been much greater than their present levels. Castor bean, for example, was expected to play a much greater role as a biodiesel feedstock (NAE, 2004). However, castor bean oil has a high opportunity cost and is a highly valued input for the cosmetic, paint, and pharmaceutical industries. In addition, the social seal criteria specify the minimum quantity of feedstock that biodiesel producers have to buy from family farms (refer to Section 3.1). In the Northeast, the inability to grow other types of biodiesel feedstock means that the social seal biodiesel companies are obliged to buy castor bean since other crops cannot be supplied by the family farms. Taking advantage of the fact that the criteria do not specify the type of feedstock that must be used for biodiesel production, biodiesel companies have made it a common practice to buy castor bean oil simply to comply with the social seal requirements. They then resell the oil to other industrial sectors without using it as a feedstock for biodiesel production (*Repórter Brasil*, 2010; Trentini and Saes, 2010).

This overwhelming use of soybean oil for biodiesel production is due to soybean's widespread availability and low cost compared to other vegetable oils. Perhaps the main reason behind this observed price competitiveness of soybean oil is the fact that its production and supply chain was already well developed by the time PNPB was launched. In fact, Brazil is the second largest soybean producer after the United States, and Brazil's current soybean productivity per unit of land is among the highest in the world (FAO, 2011). In 2009, the area under soybean cultivation amounted to almost 22 million ha, making soybean the country's top crop in terms of land use, taking up 32 percent of the total area dedicated to crops (IBGE, 2009b).

The use of soybean oil for biodiesel production came at the expense of soybean oil exports (Figure 6.6). Regressing monthly soybean oil exports with biodiesel

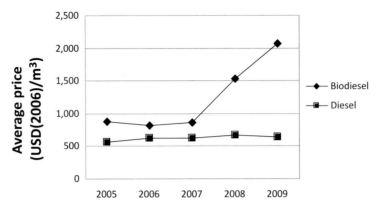

Figure 6.7. Diesel and biodiesel prices in Brazil. *Sources*: ANP (2010d, 2010e).

production indicates a statistically significant negative correlation between these two variables ($p = 0.007$ at a 95% confidence interval).

Figure 6.6 suggests a decline in soybean oil production since 2008. This is in contrast to the constant increase in soybean cultivation area and soybean production during the same period (CONAB, 2010). A possible explanation for the observed decline in soybean oil production is that the data in Figure 6.6 do not report information on the amount of soybean oil that is not directly marketed. This unrevealed information would include the soybean oil used for biodiesel production by the biodiesel companies that own their own soybean oil extraction facilities. If this hypothesis is true, biodiesel production would have kept from the market a significant amount of soybean oil, making it unavailable to consumers.

Despite biodiesel's huge production capacity (refer to Section 3.2), the variety of potential feedstocks, and the high productivity of soybean, the price of biodiesel remains noncompetitive against the price of conventional diesel (Figure 6.7).[14] Because biodiesel does not receive any direct governmental subsidies (except for the fiscal relief given to social seal–endowed companies) this price difference is paid by the consumers.[15] This lack of competitiveness is to a large extent due to the fact that Brazilian conventional diesel is among the cheapest in the world (da Silva Dias, 2007).[16] The cost of diesel is an important element of the whole industrial structure, and its low cost facilitates import-substitution industrialization. For decades, Brazil maintained a petrol price above the international market price, while keeping the price of diesel below the international average. In this regard, the introduction

---

[14] Barros et al. (2009) suggest that the cost of vegetable oil represents 78% of overall biodiesel cost.

[15] The incorporation of biodiesel to diesel is paid for by the consumer though the blending mandates.

[16] This is a result of specific transport policies that were initiated by the Juscelino Kubitschek administration. The Targets Plan (1955–1959) promoted policies that favored road cargo transit to the detriment of railways. Since then, the relative price of petrol in relation to diesel has always been managed to sustain high profits with the former and pay sufficient subsidies for use of the latter.

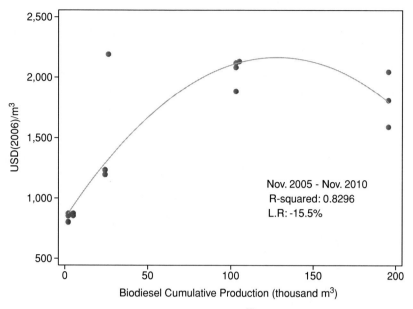

Figure 6.8. Learning curve for the biodiesel sector[17] in Brazil. *Sources*: ANP (2010a, 2010b).

of more expensive biodiesel may hinder the low-price diesel policy (da Silva Dias, 2007).

Figure 6.8 shows the biodiesel sector's learning curve since the beginning of the PNPB. Biodiesel's price, far from decreasing, as theory would predict,[18] actually increased with accumulated experience. This suggests that the learning rate is negative, seemingly implying that learning effects are absent from the sector. The PNPB was launched only a few years ago – not a long period for the development of a fuel technology.[19] Even though a potential for cost reduction exists, it is rather slim when taking into account the fact that the main cost component of biodiesel is vegetable oil. As da Silva Dias (2007) points out, research incentives for the genetic improvement of feedstock and technical innovation are essentially absent in the current format of the auction system.

---

[17] The regression was obtained by loglinear transformation of the following assumed learning function: $C = A \times CP^b$, where C is the unit cost, A is a constant, CP is the cumulative production, and b is the learning elasticity.

[18] The performance of individual technologies generally improves as organizations accumulate experience. The consequence of this accumulated experience is a decrease in unit costs. This phenomenon is called *learning by doing* or *learning by use* (Arrow, 1962; Rosenberg, 1982). The mechanisms through which learning takes place are numerous, e.g., technological advancement, economies of scale, and improvement in organizational functioning (Grubler et al., 1999). Since learning depends on the accumulation of actual experience (and not only time), learning curves assume the shape of a power function with unit costs decreasing as a function of cumulative output. The learning rate is the slope of the curve. This curve indicates the change in price per doubling of accumulated output.

[19] Sometimes, significant learning effects are possible even in such short periods of time, as observed in the bioethanol sector. Goldemberg et al. (2004a) found a decreasing learning curve only five years after the launch of the Proálcool program.

Despite the importance of the private sector, the Northeast still has few companies integrated in the PNPB. This scant business foundation limits benefits from strong competition and increases the odds for opportunistic conduct.

Finally, Petrobras plants supply 70 percent of the Northeast's biodiesel production. Considering that Petrobras is the nation's third largest biodiesel producer (after Granol and ADM), it can be easily concluded that the biodiesel program is in fact undermining one of its stated driving principles: anchoring biodiesel production in the private sector.

3.5. Social impacts

Before delving deeper into the social impacts of the PNPB, it is important to explain how family farming is understood in the Brazilian context. Generally speaking, family farming indicates a form of production in which there is unity between the worker and the factors of production. Despite the *latifundium* (large plantation) tradition in Brazil (refer to Section 2.1), a significant portion of agricultural production still rests on small farmers and their families. In Brazil, most of the so-called family farming, especially in the Northeast, cannot produce enough to allow for the farmers' subsistence. As a result, family farmers often have to seek additional income through off-farm activities. These hardships are due not only to the exiguity of the family farmers' land but also to the social and institutional context in which the farmers operate. As in several other parts of the developing world, the most impoverished segments of the rural population establish relations with highly imperfect and incomplete markets (Ellis, 1988). Such market and production conditions in the less developed rural areas are often characterized by weak information flow, weak institutions, and high transaction costs and risks.[20]

A key objective of the PNPB is to use biodiesel expansion as a means of strengthening family farming, helping especially the poorest farmers. The PNPB is, from this standpoint, the culmination of a series of policies that can trace their origins to the 1996 National Program for Strengthening Family Farming. However, a comparison of PNPB's actual results with those initially expected reveals that the program is far from achieving its highly interlinked social inclusion objectives.

It seems that only the more prosperous farmers have managed to integrate themselves into the biodiesel production chain, despite the fact that a significant share of soybean feedstock comes from family production units. For example, in the state of Rio Grande do Sul, more than 50 percent of supplied soybean comes from establishments that do not employ salaried workers. Nevertheless, these soybean producers

---

[20] The ultimate consequence of these factors is "a low level equilibrium trap" (Dorward et al., 2003: 324). If these conditions are present, then the smallholders' access to financial capital is facilitated by intermediaries. These intermediaries usually provide the inputs required for agricultural production in exchange for a right over crop production. As a result, the possibility of improving productive capacity is usually limited and depends on state intervention.

Table 6.2. *Number of family farms involved in the NPBP*

	2005	2006	2007	2008	2009	2010
Number of family farms	16,328	40,595	36,746	27,858	51,047	109,000 (est.)

Source: *Repórter Brasil* (2010b).

represent the wealthier family farmers. These wealthier family farmers have more land and, more important, better access to dynamic and competitive markets. In the Center West, for example, soybean family farmers that have joined the PNPB own properties often exceeding 200 ha.

Furthermore, feedstocks associated with low-income farmers (castor seed and palm oil) constitute only a tiny fraction of total biodiesel feedstock. Family farms in the Northeast could have profited from a competitive advantage in the production of castor bean, but the role of this feedstock has been less prominent than initially expected. More important, the markets in which small family farmers participate are rather volatile. The access of the castor bean market's main actors to bank credit is very limited, especially when compared with what happens in the soybean sector, where farmers have easier access to credit.

The Brazilian government expected the PNBP to become an important driver for the generation of jobs and income opportunities among family farmers. When the PNPB was launched, the Minister of Energy and Mines declared that biodiesel production would create 382,000 jobs (Roussef, 2004). In a report published by the Nucleus of the Presidency of the Republic for Strategic Issues (2004), it was estimated that the PNPB would generate 260,000 direct new jobs in the agricultural sector. Even though the relationship between new jobs and the number of family farms is not completely straightforward (Table 6.2), it is believed that PNPB produced much fewer jobs than initially expected. This is mainly due to the use of relatively large and highly mechanized production units, which limited the need for human labor. In fact, the amount of feedstock (soybean included) coming from the poorest regions of the North and Northeast is minimal (refer to Figures 6.4 and 6.5).

As already mentioned, biodiesel is mainly produced from soybean oil, to the detriment of soybean oil exportation (Figure 6.6) (Section 3.4). It is therefore unlikely that biodiesel has had any significant, direct effect on the availability of food within Brazil. However, the impact of biofuel production on food security can sometimes be more complicated (see also Chapters 1 and 2). As in the case of sugarcane (see Section 2.4.4), biodiesel feedstock production can generate more income for small producers and farmworkers, which, in turn, can have a positive impact on these farmers' and producers' food security. In any case, more research is required to unravel the true impact of biodiesel production on national food security.

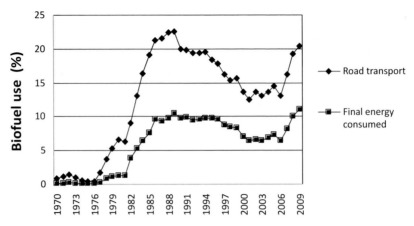

Figure 6.9. Bioethanol use in road transport and in the national energy mix.
Source: MME (2011).

4. Discussion

Table 6.3 provides a summary of the main drivers, policies, and impacts of the two programs. Bioethanol and biodiesel production most certainly have very different origins and impacts, not the least because of the different production practices and land areas involved. Ethanol production takes place mostly in São Paulo State, the richest state in Brazil, while biodiesel production is located in other regions (South and Center West). The very different environmental, socioeconomic, and institutional features of these regions play a significant role in shaping the two programs' impacts and successes but also make the comparison and joint analysis of the two programs problematic.

Energy security has been the main policy concern behind the Brazilian bioethanol program. Ethanol's share in the national energy mix has been consistently high since the mid-1980s, despite signs of decrease in the 1990s (Figure 6.9). More important, ethanol's share in the energy mix has been consistently increasing over the past six years. This high penetration of ethanol in the national energy mix is a combination of the 20 to 25 percent blending mandate of ethanol into gasoline and the large-scale production of FFVs.[21] This mandate is by far the highest in the world and has resulted in bioethanol constituting 20.4 percent of the total energy consumed in the transport sector in 2009 (MME, 2011). These figures suggest that Brazil is currently the only economy in the world in which biofuels play such an important role in the energy mix, with an estimated 11.1 percent of total final energy consumed in 2009 (MME, 2011). In this respect, the Brazilian bioethanol plan can be seen as a huge success that other developing countries attempt to emulate (refer to Chapter 12).

---

[21] The latter has been the result of a strong coalition between the Southeast sugarcane sector (*usinas*) and the automobile industry that aimed to avoid the decreases in bioethanol production following the first period of the Proálcool.

Table 6.3. *Drivers, policies, and impacts of the Brazilian bioethanol and biodiesel programs*

	Bioethanol program	Biodiesel program
Time line	• 1975 to present	• 2004 to present
Blending mandate	• E20–E25 • FFVs allow the use of pure ethanol	• B5
Feedstock	• Sugarcane	• Soybean (main) • Beef tallow and minor contribution from cotton seeds, sunflower seeds, and castor beans • Palm oil (emerging)
Main producing areas	• Southeast (mainly São Paulo State) • Northeast • Center West	• Center West • South • Southeast • Small contribution from the North and Northeast
Drivers	• Energy security (initial) • Create an international ethanol market (emerging)	• Energy security • Rural development • Develop an economically competitive and renewable fuel
Environmental impacts	• High EROI • High GHG savings • Water pollution problems • Inconclusive evidence regarding its impact on indirect LUCC, which potentially results in deforestation (in Cerrado and Amazon) and biodiversity loss • Atmospheric pollution from sugarcane burning; inconclusive evidence regarding ambient air quality improvement from bioethanol combustion	• Low EROI • Moderately high GHG savings if LUCC is disregarded • Possibly responsible for indirect LUCC, which potentially results in deforestation and biodiversity loss • Inconclusive evidence regarding ambient air quality improvement
Economic impacts	• Economically competitive and renewable fuel • Significantly positive effect on trade balance and foreign exchange reserves • Agent for innovation in the Brazilian economy	• Production still depends on government protection (e.g., credit and social seal) • Might trigger potential increases in fuel costs due to the fact that the price of biodiesel is higher than the price of conventional diesel

(*continued*)

Table 6.3 *(continued)*

	Bioethanol program	Biodiesel program
	• Major employer in the agricultural sector • Does not depend on government subsidies • Attracts significant domestic and international investment • Is becoming an internationally traded commodity • Integration of the sugarcane industry with the chemical industry for the production of organic plastic • Cellulosic ethanol (from bagasse) will elevate productivity in the sector but might compete with electricity generation (from bagasse combustion)	• New source of income for the soybean oil industry • Learning effects are still taking place • National companies are mainly involved in biodiesel production currently • Potential for exportation depends on traceability and certification schemes
Social impacts	• Provides relatively higher income when compared to similar agricultural activities. This income is not usually enough to allow sugarcane cutters to escape from poverty • Bad working conditions • Land concentrated among a few powerful actors • Limited signs of food–fuel competition • Mechanization will eliminate bad working practices and reduce environmental stress; however, it will be responsible for several thousand people losing their jobs • No signs of conflict with food production or food security	• Designed to promote social inclusion, but the participation of family farmers is rather poor and lower than expected • Family farming participation depends on the agricultural sector's general characteristics • Benefits the most prosperous grain-farming sector (soybean) • Farmers' organizations play the role of mere grain producers. There are no provisions for their further involvement in the biodiesel chain • Potential conflict with food production as the blending mandate increases

An emerging strategy of the Brazilian government attempts to take advantage of Brazil's competence in bioethanol production by making the fuel an internationally traded agricultural commodity. It has been suggested that the worldwide interest in renewable liquid transport fuel can boost the creation of an international ethanol market (Hira, 2011; Rosillo-Calle and Walter, 2006). Brazil and the United States are the largest producers of bioethanol. In some cases, Brazilian sugarcane ethanol is more competitive than U.S. maize ethanol (Crago et al., 2010), which would allow Brazil to benefit significantly from such an international ethanol market (Farinelli et al., 2009). However, the lack of mutually agreed-on fuel standards (coordinate international ethanol trade, investment, and disputes), a commodity exchange, and protectionist policies in the United States have hindered the development of an international ethanol market (Hira, 2011).

The first important step toward the creation of an international ethanol market happened during the 2007 Brazilian visit of G. W. Bush in his capacity as president of the United States. During that visit, the term *ethanol diplomacy* was coined, and an agreement was signed with an aim to "enlarge the ethanol market, including the infrastructure of production, distribution, and consumption, at the hemispheric scale and beyond" (Hollander, 2010: 707).[22] Since that visit, there have been signs that an international ethanol market is in its formative steps. For example, Brazilian sugarcane mill managers are lobbying the government to allow the import of ethanol from the United States during the periods between sugarcane harvests.[23] At the same time, there is a growing opposition within the United States toward the extra tax levied on Brazilian ethanol. In addition, the pace of mergers in the Brazilian ethanol sector began accelerating after an agreement was struck between Shell and the largest Brazilian ethanol company, COSAN. Finally, UNICA and the Brazilian government are lobbying the European Union, not unsuccessfully, to drop tariff barriers, increase the GHG saving requirement of European Union biofuel legislation, and promote sugarcane ethanol as the best choice to meet carbon-saving and environmental objectives (Franco et al., 2010). The current increase in certification initiatives serves as a further indication of a gradually consolidating global ethanol market (see Chapter 7). In 2009, Brazil exported 3,296 billion L of ethanol, mainly to the United States (through Caribbean countries), the European Union, South Korea, and Japan (MDICE, 2011). It should be noted that the increased international demand for Brazilian ethanol is expected to drive significant land use changes in Brazil (Sparovek et al., 2007) (see also Chapter 8).

For ethanol to become an internationally acknowledged commodity, two obstacles need to be surmounted. First, the current American–Brazilian ethanol production

---

[22] It is worth noting that the reduction of the U.S. tariff on Brazilian ethanol was not properly addressed (Hollander, 2010).

[23] Imports from the United States during the months between harvests can decrease energy security.

duopoly should be broken, to allow for a greater diversification of producing countries. In fact, this constitutes one of the key provisions of the new ethanol diplomacy. Since 2007, the Brazilian government has facilitated the transfer of relevant knowledge to countries in Latin America (Honduras, Nicaragua, Costa Rica, the Dominican Republic, Haiti, Colombia) and Africa (Ghana, Angola, Mozambique, Kenya) (Almeida, 2009; Franco et al., 2010). The second challenge consists in achieving ethanol expansion using only nonpredatory forms of land occupation and nondegrading labor conditions. Actually, one of the main reasons for Brazilian sugarcane's high economic efficiency (and, consequently, the competitive price of ethanol) lies in the economies of scale of the cultivation and postharvest operations. These economies of scale are achieved thanks to extensive sugarcane properties, favored by the specifically Brazilian land distribution structure: the *latifundium*. In our opinion, if the ethanol expansion schemes that are planned outside of Brazil (see Chapter 12) adopt the Brazilian land concentration model, then any economic and energy-related advantages that could have resulted will be eclipsed by the ensuing negative social impacts.

Bioethanol certification will be an important tool toward the development of an international ethanol market (see Chapter 7). The participation of stakeholders that represent social concerns in these certification processes would only benefit the biofuels market, reducing its negative effects, making it more transparent and increasing investor security.

The Brazilian biodiesel program has on the other hand a much shorter history. Nevertheless, some tentative conclusions can be drawn regarding its successes and impacts. As a reminder, the PNPB seeks to accomplish three main objectives:

1. To set up a biodiesel production system that will guarantee competitive supply, quality, and price
2. To diversify biofuel feedstocks and to further promote energy security
3. To strengthen family farming

Regarding the first objective, the PNPB was designed when the price of agricultural commodities was low. It was conceived as a means of stabilizing prices and guaranteeing farmers' incomes. However, the recent increases in the price of internationally traded agricultural commodities suggest that PNPB's original vision cannot be materialized. First, high agricultural commodity prices render the blending of biodiesel into fossil fuel more difficult, given the potential impacts on the final consumer's pocket. Moreover, higher feedstock prices increase the profitability of trading farm products and push up the opportunity cost of biodiesel production.

In Brazil, biodiesel commercialization and distribution are completely centralized, and the use of locally produced biodiesel is not allowed. This is because only two companies are allowed to distribute biodiesel. One of these two companies, Petrobras, distributes approximately 90 percent of total biodiesel production. In addition, the current biodiesel sale system based on large-scale auctions creates strong entrance barriers for smaller companies. As a result, only a few companies auction most of the biodiesel produced. If biodiesel were to be consumed at the local level (near the place of production), its energy balance would be improved. Moreover, distant and/or dispersed production points would benefit from reducing the transport cost of biodiesel and from allowing local production companies to set up sales arrangements with local transport companies. Finally, given the higher cost of diesel in the northern region of Brazil, the use of locally produced palm oil biodiesel could render this biodiesel the only economically viable alternative to fossil diesel in the region (de Camargo Barros et al., 2009). However, the Brazilian experience with palm oil production is very limited (refer to Section 3.2).

Regarding the second objective (diversifying biofuel feedstocks to further promote energy security), to date, soybean, and to a lesser degree, beef tallow, have been the main biodiesel feedstocks. As we discussed in the previous section, this trend is not going to change in the near future, which causes serious concerns about the PNPB's ability to meet this second objective. Feedstock diversification should be included in Brazil's general agricultural policy alongside other issues such as credit and technical assistance, research, road infrastructure, and the development of input markets and alternative output markets. As long as these agricultural policy components are not suitably managed, biodiesel's production and transaction costs will be too high. Additionally, farmers considering planting jatropha or castor beans may fear the so-called holdup problem[24] and limit as a result their engagement as vegetable oil suppliers. Finally, the low net energy ratio of Brazilian soybean biodiesel (see Table 6.1) implies that instead of contributing to net energy delivery, biodiesel could cannibalize other energy sources. Thus soybean biodiesel does not effectively contribute to energy security.

As for the third objective, the involvement of family farmers is much lower than initially expected. Castor bean's contribution (a feedstock that was expected to facilitate family farm involvement) is essentially negligible. This is due to its high opportunity costs, which make it uncompetitive with other biodiesel feedstocks. Ambitious biodiesel production targets that fail to address the structural problems faced by family farms may cause the biodiesel experience to be similar to that of the ethanol program, which ended up promoting only large-scale agriculture. At this point a parallel can be drawn between Proálcool and PNPB. When Proálcool was drafted, it included certain provisions designed to boost income opportunities for smallholders that used cassava

---

[24] Once a farmer has invested in planting an energy crop whose produce has value for only one buyer, the farmer finds himself locked into a relationship with the buyer (Williamson, 1979, 1985). The holdup problem refers in this example to the buyer behaving opportunistically and exploiting his or her stronger bargaining position to extract rents (e.g., demanding higher prices than those previously agreed). This may be common since there are only very few companies with the necessary facilities to extract jatropha and castor bean oil. By the same token, the processor, after having invested in expensive processing equipment, is vulnerable to decisions by local producers to stop harvesting.

as a feedstock. However, this proved to be economically unviable, so subsidies ended up favoring large-scale sugarcane production (Lehtonen, 2009).

5. Conclusions

The Brazilian bioethanol program has been a major success in the sense that it has managed to strengthen national energy security by substituting significant quantities of imported petroleum products with biofuels. Building on this success, the Brazilian government is now promoting ethanol as an internationally traded commodity. To achieve both these objectives, the ethanol sector had to undergo important technological and organizational transformations, which have altered the very physiognomy of the sector's corporate structure. This is particularly evident from the recent attraction of international capital and large national investors usually not involved in the agroindustry. However, big investments in agroenergy threaten the patrimonial nature of Brazil's agrarian system (especially in the Cerrado) because investment funds and international companies replace the traditional farm owner.

Brazil's historical land concentration legacy has resulted in today's extraordinary concentration of wealth and power held by only a few actors of the sugarcane sector.[25] Mechanization – a positive development that reflects technical progress as well as the social pressure to end slash-and-burn agriculture and degrading labor practices – will result in tens of thousands of laborers being laid off from the ethanol sector. Despite the training programs put in place by private sector actors, most of these low-skilled workers will not be reabsorbed by the industry. Addressing these workers' social plight will require policies that are designed to economically develop, in an environmentally friendly manner, the workers' home regions. The big challenge of the Brazilian government is to promote policies that will ensure the competitiveness of its bioethanol in the world market, all the while improving the working conditions of ethanol sector employees and avoiding directly or indirectly compromising ecosystems and biodiversity in the state of São Paulo and beyond (Cerrado, Amazon).

Concerning the PNPB, the program easily achieved its production target but failed to reach its feedstock diversification objective. In addition, the overall participation of family farming seems to be low and limited to the most prosperous sector: soybean.

However, environmental criteria are blatantly missing. The environmental and energy performance of biodiesel depends to a large extent on the type of feedstock used. As long as soybean remains the main feedstock, biodiesel's contribution to GHG displacement will be quite reduced, given soybean biodiesel's low net energy ratio and its high LUCC potential. It is therefore important that the social seal be upgraded to a socioenvironmental seal that would include tracking as well as public disclosure

---

[25] It is worth stressing the scarcity of social proposals and public debates regarding the appalling concentration of land and wealth in the Brazilian ethanol sector.

of the production process's impacts on the environment. This would perhaps be a first step toward feedstock diversification and the strengthening of family farming.

Acknowledgments

This chapter was written while A.G. was based at the Institute of Advanced Studies of the United Nations University (UNU). A.G. would like to acknowledge the financial support of the Japanese Society for the Promotion of Science (JSPS) through a JSPS-UNU Postdoctoral Fellowship. M.B. is thankful to all the members of the PENSA group of the University of São Paulo for their help orientating among Brazilian data sources and for the inspiring discussions.

7

Power, social impacts, and certification of ethanol fuel: View from the Northeast of Brazil

MARKKU LEHTONEN
University of Sussex, Brighton, UK

Abstract

International discussions on the sustainability of Brazilian ethanol biofuel and efforts to develop biofuel sustainability certification have, until recently, concentrated on the environmental effects – notably deforestation and the indirect land use impacts – of the expected expansion of sugarcane cultivation. The social impacts of large-scale sugarcane cultivation have received major attention in the debate only during the past few years. However, this attention has primarily concerned the impacts in São Paulo State, currently the main sugarcane-producing area of the country, and in the Center West, the expected main area of expansion for sugarcane cultivation. This chapter brings into focus the socioeconomic situation and potential impacts of further biofuel expansion in the coastal area of the poor Northeast of Brazil, whose economy and society have been fundamentally shaped by sugarcane cultivation since the late sixteenth century. In particular, the chapter starts from the assumption that the highly unequal power relations in the Northeast crucially condition the impact of biofuel expansion in this region and that the various forms of exercise of power should be given greater attention when designing biofuel policies, notably international sustainability certification. In the context of the highly polarized social relations and competing development paradigms, a crucial question concerns the degree to which sustainability certification can indeed help profoundly transform rather than consolidate the prevailing unequal social and power relations.

Keywords: Brazil, certification, power relations, social sustainability

1. Introduction

While the environmental and economic dimensions of sustainability have received plenty of attention in biofuel policies and assessment, only recently has the social pillar gained increasing weight, demonstrated, for example, by debates over dilemmas such as food versus fuel and large-scale versus small-scale biofuel production (e.g., Eide, 2009; von Braun and Pachauri, 2006; see Chapter 1). Perhaps to an even greater degree than environmental sustainability, decisions and debates concerning the social aspects of sustainability are intricately linked with the different forms of exercise of power in society. In biofuel policies, the importance of power relations manifests itself, for instance, in the unequal access of different actors to economic resources and political decision making, including unequal structures of landownership (e.g., Cotula et al., 2008). Inequality also concerns the discursive level in that different actors have highly varying degrees of influence on the ways in which problems are framed, on whose knowledge of the problem is considered authoritative, and on which societal visions are perceived as desirable.

Such unequal power relations crucially influence policies in the emerging context of hybrid or multilevel governance in the area of biofuels (see, e.g., Swyngedouw, 2005). The definition of standards, meta-standards, labeling schemes, and codes of conduct (often in collaboration between supranational organizations, governments at national and subnational levels, industry, and civil society organizations) is a central part of such governance. The multiplication of sustainability certification schemes at different levels has prompted attempts at harmonization, for instance, through the creation of meta-standards or internationally harmonized certification systems, notably within the European Union. In parallel, there is an emerging academic literature on the subject, including suggestions on how an appropriate certification system should look (for suggestions concerning Brazil, see, e.g., Delzeit and Holm-Müller, 2009).

This chapter starts from the assumption that the consequences – the success – of a sustainability certification scheme crucially depend on the underlying sociopolitical and institutional conditions prevailing within the sector. The chapter therefore examines the social and political conditions prevailing in the sugar and ethanol sector, with particular attention given to the sugarcane-growing zone in the Northeast region of Brazil. It draws some tentative conclusions on the ways in which such conditions affect the ability of sustainability certification to alleviate the harmful social impacts of ethanol in the Northeast.

The case of Brazil is relevant not only because of the country's position, since the launching of the national alcohol program in 1975, as the world leader in the production of sugarcane ethanol but also because of Brazil's ambitions to turn ethanol into an internationally tradable commodity (see Chapter 6). In this task, the establishment of a credible international sustainability certification scheme is expected to play a central role (Huertas et al., 2010). Furthermore, certification has been suggested as a key mechanism for introducing social aspects into a biofuel debate hitherto largely dominated by environmental concerns (Abramovay, 2008). Indeed, a broad consensus prevails today concerning the position of Brazilian sugarcane. Among different biofuel practices, Brazilian biofuel is considered a special case and as probably the most environmentally sustainable (or at least one of the least unsustainable) among the currently available first-generation biofuels (see, e.g., RFA, 2008; Wilkinson and Herrera, 2010). Debates and controversies still continue over the exact amount of GHG emission reduction, impacts of the expansion of sugarcane cultivation on Cerrado shrublands, the indirect land use impacts, and the wider environmental impacts of Brazilian sugarcane ethanol (e.g., on biodiversity, water use, and water pollution; see Chapter 6). Yet the overall environmental impacts of sugarcane cultivation are deemed comparable to or slightly less harmful than those from the large-scale cultivation of other comparable cash crops (e.g., Amaral et al., 2008). While not negligible, the pollution problems are usually considered as manageable, for instance, by using the effluents from sugar and alcohol mills for ferti-irrigation in the fields and by mechanizing sugarcane harvesting to avoid the need to burn sugarcane fields prior to harvesting (Amaral et al., 2008).

By contrast, the social impacts of ethanol are much more ambiguous. Key issues of concern include the model of large-scale monoculture cultivation inherent to sugarcane production, a highly unequal distribution of landownership, often appalling working conditions, frequent violations of worker legislation, and the serious health hazards that inappropriate use of agrochemicals poses to sugarcane field workers (e.g., Alves, 2006; Noronha et al., 2006; Plataforma BNDES, 2008). Such social problems are prevalent in all sugarcane-producing areas of the country, yet this chapter highlights the Northeast as a largely forgotten area in debates concerning Brazilian bioethanol. The chapter starts from the assumption that despite the limited contribution of the Northeast to the country's ethanol supply, the sociopolitical conditions in the region may be crucial for the success of sustainability certification. Furthermore, the evolution of the ethanol sector will have crucial impacts on the socioeconomic well-being of the region.

The social impacts of ethanol in Brazil are conditioned by two profound and partly contradicting historical tendencies in the country's sugarcane ethanol sector. The first, eagerly put forward by the country's political and industrial elite, is the sector's position as the spearhead of Brazilian modernization and technological innovation (e.g., Bound, 2008; Furtado et al., 2011). The second, conversely, is frequently underlined by the critics of the current large-scale biofuel-based development model: the centuries-long burden of large-scale plantation agriculture based on slave labor (e.g., Andrade, 1988; Dabat, 2007; Porter et al., 2001; Rogers, 2005; Wolford, 2004). Until the recently begun internationalization of biofuel politics (see, e.g., Hollander, 2010), this dualistic nature of the Brazilian ethanol sector could conveniently be expressed in terms of center–periphery relationships (Lehtonen, 2011). Such relationships were manifest both between the regions, notably between the prosperous and dynamic São Paulo and the poor and stagnant Northeast, and between social classes within these regions. Especially in the traditional sugarcane-growing coastal zone of the Northeast, such social inequities have been pervasive since sugarcane cultivation was brought to the region in the sixteenth century. However, with the emergence of global biofuel markets and politics, the historically unequal power relations and center–periphery relations are being embedded in, and transformed by, a far more complex setup involving the entry of new players onto the field; the creation of new network structures; and an increasing interplay between governments, industry, independent experts, and nongovernmental organizations (NGOs) across national boundaries (Saturnino et al., 2010).

This chapter argues that for certification to contribute to social sustainability of bioethanol in the Northeast of Brazil, the entrenched center–periphery relationships in the sugar and alcohol sector would need to undergo a profound transformation. The Northeast, especially the state of Pernambuco, continued to be the major sugarcane-producing region of Brazil until the early 1950s (Andrade, 1988). However, drawing benefit from more favorable natural conditions, and thanks to its more entrepreneurial sugar and alcohol sector, São Paulo has progressively established its hegemony as the country's absolute leader in sugarcane, sugar, and ethanol production. Today the Northeast accounts for less than 15 percent of national production (DIEESE, 2007). Despite this minor participation of the Northeast in the national ethanol supply, the evolution of the sugar and ethanol sector crucially shapes the socioeconomic conditions in the coastal sugarcane zone, especially in the states, such as Alagoas and Pernambuco, that are economically most dependent on sugarcane.[1]

While recognizing the importance of the concrete manifestations of inequality, in terms of access by different groups in society to economic and political power, this chapter focuses in particular on the discursive dimensions of power and the role of different paradigms or worldviews in guiding actors' attitudes toward biofuels in general and sustainability certification in particular. In a context in which sustainability has become a dominant discursive frame for biofuel debates, and the development of certification systems is perceived as a major tool for operationalizing sustainability, the key question concerns the discursive struggles over the definition of the terms and framing of the debate. In Brazil, such discursive battles are conducted in the context of concerted and long-standing efforts:

1. By the government and the ethanol industry, with support from numerous Brazilian and foreign experts, to define ethanol as a modern, clean, and environmentally benign alternative to petrol (e.g., CTC, 1989; Goldemberg et al., 2008; Zuurbier and van de Vooren, 2008)
2. By the Northeast ethanol industry to secure state support in the name of social sustainability (e.g., Bacoccina, 2007)
3. By Brazilian and international NGOs to resist the development of biofuels, seeing them as a source of social injustice and environmental destruction (e.g., Mendonça, 2006; Noronha et al., 2006; Plataforma BNDES, 2008; *Repórter Brasil*, 2010c)

---

[1] In the mid-1990s, 75.4% of the gross value added in agriculture in Alagoas and 36.1% in Pernambuco came from sugarcane cultivation. Moreover, in many municipalities in the Zona da Mata, sugarcane cultivation was virtually the only source of employment for local workers (Lima and Sicsú, 2001).

The chapter is structured as follows: Section 2 briefly describes the evolution of the sociopolitical structure of the Brazilian sugar and alcohol sector, followed in Section 3 by a brief summary of the main social issues related to sugarcane ethanol in Brazil. Section 4 then describes the situation in the coastal sugarcane-growing zone of the Northeast of Brazil and considers the potential impacts of the ongoing biofuel boom on the social relations in the region. Section 5 considers the potential and limits of biofuel certification against three key principles of successful certification. Finally, Section 6 situates the debates on sustainability certification within a broader framework of competing paradigms and the increasingly globalizing biofuel politics.

2. Brazilian sugarcane ethanol and power: History and structures

Sugarcane cultivation in Brazil started in the country's Northeast coastal zone (Zona da Mata) in the mid-sixteenth century. Until the abolishment of slavery in 1888, cultivation was based on slave labor, which has left its marks in the highly unequal worker–employer relationships, antiquated production methods, and overwhelming dominance of sugarcane in the economy and society of large areas of the Zona da Mata. The type of center–periphery relationships observed in the sector are, nevertheless, more complex than a mere opposition between the industrialists and farmworkers. Indeed, the sugarcane planters (*fornecedores*) have played a historically central role in the Northeast sugar and ethanol industry (e.g., Andrade, 1988; Porter et al., 2001).

Furthermore, the impacts on discourses and social identities in the Northeast have been at least equally significant. The complexity and the disputed character of the social relations and their influence on public imagery are highlighted by Rogers (2005: 2), who argues that even the renowned sociologist Gilberto Freyre, despite his crucial contribution in producing the first critical cultural history of the Northeast slave economy and in revealing the deeply rooted structures of domination in northeastern society, paradoxically also helped perpetuate the elite discourse that naturalized this very domination. In the same vein, Dabat (2007: 219) describes the kind of naturalization and determinism in Freyre's argumentation, which portrayed both landowners and slaves as victims of a *force majeure*, of natural, environmental, and historical phenomena, thereby essentially freeing the actors from any responsibility for the prevailing social conditions.

Part of the Northeast's self-identity is the history of oppression by the dominant landowning classes and the continuous economic decline of the region, which was paralleled by the Southeast's modernization (e.g., Andrade, 1988; Porter et al., 2001).[2] This decline became manifest in the sugar and alcohol sector as the Northeast gradually lost its leading position to São Paulo, as part of the modernization of the sector that

---

[2] The move of the capital from Salvador to Rio de Janeiro in the eighteenth century and the rise of São Paulo as the leading industrial area in the early twentieth century were signs of this modernization.

began in the early twentieth century. The Northeast state of Pernambuco was the country's largest producer of sugar until the early 1950s, but by the mid-1970s, the eve of Proálcool, São Paulo had taken Pernambuco's place as the incontestable leader in the sugar and alcohol sector (Andrade, 1988: 516; see also Chapter 6).

As discussed in chapter 6, the Proálcool program was launched in November 1975, mainly in response to, first, the oil crisis, which threatened to put an end to the Brazilian economic miracle, and second, the rapid collapse of sugar prices in the world market. Lobbying from the sugar and alcohol sector was elemental in bringing about the program (e.g., de Castro Santos, 1987; Demetrius, 1990). The program laid the basis for Brazil's hegemony as the leading nation in sugar and alcohol–related technology and know-how and for a continued improvement of the average productivity and economic viability of ethanol production in the country (e.g., Goldemberg et al., 2004a). However, in doing so, it also supported the economically unprofitable and both technologically and culturally backward sugar and alcohol sector in the Northeast coastal zone. The generous subsidies protected the Northeast sugar and alcohol sector from competition and allowed sugarcane to expand to marginal and less productive lands, to new states, and to areas lacking previous tradition in and suitable natural conditions for sugarcane cultivation. In consequence, food crops were increasingly pushed to marginal lands, thereby strengthening the domination of sugarcane as the hegemonic crop in the Northeast's coastal zone, monopolizing the utilization of newly opened lands, and increasing the concentration of landownership (e.g., Grenier, 1985). In summary, Proálcool helped inefficient producers in the Northeast stay in business, prevented structural change toward more egalitarian power relations, and impeded the diversification of agriculture in the region (e.g., Demetrius, 1990; Lehtonen, 2011; Puppim de Oliveira, 2002; Sandalow, 2006). Proálcool has therefore been described as yet another expression of the centuries-long continuity and inertia in the social and power relations within the Northeast sugarcane-producing region (e.g., Andrade, 1988; Dabat, 2007: 18–20).

The phasing out of subsidies to the sugar and alcohol sector and the virtual disappearance of the Proálcool program started in 1986. Factors contributing to the demise of the program included the drop in international oil prices, the increase in domestic oil production, record-high international sugar prices, and liberalization policies aimed at solving the acute debt and economic crises (e.g., Brilhante, 1997; Novo et al., 2010). The drastic reduction of direct government support initiated a rapid decline in the share of ethanol in transport fuel supply and a process of reorganization in the sugar and ethanol sector. However, the abolishment of subsidies forced the sugar and ethanol industry, especially in São Paulo, to seek technological innovation (Puppim de Oliveira, 2002) and led to the creation of a strong sectoral innovation system, driven by private sector R&D (Furtado et al., 2011). In the Northeast, the loss of government support led to yet another crisis in the sugar and alcohol sector but stimulated reorganization of the industry, prompting, notably, the minority of the more entrepreneurial Northeast sugar and alcohol producers to invest in production facilities in São Paulo and in the Center West (particularly in Minas Gerais) to benefit from the more favorable agroecological conditions for production in the Southeast (e.g., Porter et al., 2001: 851). Many producers also diversified into a wide range of other activities across the country, including ceramics, car dealerships, cement, textiles, and hotels (Porter et al., 2001: 833). This period also saw a reorientation of the sugarcane crop toward the production of sugar, to the detriment of ethanol (Novo et al., 2010).

The reorganization of production in the Northeast opened a window of opportunity with the potential to reduce the dominance of sugarcane in the regional economy, and thereby to help diversify the agrarian structure and alleviate the polarized social relationships. However, largely because of the rigid social structures in the region, the restructuring remained only a promise, consisting essentially of a reorganization of production in a more efficient manner for the sugarcane industry (Wolford 2004: 159). Importantly, the conditions for restructuring and diversification were again compromised by the ethanol boom, which followed the introduction in the national market in 2003 of the flex-fuel vehicles, capable of running on either petrol, ethanol, or any combination of the two. As a result, sugarcane cultivation began to expand, especially in states outside the traditional sugarcane regions of São Paulo and the Northeast, such as in the states of Goias, Mato Grosso do Sul, and Minas Gerais. In recent years, the cultivated area in these states has increased by between 20 and 50 percent (*Repórter Brasil*, 2010c: 7). The Lula government took a number of steps to facilitate the expansion of ethanol production, resulting in the installation of more than 100 new mills since 2006, with an estimated production capacity of one-third of the total existing production in the country.[3]

Despite the expansion of sugarcane cultivation into new areas, the state of São Paulo still produces about 60 percent of Brazilian sugarcane, while the share of the Northeast has continued to decline, being currently less than 14 percent (DIEESE, 2007; Chapter 6). The declining share of the technologically stagnant Northeast in the country's ethanol supply has also helped to increase the average productivity in the ethanol sector as well as to make the sector economically independent of direct state subsidies. In practice, numerous forms of support are still in place, including favorable tax treatment to ethanol, blending mandates, tax reductions and exemptions for flex-fuel vehicles, and, most notably, support for investment and R&D (e.g., Gauder et al., 2011: 673; Novo et al., 2010; Plataforma BNDES, 2008). Indeed, a major part of the expansion in the sugar and alcohol industry is funded by public resources through the Brazilian economic and social development bank (BNDES), Banco do Brasil, and other government agencies (FIAN, 2008: 23).

---

[3] During the previous presidential election campaign, Lula had been described as the president with the most favorable policy ever toward the sugar and ethanol sector in Brazil. Lula indeed went as far as calling mill owners "heroes" (Valor Econômico, 2010).

The transformation of São Paulo into not only the country's industrial core but also an agricultural powerhouse laid the early bases for Brazil's current agroindustrial development model, of which the sugarcane ethanol industry is an integral part (e.g., de Miranda et al., 2007; Hall et al., 2009; Wilkinson and Herrera, 2010). Hence the Lula government's so-called ethanol diplomacy, that is, its efforts to turn ethanol into an internationally tradable commodity and to thereby create international markets for ethanol and ethanol-related know-how and technology, must be seen against the central place that ethanol occupies in this agroindustrial development model (see Chapter 6). An essential objective of the ethanol diplomacy is the expansion of the number of producing countries to reassure potential buyers about the security and stability of supply (e.g., Hollander, 2010; *Repórter Brasil*, 2010c: 9; see Chapter 6). Sustainability certification, in turn, is a vital element in attempts to ensure that Brazilian ethanol enjoys sufficient legitimacy and credibility in the eyes of buyers.

3. The social virtues and vices of ethanol in the Northeast

As noted earlier, the environmental impacts of Brazilian ethanol are in general considered less harmful than those engendered by other first-generation biofuels and are comparable to those from other large-scale agricultural crops. Numerous efforts have, indeed, been made, especially in São Paulo, to reduce the harmful environmental impacts of ethanol (see Chapter 6). These measures range from economically motivated win-win measures, such as ferti-irrigation, that is, the process of irrigating sugarcane fields with the high-nutrient effluents from the ethanol factories, to legislation and government-induced agreements fostering environmentally responsible farming and production practices (Amaral et al., 2008).[4]

Conversely, certain negative social impacts are still pervasive in the sector (see Chapter 6). In fact, the political influence of the sugar and ethanol elite has decisively helped to perpetuate the highly unequal social relations in the sector. These social relations have been, throughout history, particularly polarized in the Northeast sugar and ethanol sector (e.g., Porter et al., 2001). While the social problems in the state of São Paulo are largely associated with the increasing pressure to improve productivity, the roots of inequality in the Northeast are more diverse. In particular, the region's sugar and alcohol elite has been frequently criticized for its lack of entrepreneurial spirit, manifest in the landowners' and industrialists' major preoccupation with retaining their political power rather than productivity and performance improvement (e.g., Demetrius, 1990; Mendonça, 2006; Porter et al., 2001).

---

[4] The agreements include the Green Ethanol Program, one of São Paulo Environment Secretariat's strategic projects, whose aim is to reward good practices in the sugar and alcohol sector; the Brazilian Sugarcane Industry Association, UNICA, Sustainability Report, based on Global Reporting Initiative (GRI) guidelines; and the agroecological zoning (ZAE) for sugarcane, introduced in 2009 by President Lula, which forbids the expansion of sugarcane cultivation and the installation of new mills in environmentally sensitive areas (Decreto, 2009; Schaffel and La Rovere, 2010).

The federal government and the sugar and ethanol industry have in recent years undertaken a number of measures to reduce the social inequalities and improve the working conditions in the sector (e.g., FIAN, 2008; Schaffel and La Rovere, 2010).[5] While the motivations behind such efforts are certainly diverse, they can be seen as essential elements in the attempts by the industry and the government not only to expand the production of ethanol but also to ensure international buyers that Brazilian ethanol is produced "sustainably" (*Repórter Brasil*, 2010c: 35). Hence the interpretations concerning the efficacy and sufficiency of such efforts are highly contrasted. On one side are those who see these industry efforts as little more than palliative measures, a mere fig leaf aimed at legitimizing the continued exploitation of the workers and smallholders; on the other side are those who see in such efforts welcomed signs of the ecological modernization of the sugar and ethanol sector (Fernandes et al., 2010; Schaffel and La Rovere, 2010). These views will be examined in more detail in Section 6.

4. The forgotten Northeast

Most of the debate on Brazilian bioethanol has thus far tended to concentrate on the potential impacts of sugarcane cultivation in São Paulo and the new production regions, while the Northeast tends to receive little attention, both in the national media and in research concerning the scenarios of increasing ethanol production in the country. At least three reasons can help to explain the relative neglect of the Northeast. First, given the declining share of the Northeast in national sugarcane and ethanol supply, the area is simply considered "not to count" for much in terms of the sustainability of the country's bioethanol sector. Second, since most favorable production sites in the coastal area of the Northeast (as well as in the state of São Paulo) have already been taken into sugarcane cultivation, the potential for expansion in the area is limited. The main areas of expansion are instead situated in the Center West of the country, in the states of Goias, Minas Gerais, and Mato Grosso do Sul (see, e.g., WWF Brasil, 2009). Third, the reputation of the Northeast sugarcane sector, notably the image of the landowning elites as a technically and culturally backward, corrupted class dependent on continuous state support (e.g., Lehtonen, 1993), means that the

---

[5] These efforts include, notably, the RenovAção program for retraining of workers delocalized as a result of mechanization of sugarcane harvesting, and the National Commitment to Improve Labor Conditions in the Sugarcane Cultivation (a state-led three-partite agreement signed by federal and state authorities, including five ministries, as well as by labor market organizations at national and São Paulo levels). In 2006, a protocol of intention was signed between UNICA and the Federation of Rural Wage Workers of the State of São Paulo (FERAESP), foreseeing the elimination of intermediaries by 2011 (FIAN, 2008: 42). The previously mentioned bill for agroecological zoning (ZAE) for sugarcane forbids the expansion of sugarcane cultivation and the installation of new mills in areas where sugarcane might threaten food security (Schaffel and La Rovere, 2010). Further measures by the Lula government include the adoption of a National Plan for the Eradication of Slave Labor and the establishment of the National Commission for the Eradication of Slave Labor (FIAN, 2008: 39).

Northeast is frequently perceived as a threat to the image of the Brazilian ethanol sector in general and to the credibility of sustainability certification in particular.[6]

4.1. The continuous decline of the Northeast sugar and ethanol sector

Trends in the number of sugar and ethanol mills in different areas of Brazil illustrate the continuing decline of the Northeast. In 2006, 84 out of the country's 334 mills were located in the Northeast, and no new mills were under planning or construction, while the Southeast hosted almost two-thirds (205) of the mills (NAT, 2007). As many as 113 new mills were expected to be constructed throughout the country between 2005 and 2014, of which 116 were in the Southeast, 41 in the state of São Paulo alone (Nastari, 2006). The Northeast sugar and alcohol sector crisis reached its deepest point at the end of the 1990s, when, for example, in Pernambuco, according to media reports, only one-third of the 45 mills that processed cane in the late 1970s were still in operation, others having gone bankrupt or having ceased to operate (Gusmão, 1998). Between the harvest years 1982–1983 and 1994–1995, the share of the Northeast in Brazil's sugarcane production declined from 30 to 18 percent (Lima and Sicsú, 2001). In 2006, the number of Pernambuco mills in operation had increased again to 30 (NAT, 2007). Yet local newspapers reported a total of 18 mills having been closed in the state of Pernambuco between 1992 and 2007, while 30 new mills were set up in São Paulo between 2005 and 2007 alone (Bahé, 2007).

The sugarcane-growing area in the Northeast is characterized by extreme inequality, be it in terms of income, wealth, or education. Pernambuco has the fifth lowest Human Development Index (HDI) in the country, with the coastal area populations often enduring the poorest living conditions and lacking access to most elementary social infrastructure such as sanitation, medical care, or safe drinking water (e.g., Sicsú and Silva, 2001). About 70 percent of the working population in the Northeast coastal zone earns the minimum wage or less, and only 2.55 percent have a higher-level educational degree (Saldanha, 2005). According to the peasant farmers' movement CPT (Comissão Pastoral da Terra), in the municipalities of the sugarcane-growing coastal zone in Pernambuco, the concentration of landownership reaches figures as high as 0.9, as measured by the Gini index[7] (Junior and Albuquerque, 2010). Sugarcane field workers have a particularly low educational level, with one-third of sugar workers in the Northeast being illiterate against 4 percent in the Center South (Krivonos and Olarreaga, 2006: 96, 102). Almost 40 percent of them lack guaranteed labor rights,

---

[6] This image and the Southeast–Northeast feud surfaced again in 2008 in a dispute following accusations by environment minister Carlos Minc, who called the northeastern *usineiros* outlaws and the state of Pernambuco a "disaster of a disaster" (Éboli, 2010).

[7] Gini index measures the extent to which the distribution of income (or, in some cases, consumption expenditure) among individuals or households within an economy deviates from a perfectly equal distribution. An index of 0 means perfect equality, whereas an index of 1 implies perfect inequality.

and their salary reaches only 32.6 percent of the average wage of an ethanol factory worker in the same region (DIEESE, 2007: 21). Again, in the Center South, a field worker earns almost 50 percent of the average salary of an ethanol factory worker (DIEESE, 2007: 21). The crisis in the Northeast sugar and ethanol sector in the 1990s was accompanied by a similarly serious and "deepening crisis for labor," especially in the older sugar zone of the southern Zona da Mata (Porter et al., 2001: 843).

The modernization of the sugar and alcohol sector and the rise of São Paulo forced the Northeast sugar and ethanol industry to seek greater competitiveness through modernization of its production systems. Since the Second World War, a small number of large landowning industrialists (*usineiros*) gained greater control over the sector both in the Northeast and in São Paulo. In the Northeast, this process led to the gradual loss of markets and status among the once dominant small- and medium-sized sugarcane farmers (*cultivadores* and *fornecedores*), as the industrialists increasingly cultivated their own cane, instead of buying it from the *fornecedores* and *cultivadores* (e.g., Andrade, 1988).[8] For sugarcane field workers, this decline of the small and medium producers has been a double-edged sword. On one hand, the small- and medium-sized planters (sugarcane producers that supply cane to the industry) are major employers, as their production methods are highly labor intensive. On the other hand, it has been suggested that the working conditions on their farms are often even worse than on the large industrial farms (Andrade, 1988; Porter et al., 2001: 837). Yet some contest the evidence, especially labor movement activists, who see such claims as merely attempts by the sugarcane industrial elite to create a false opposition between the "modern" *usineiros* and the "backward" planters (Porter et al., 2001: 837). In any case, dichotomies opposing the good small-scale ethanol and bad large-scale agribusiness should be viewed with caution and seen as simplified representations of a far more complex reality.

The Northeast production mode is highly labor intensive, reflected in the fact that the region provided in 2006 about 25 percent of the jobs in the country's ethanol sector, even though its share of total ethanol production was less than 10 percent (Compéan and Polenske, 2010).[9] The social and employment benefits of sugarcane in the Northeast are one argument used by the region's sugar and ethanol elite in favor of state support and of social certification for sugar and ethanol from the region (Bacoccina, 2007). Most of the jobs are, however, low-quality, hard manual labor because the high declivity of terrain in the coastal zone makes mechanized harvesting impossible in many areas of the region (e.g., Compéan and Polenske, 2010). Moreover, there has been a long trend of proletarization of the workforce, which has reduced the

[8] The state of Pernambuco had, in 2007, an estimated 13,000 *fornecedores* and *cultivadores* (most of them family farmers) producing between 1 and 1,000 tons of sugarcane per year. The proportion of sugarcane provided by these producers of the entire amount of sugarcane processed by the mills in Pernambuco had declined to between 20% and 30% from the all-time high of 70% (Fritz, 2008: 11; *Jornal do Commercio*, 2007a).

[9] The share of the Northeast of employment in sugarcane cultivation is probably even greater (see Guedes, 2010).

opportunities of the farmworkers to cultivate subsistence crops for their own use. The mill owner may still provide land, but the plots have become smaller and are often situated farther away from home (Porter et al., 2001).

The expansion of biofuel production in the Southeast has also indirectly influenced the social conditions in the Northeast, a region which provides a large share of the migrant seasonal labor force to São Paulo's sugarcane plantations. Ramos (2006) estimated that almost 120,000 migrants were cutting cane in São Paulo, most from the Northeast. These workers are frequently nonunionized and often end up being exploited by the intermediaries recruiting workers for the mills in the Southeast (e.g., Fritz, 2008: 14; see Chapter 6). Allegedly, many mill owners in São Paulo prefer using migrant workers because these are "more vulnerable and thus tend to accept intense levels of exploitation and low wages without complaint" (FIAN, 2008: 42). However, a similar trend toward an increasing use of temporary contracts has been observed in the Northeast, which, according to Porter et al. (2001: 840), is a sign of "employer efforts to divest themselves of social attachments and thus achieve a more 'competitive' labor market."

4.2. Political, economic, and cultural hegemony of the Northeast usineiro

A major obstacle preventing the improvement of social conditions in the Northeast is the historically highly unequal distribution of power between the sugar barons, on one hand, and the poor and uneducated farmworkers, on the other. Unequal power relations and poor labor conditions are persistent problems in many sectors of Brazilian agriculture. These problems are specific and arguably particularly complex to solve in the sugar and ethanol industry, for two main reasons: first, because the history of sugarcane cultivation was characterized by the exploitation of slave labor, and second, because of the dominance of the sugar and ethanol sector in the regional economies in the northeastern coastal zone. Any solutions to the social problems in the sugarcane zone of the Northeast are complicated by what Porter et al. (2001: 829) call the "historically embedded mistrust, on all sides, among players in the Zona da Mata," that is, in the sugarcane-growing coastal zone (see also Wolford, 2004). Furthermore, social inequality is perpetuated by the intimate links between economic and political elites in the region, as the large-scale landowners and industrialists in the sugar and alcohol sector traditionally used either to choose among themselves a candidate for state elections or support a common candidate (Demetrius, 1990; Johnson, 1983). The consequences of this political inequity, the almost caricatural center–periphery relationships between social classes in the Northeast, manifest themselves in multiple ways, with more or less direct consequences for the social conditions in the area.

Views diverge concerning the role of these sociopolitical conditions in explaining the reasons for the continuous decline, recurrent crises, and low labor productivity in the Northeast sugar and alcohol sector. On one hand, natural conditions (the hilly landscape) and the end of government subsidies following the abandonment of the Proálcool program in the late 1980s have been suggested as the major reasons for the latest crisis of the sector in the 1990s. However, such an explanation has been called into question for a number of reasons. First, the willingness of the state to act as the ultimate guarantor and bail out the debt-ridden Northeast producers in times of trouble can have led to a low uptake of technology and the absence of productivity improvement (Sicsú and Silva, 2001). Second, at the national level, the sector continued to expand even in the 1990s, despite the abolition of subsidies (Moura et al., 2004: 80). Porter et al. (2001: 832, 850) seek explanations in the history of the region and argue that the ways in which the global production relations in the sugar and alcohol industry are embedded in national and local processes work mostly to the disadvantage of the Northeast. Hence, in the Southeast, labor shortages since the mid-1850s meant that European immigrants worked alongside slave labor, the region was consequently open to the introduction of labor-saving machinery, and wage flows led to growth in demand and capital accumulation. By contrast, the Northeast was characterized by a concentration of income through slavery and the consequent curb on economic activity (Jatobá, 1986). Thus, in the Center South, the sugar and alcohol sector is less embedded in disadvantageous and historically uneven social contexts. Furthermore, Porter et al. (2001: 849) argued that even during the crisis years in the 1990s, "the members of the landowning elite continue[d] to exert remarkable pressure, despite their economic difficulties and the high degree of rural unionization."

A particularity of the Northeast is the coexistence of the old paternalistic tradition alongside the strong labor movements, which arose as a reaction to the modernization and proletarization of the workforce in the 1950s (Porter et al., 2001). While the trade unions were repressed in the 1964 military coup, they nevertheless continue to mark identities and memories in the region, without nevertheless being able to dismantle the remarkably persistent paternalism (Garcia, 1988; Scheper-Hughes, 1992: 90, cited in Porter et al., 2001). Such coexistence is illustrated in the varying explanations for the poor productivity in the Northeast sugar and alcohol sector. On one hand, especially the sugar and alcohol sector elites emphasize so-called cultural explanations and high union activism (e.g., Lehtonen, 1993; Porter et al., 2001: 833, 843). Porter et al. (2001: 833) refer to a representative of the Northeast sugarcane employers' organization, who argued that laborers were simply lazy, as they "work only three to four hours per day, despite the wage incentives provided by some millers to improve performance." Despite clear evidence of employers' desire to cut the old semifeudal linkages to labor (to depersonalize relationships) to achieve a more competitive labor market, landowners and industrialists, conversely, hesitate to adopt the "hard position on labor" necessary for breaking the cultural habits (Porter et al., 2001: 834–835). To alleviate the problems of low labor productivity, some companies reputedly even hire labor from the state of São Paulo to "show and motivate" local people to increase their workload (Porter et al., 2001: 834–835).

By contrast, worker unions explain the low productivity through the nature of the Northeast's terrain (especially in Pernambuco) as well as through the climate and associated technological difficulties rather than through worker deficiencies. Moreover, Porter et al. (2001) refer to diverging economic rationalities as a fundamental explanation for low labor productivity. Not only are cutting requirements lower in the Northeast than in São Paulo, but instead of maximizing their daily cutting to maximize their income, workers place a higher priority on allocating time for working in other activities, notably subsistence farming. Farmers have introduced, in vain, a number of incentives (bonus systems, prizes, and midday meals) to persuade workers to stay on and cut a second day's workload in the afternoon for a second wage. Labor unions, in turn, argue that their workers generally cut far more than one day's "task" (Porter et al., 2001: 844).

In summary, Porter et al. (2001: 826) observe that "historical antagonisms between (white) landowner classes and (black) labor, and their reworking in recent years, complicated by competing economic rationalities between these groups, make progress towards viable, sustainable socioeconomic improvement difficult." Hence Moura et al. (2004) call for an "attitude change" in the sector as a necessary condition for productivity improvement, achievable through the formation of producer cooperatives to foster the research and adoption of new production methods, sugarcane varieties adapted to the needs of the region, and better use of subproducts.

While São Paulo sugar and ethanol industrialists are clearly in a leading position in decision making within the sector nationally, the Northeast sugar and alcohol elite has, at least until recently, managed (thanks to close ties with political decision makers both at national and regional levels) to retain its privileges and exert an influence disproportionate to its contribution to domestic sugar and ethanol output. This continuing hegemony of the sugar and ethanol elite in northeastern society has also helped perpetuate the historically unequal relations between the social classes. At the national level, the sector seemed to lose some of its direct access to political power due to the drastic reduction of state support to the sector since the late 1980s, but the recent biofuel boom has placed sugar and ethanol production again at the heart of the Brazilian development model (e.g., Bound, 2008; Furtado et al., 2011), in which export-led agriculture plays a crucial role (e.g., de Miranda et al., 2007; Wilkinson and Herrera, 2010).

Two enduring features of the Northeast sugarcane economy, and an illustration of the political influence of the Northeast sugar and alcohol sector elite, have been the chronic indebtedness of the sugar and alcohol producers in the region and the state's tradition of rescuing the industry from its repeated economic crises. While the sugar and alcohol sector in general has held close links with political power, the northeastern *usineiros* have been highly successful in securing continued state support in the recurrent situations of crisis. The São Paulo producers have direct access to political power at the federal level through the state's sugar and alcohol producers cooperative, UNICA (until 1995, Copersucar), whereas the Northeast sugar elite used to have two main avenues for exercising political power. First, the Sugar and Alcohol Institute was, from the time of its creation in the early 1930s to its abolishment in 1990, a major defendant of the interests of especially the independent sugarcane growers in the region, for instance, by developing cane varieties adapted to the climatic and soil conditions of the Northeast (Lima and Sicsú, 2001). Second, the northeastern sugar barons have traditionally had indirect access to political power at the national level, through their close links with the regional political elites.

The strong ties between the sugar and alcohol elite with regional decision makers that have ensured continuous economic support to the sector have also helped prevent the diversification of agriculture and economic activities, recognized for a long time as essential for harmonious development in the cane-growing coastal zone (e.g., Cavalcanti et al., 2002; FIPE, 1991). Some progress has taken place, partly thanks to efforts by the regional authorities to promote the cultivation, in the coastal zone, of local fruit crops (e.g., Cavalcanti et al., 2002). Yet, for example, the state of Alagoas still depends to more than 90 percent on sugar, ethanol, and molasses for its export revenue (Saldanha, 2005). Again, Porter et al. (2001: 845–846) note that "it is the planter group, tenuously clinging to its rural power base, which seems least attracted to diversification and most keen to retain labor-intensive agriculture." They evoke, again, the cultural traditions and entrenched power relations as potentially significant explanations of such reluctance to engage in further diversification. Moving out of cane production would appear to some farmers as "the end of an era," perceived as a threat of loss of the long-standing power and authority over the region (Porter et al., 2001: 845–846).

The political clout of the northeastern sugar barons has also been a major reason for the weak implementation and enforcement of the slowly emerging labor and environmental regulation in the sector. Both burning of the fields prior to harvesting and the discharging of distillery effluents untreated to watercourses still represent serious problems in the Northeast, admittedly due in part to economic reasons (Abramovay, 2008). The use of child and slave labor is probably an even more acute problem in the Northeast than in the Southeast. Indeed, in 2009, 4 out of the total of 16 mills accused of using slave labor in the country were from the northeastern state of Pernambuco (*Repórter Brasil*, 2010c: 16).

4.3. The biofuel boom and the future of Northeast sugar and ethanol sector

In light of the highly unequal sociopolitical situation described earlier, and in the context of the presently ongoing biofuel boom, one must ask, what are the prospects for northeastern Brazil's sugar and ethanol sector in general and its social sustainability in particular? Could the rapid expansion of the global demand for ethanol help revive "derelict sugarcane industries," or is such an expectation unrealistic (as argued by Hollander, 2010: 716) in a situation where most investments go to greenfield enterprises in regions without previous experience in large-scale sugarcane cultivation? And furthermore, even if such a revival takes place, would this reduce social inequality in the Northeast? From a pessimistic perspective, one could fear that the current ethanol boom will not benefit the northeastern sugar and alcohol sector simply because poorer agroecological conditions make the region uncompetitive. Some have even argued that the Northeast has already "given what it can" in terms of sugarcane because the best areas for sugarcane are already in cultivation, and there is not much room for productivity improvement in the area (e.g., Gusmão, 1998). According to this argument, not only do sociopolitical and cultural factors work against productivity improvement but so do the topography – which makes mechanization difficult, the concentration of the areas most suitable for sugarcane in a relatively small coastal area, and the four decades of monoculture that have exhausted the soil (e.g., Compéan and Polenske, 2010; Smeets et al., 2006: 36–37). And yet, it can equally be argued that the potential for expansion is in fact more limited in the Southeast (especially in São Paulo), where almost all suitable lands have already been taken into sugarcane cultivation (see Frizzone et al., 2001: 1131). The Northeast, in turn, still has potential areas for expanding sugarcane cultivation, especially in irrigated areas.

Sicsú and Silva (2001) have evoked further comparative advantages of the Northeast, notably the following:

- The different timing of harvesting periods in the Northeast and in the Southeast, which could lead to complementarity rather than competition between the two regions
- The possibility to concentrate production in the areas best adapted to sugarcane, notably the "várzeas," where productivity can reach 140 metric tons/ha, under irrigation
- Preferential access of northeastern sugar to the American market (quotas and guaranteed prices above the world market price)
- While 30 to 40 percent higher than in the Southeast (e.g., Compéan and Polenske, 2010), production costs are low in the Northeast compared to the international average, largely because of low labor costs

To benefit from the ethanol boom, many state governments in the Northeast have sought to give a new boost to the sugar and ethanol industry. For instance, the Pernambuco state government created, on May 30, 2007, a special commission (Câmara Setorial da Cana-de-Açúcar, Açúcar e Álcool do Estado) to promote collaboration between the relevant actors in the sector (SDE, 2011). A specific strand of initiatives consists of efforts to expand sugarcane cultivation into new frontier areas, notably on what are called *degraded lands* – lands that have long been logged and transformed into pastureland and are now idle (Wilkinson and Herrera, 2010: 754).

Three suggestions, aimed at ensuring that the poorer areas of the Northeast actually benefit from the ethanol boom, deserve attention. First, a project is under way to expand sugarcane cultivation to the semiarid inland Sertão region. By 2026, the project is expected to generate as many as 333,000 jobs in 17 municipalities and benefit a total of 640,000 persons (Peres, 2009), but subsidies from the federal government would be needed to realize the needed USD 2 billion investment and to install the necessary irrigation canals (Bahé, 2007).[10] One can hardly avoid drawing parallels with frequent past efforts to solve the chronic drought-related poverty problems in Sertão with projects whose main outcome has been to create a lucrative "drought industry". In fact, the situation of those suffering from the consequences of drought has remained as difficult as ever. Second, mini-distilleries have been constructed, partly by financing from BNDES, to replace old and abandoned mills in areas that had become pockets of major poverty, for instance, in the coastal area of the state of Pernambuco (*Jornal do Commercio*, 2007a). Yet past experience from the Proálcool program suggests that making mini-distilleries economically viable in a market dominated by large players is not easy (e.g., Demetrius, 1990; Furtado et al., 2011: 158). Third, there has been a revival of the old plans to produce both ethanol and biodiesel from manioc, predominantly cultivated by smallholders and more suitable for small-scale production in micro- and mini-distilleries for local demand (Peduzzi, 2009; Peres, 2009). Again, experience from the recent efforts to promote manioc cultivation for biodiesel suggests (e.g., Wilkinson and Herrera, 2010: 764) that the conditions for success of manioc biofuel may not be any greater today than they were during the early years of Proálcool, when the small-scale manioc producers failed to compete with the large-scale sugarcane producers (e.g., Demetrius, 1990).

One may ask whether the recent biofuel boom might not turn out to be a curse rather than a blessing, preventing the necessary reorganization and diversification of the economic (notably agricultural) activities in the coastal sugarcane area. The crisis of the late 1990s may have been an opportunity for a gradual normalization of the sector and the entire economy in the Northeast. However, this opportunity was missed once the ethanol boom again improved the competitiveness of the sector. In particular, the idea of a more harmonious coexistence between small- and large-scale producers in the region, as part of an agrarian reform, may have proved an illusion. Indeed, many NGOs mention the conflicts with agrarian reform as a major problem of the current agroindustrial model within which the biofuel production is embedded in Brazil (e.g., Fernandes et al., 2010; Pinto et al., 2007). Hence the prospects for further diversification of the economy and for the development of small-scale production will be limited to the extent that the sugar and alcohol elite manages

---

[10] In principle, increasing the area of sugarcane cultivation under irrigation goes against objectives of the current national sugarcane policy. The current policy allows cane cultivation to expand only in areas where full irrigation is not needed, and even so, as only "complementary irrigation" in exceptionally dry periods (Decreto, 2009). While the project benefited from the support from the half state-owned oil company Petrobras, Companhia de Desenvolvimento do Vale do São Francisco (Codevasf), the state and federal governments, and the government of Japan (*JornalCana*, 2007), the northeastern politicians were concerned about the possibility that the new sugarcane zoning policy would put the project in danger (Folha de Pernambuco, 2009).

to retain its privileges and as long as the widespread and long-standing belief persists that sugarcane represents the only viable agricultural crop in the region.

Some recent events have further consolidated the widespread perception in the Northeast that the large landowners and mill owners in the sugar and alcohol sector are systematically privileged at the expense of other sectors of agriculture.[11] A news article published in the major daily newspaper *Folha de São Paulo* (Souza, 2007) and subsequently reproduced on various websites[12] (Nordeste, 2007) reported that in January 2007, while the cane growers were experiencing one of the most lucrative periods in history, Banco do Brasil granted a debt relief of more than R$ 1 billion to at least 20 mills, most of them located in the Northeast. The politicians and the sugar elite in the Northeast continue facing accusations of nepotism, fiscal fraud, and clientelism, often manifest in vote buying (Extra Alagoas, 2008). Critics have also claimed that the very same mill owners that are highly competitive in the Southeast are among the first to call for state subsidies, which would permit them to continue producing in the Northeast. Indeed, the Northeast producers have requested for the region a privileged status similar to the one accorded to the Tax Free Zone of Manaus and to the car industry (Bahé, 2007).

The recent internationalization of the sector has affected also the Northeast, with multinational groups investing in mills in the region (e.g., Wilkinson and Herrera, 2010). Some observers have seen the arrival of foreign capital as an opportunity to improve the socioenvironmental management in the sector, as multinationals presumably would need to pay more attention to social equity and environmental protection (*Jornal do Commercio*, 2007b). Others, by contrast, take a more radical view, rejecting foreign ownership as an integral element of the inherently undesirable agrocapitalist model of development. For the civil society movements, the penetration of foreign capital therefore presents a dilemma, with potential to create internal division within the movement (e.g., Fernandes et al., 2010).

On a more positive note, the Brazilian government's antipoverty political agenda, together with international concerns for social sustainability and the certification schemes under development, could help small-scale sugarcane production. Recent events in the Northeast's regional and local politics may suggest that change is under way. Hence, for the first time in the history of the state of Alagoas, two mill owners were disputing for the position of state governorship. This might be an early sign of the normalization of the politics in the region. Political divisions would enter the sugar and alcohol sector, hence replacing the old tradition whereby the mill owners would not directly enter politics but would unite (and finance) a

---

[11] As reported by Novo et al. (2010), in other regions of the country, notably in São Paulo, the sugar and alcohol sector also enjoys a similar privileged position within agriculture. However, the agricultural sector in São Paulo is more diversified than its northeastern counterpart, and therefore such privileges may not generate equally dramatic impacts.

[12] E.g., Bancada do Nordeste (2007).

single candidate to represent their cause. A different reading of the situation would be to consider such a so-called normalization as little more than another step in the complete integration of peasant agriculture into capitalist relations of production, a perspective strongly resisted by a number of civil society actors, as will be discussed later.

5. Criteria for successful certification and power relations in the Northeast

Numerous, mostly voluntary sustainability certification schemes and multistakeholder initiatives worldwide are in use or under development to ensure that biofuels are pro-duced in a sustainable manner (see Chapter 1). Van Dam (2010) identified more than 50 schemes worldwide that attempted to define some form of sustainability criteria for energy derived from biomass, both for transport fuel and other forms of energy consumption. In addition, numerous schemes exist for certifying the cultivation of major biofuel crops (soy, sugarcane, palm oil) and forestry and agriculture certifica-tion schemes with potential relevance to biofuels production (van Dam et al., 2008). Apart from the voluntary schemes, one of the most important schemes for the Brazil-ian ethanol industry is the European Union scheme under the Directive on Renewable Energies, which is mandatory and potentially very powerful. In addition to the inter-national schemes, pilot programs have been introduced in different regions of Brazil, such as in São Paulo (Azevedo, 2009) and Alagoas (Delzeit and Holm-Müller, 2009: 665). At the national level, a pilot scheme was launched in 2008, largely in response to the international concern for the sustainability of Brazilian ethanol and for the needs of the export industry. The scheme is being developed under the leadership of the national standardizing organization Inmetro (Instituto Nacional de Metrolo-gia, Normalização e Qualidade Industrial). Selected participating mills were included from São Paulo, Center West, Northeast, and Paraná. The criteria were developed in collaboration with foreign experts, notably from Germany, in view of a harmoniza-tion with future international umbrella criteria (e.g., Wilkinson and Herrera, 2010). Finally, Brazilian companies have signed several bilateral agreements with importing countries, including provisions concerning sustainability criteria and their verification (e.g., Wilkinson and Herrera, 2010: 755).

5.1. Certification should be based on theory

While the intention in this chapter is not to conduct a detailed analysis of the potential and limits of the various certification schemes, the preconditions of certification in the Northeast can be examined in light of the three overarching criteria for successful certification described by Delzeit and Holm-Müller (2009: 663). First, a certification scheme (notably the certification criteria) should be based on theory, that is, on a scientifically based analysis of the potential impacts of biofuel production. The preceding analysis of the social relations in the sugar and alcohol sector in Brazil in general and in the Northeast in particular raises a number of critical issues concerning the potential of this criterion being fulfilled. On a "technical" level, the science concerning biofuels, supposed to underpin certification efforts, is highly contested, and the knowledge in many impact areas is patchy at best. For instance, the life cycle analyses (a key source of information for certification) are contested and contestable. There are few peer-reviewed studies in the area, and even these generate debate and controversy. Furthermore, certification schemes must decide how to deal with the stark differences in, for example, productivity figures between the Northeast and São Paulo. Using national productivity averages would provide a highly flawed picture of the situation, especially in the Northeast.

Beyond the knowledge gaps and technical uncertainties, the political and social dimensions of uncertainty are probably even trickier. An essential question here is, who has the power and ability to define the type of authoritative evidence base that should underpin certification? Hence the problems generated by scientific uncertainty are compounded by the asymmetries in knowledge and expertise between different stakeholders. Such asymmetries play a role also at the international level, as demonstrated by the influence that the Brazilian industry and government have exerted, for instance, on the German biofuel certification criteria. Furthermore, the space that scientific uncertainty leaves for power play highlights the crucial role of experts (both Brazilian and international) in defining the criteria of sustainability. To get an idea of the potential power of experts, one need only look at the tremendous impact that individual studies have had on international debate (e.g., Fargione et al., 2008; Patzek and Pimentel, 2005; Searchinger et al., 2008, at the international level; Macedo et al., 2004; Smeets et al., 2008, on Brazilian ethanol specifically). In Brazil, the academic expert community includes highly influential academics traveling internationally with Brazilian government and industry delegations to promote "sustainable Brazilian ethanol" and correct the "erroneous claims" made especially about the environmentally harmful impacts of ethanol in the country. Other academics, in turn, hold close ties with civil society movements that criticize biofuels on the grounds of their deleterious impacts, notably on peasant agriculture, and resist the biofuel expansion, which they see as part of the development of capitalist agriculture (e.g., Fernandes et al., 2010).

In light of the sociopolitical context in the Northeast, the criterion concerning the "theory-based" nature of certification raises questions. These questions concern, notably, the ways to ensure that certification is based on a solid evidence base in a context of highly unequal access to power and authority by various participants and means of ensuring that the Northeast sugar and ethanol elite do not capture the certification system so as to make it a marketing tool designed to give ethanol a social label. The conventional answer to these questions would be to ensure the widest possible participation in decision making by the affected stakeholders. However, the asymmetries of power make such participation highly problematic and potentially even counterproductive to the objective of improved social equity. Hence numerous local-level NGOs and civil society movements have rejected the opportunity to participate in the allegedly participatory roundtable processes, refusing to take part in what they see as mere attempts to give a green label to the inherently harmful biofuels.

5.2. Certification should be relevant to key stakeholders

The second criterion put forward by Delzeit and Holm-Müller (2009) is in principle less problematic, as it calls for the certification to be relevant to key stakeholders. However, achieving relevance is not an easy task in view of the highly contrasting expectations on certification. On one hand, the industry and government actors see certification as a tool in the efforts to turn biofuel into an internationally tradable commodity, while on the other hand, the environmental and social sector players emphasize the primary role of certification in guaranteeing the environmental and social sustainability of biofuel. The relevance criterion obviously is intimately linked with the question of participation. Who are the stakeholders whose concerns for relevance should be taken into account? How should these groups be integrated into the decision-making system? How can we ensure that participation takes place on a more or less level playing field? One would imagine that an appropriate certification system should integrate the concerns of the peasants and sugarcane field and factory workers in the Northeast and give these groups a voice in decision making. However, the historical legacy of deep inequality, conflicts of interest, and outright hostility between the ruling landowning elite and the poor uneducated working classes makes any participatory decision-making arrangement highly challenging.

How could Northeastern smallholders participate on a level playing field with the multinational ethanol industry, given the formidable difficulties in guaranteeing equitable participation even for representatives of governments of developing countries, for example, in forums such as the ISO? Could the smallholders ever have the required resources and expertise for such participation? Furthermore, ensuring participation by all relevant stakeholders faces a host of more technical and practical challenges, familiar from participatory practice and theory more generally (e.g., Parkinson, 2003). Organizing such a process is particularly tricky when relevant actors range from farmworkers in the Northeast of Brazil to the state authorities at different levels (both within Brazil and internationally), up to the multinational agribusiness, car, and energy companies, and ultimately, to the consumers and car owners in developed countries. In the traditional model of representative democracy, participation takes place through elected representatives and organized interest groups. However, there is a strong argument in favor of complementing representative democracy with a model of deliberative democracy, with the ambition of giving all of the affected groups a more direct voice in public deliberation concerning the problems and the appropriate solutions (e.g., Gutmann and Thompson, 2004). Implementing the principles of such deliberative democracy in the context of multilevel governance is challenging, to say the least.

5.3. Certification should be implemented and enforced

The third and last principle guiding good certification is verifiability. To put it simply, adequate procedures and mechanisms must be in place to ensure actual compliance. This is where the power relations in the Brazilian sugar and alcohol sector pose perhaps their greatest challenges. The sector in general and the Northeast in particular are characterized by a chronic problem of insufficient enforcement of labor and environmental legislation (e.g., Silva et al., 2005: 895; Smeets et al., 2006). Not only has the Northeast sugar and alcohol elite sought, via its close links to political power, to systematically oppose new legislative measures that might limit its freedom, but it also has a notoriously poor record in respecting whatever environmental and social legislation has managed to be enacted (e.g., Porter et al., 2001: 851). Today this systematic noncompliance with extant legislation is seen as a major hazard for Brazil's endeavor to build the image of its ethanol as a sustainable biofuel. This has been demonstrated, for instance, by the recent polemics between the environment minister and the Northeast sugarcane industrialists mentioned in Section 4.

The question of verifiability further raises the question of the concrete ways in which verification and enforcement should be organized, including the necessary resources and institutional capabilities. While the decentralization of the verification procedures to producer level could be a viable solution capable of distributing the administrative burden involved, it would be highly questionable and prone to regulatory capture by the northeastern elite. As for the institutional capabilities, the prospects are not much brighter, given that the state authorities in the Northeast lack sufficient resources for enforcing the most basic social and environmental regulation. It is highly unlikely that enforcing a biofuel sustainability scheme would feature among their top priorities and obtain anything near to sufficient attention.

6. From center–periphery relations to global biofuel assemblages – competing paradigms, biofuels, and certification

The debate on Brazilian ethanol, the Northeast, the social impacts, and sustainability certification can be situated within a broader context of global biofuel politics, on one hand, and the different intellectual paradigms, notably those relating to the desirable societal development models, on the other. The center–periphery relations that prevailed for a long time in the Brazilian sugar and alcohol sector (Galtung, 1971; Lehtonen, 2011) are increasingly becoming embedded within the globalized biofuel politics. This involves the entry of new players onto the field and an increasing

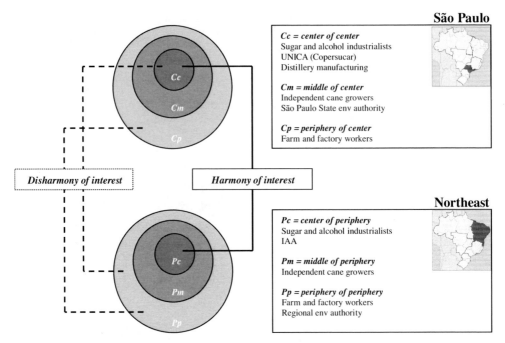

Figure 7.1. Center–periphery relationships in the Brazilian ethanol sector.

creation of networks, alliances, and "biofuel assemblages" (Hollander, 2010; Saturnino et al., 2010).

In a schematic manner, these center–periphery relationships could be described as a dual system of Southeast "Center" and Northeast "Periphery," with both Center and the Periphery having their own centers (the sugar and alcohol elites in the respective regions) and peripheries (the poor sugarcane field workers and alcohol factory workers). While the exchange relations on the whole worked to the benefit of the Center (the inexorable strengthening of the hegemony of the Southeast sugar and alcohol sector serving as a typical example), the elites in the periphery have continued to operate as a bridgehead of the Center, to the detriment of the periphery (i.e., workers) in their own region. Figure 7.1 describes schematically the power relations in the Brazilian sugar and alcohol sector, drawing on the center–periphery model of Galtung (1971). Here the center of this industry sector is São Paulo, whereas the Northeast represents the periphery. Following Galtung's model, both regions have their own center and periphery, with the center composed essentially of the sugar and ethanol industrialists and their organizations, large landowners, and equipment manufacturers. Farmworkers and factory workers, in turn, constitute the periphery, while the independent cane growers can be located in the middle between center and periphery in both regions (for further details, see Lehtonen, 2011).

International capital has entered the Brazilian biofuel industry, and today controls about 25 percent of the sugarcane processed for Brazilian sugar and ethanol production. Of the five largest sugar and alcohol producers in the country, only the Santa Terezinha company is still fully in national ownership (Ramalho, 2010; Unisinos, 2010). Yet, despite the increasing concentration, the Brazilian sugar and ethanol industry remains highly fragmented (Wilkinson and Herrera, 2010: 751), with the five largest companies controlling only about 20 percent of the market. Whether the consolidation trend (i.e., mergers and acquisitions) will continue is uncertain, and industry experts hold diverging views on the issue. Some predict further consolidation and penetration of foreign capital (e.g., Simões Lopes, 2009; Wilkinson and Herrera, 2008b), whereas others, such as the president of UNICA, Marcos Jank, expect the trend to slow down once the economic prospects improve and sugar and ethanol prices increase (Brasil Energy, 2010). Moreover, large companies may find greenfield investments more profitable than the purchase and renovation of small- and medium-sized mills, which often have relatively outdated equipment (Ramalho, 2010).

Networking and internationalization are central features also in the processes of certification, operation of experts, and NGO activities. Importantly, they have contributed to the introduction of a sustainability discourse in Brazil and the adoption of such a discourse by major actors. The biofuel boom has helped to connect the numerous local grassroots-level battles, such as those fought by local trade unions in the Northeast of Brazil, with the activities of various movements worldwide critical of capitalism. The debate on the role of biofuel in general, and of sustainability certification in particular, has led to internal splits within both the NGO and expert communities in Brazil. The majority of the NGOs, peasant movements, and researchers close to these groups oppose any engagement by peasantry with biofuel projects controlled by large-scale agribusiness, while others, such as the WWF (see, e.g., WWF-SNV, 2008; WWF Brasil, 2009) and the Brazilian National Confederation of Agricultural Workers, consider that even large-scale biofuel projects can indeed provide benefits to the environment, family farmers, and rural workers (Fernandes et al., 2010).

Fernandes et al. (2010) have described such divisions in terms of two competing paradigms. The first one, the *agrarian capitalist paradigm*, prevails among policy makers and academic experts, most notably economists and political scientists. It sees development as an almost inevitable process of capitalist state formation, in which peasantry would gradually be transformed into urban workers or capitalist family farmers. The alternative, the *agrarian question perspective*,[13] advocates resistance against the encroachment of the capitalist agriculture and the accompanying integration of peasantry into capitalist relations of production because it considers such integration as a form of subordination. While the agrarian capitalist paradigm sees the social inequalities in agrarian society as outcomes of conjunctural problems, resolvable

---

[13] In fact, Fernandes et al. (2010: 799) note that the agrarian question perspective is not a fully fledged paradigm because of its inability to provide a coherent vision of the future development path.

through policies aiming to integrate the peasantry within capitalist markets, the agrarian question perspective considers that "power relations that destroy and recreate the peasantry are inherent structural qualities of capitalism" (Fernandes et al., 2010: 797). In consequence, while the agrarian capitalist paradigm treats biofuels as the answer to a number of economic, environmental, and social problems, "including the liberation of peasants from oppressive labor in the cane fields," the agrarian question perspective refers to empirical evidence showing that sugarcane has increased the number of enslaved workers, aggravated water and air pollution, and reduced the land available for agrarian reform (Fernandes et al., 2010: 807). In this vein, for instance, Hall et al. (2009: S81) suggest that increased international trade in biofuels will foster economic efficiency and industry concentration at the expense of greater social inclusion.

Given the sharply contradicting perceptions concerning the sustainability of Brazilian ethanol, it is hardly surprising that the two paradigms have diverging views on biofuel certification. The varying views on certification can hence be placed within a broader context. Portrayed by Fernandes et al. (2010) as one of the key proponents of the agrarian capitalist paradigm, Abramovay (2008) has argued that international pressure through sustainability certification could be one of the few possibilities of integrating social concerns into decision making in the Brazilian biofuel sector, thereby helping to alleviate the drastic inequalities in the sector. The agrarian question perspective, by contrast, sees certification as little more than a smokescreen disguising the continued exploitation of workers by the sugar and ethanol sector (Fernandes et al., 2010).

A similar prism for analyzing the role of certification is suggested by Laschefski (2008), who distinguishes between four attitudes toward biofuels: (1) ecological modernization, (2) the reconstruction of the urban–industrial–capitalist society, (3) a model underlining the heterogeneity of cultures, and (4) the environmental justice movement. To a large extent, these four perspectives represent a translation of Fernandes et al.'s (2010) dichotomy into the debate on environmental policy, with the ecological modernization theory representing the former, "reformist" view of working within the capitalist system, while the three other theories are variants of the contesting, "radical" discourse, rejecting to varying degrees and in various ways the dominant development model based on capitalism.

Against this background, a fundamental question is whether, and under which conditions, certification could actually fit within the other perspectives than its natural home base – ecological modernization. The response by numerous civil society groups in Brazil and internationally has been to reject certification as an inherently flawed mechanism of "market environmentalism." Against this background, it is easy to see why the more radical NGOs refuse to consider certification as a serious means of achieving the objective, that is, a complete revamping of the current agricultural development model in Brazil toward a system based on autonomous small-scale production for local demand. Rather than representing a means to achieve positive change, certification is seen as an instrument that reinforces the existing capitalist relations, incapable of generating the necessary, more profound sociotechnical change. Such highly skeptical views on the positive potential of certification reflect what Porter et al. (2001: 843) refer to as the "low trust dynamics" in Brazil's Northeast sugar and alcohol industry. They further characterize the Northeast as a "bad faith economy" (Scheper-Hughes, 1992: 111), in which the mutual suspicion between management and workers, stemming from "historical antagonisms bred in a slave economy and nurtured by continued repression, leads to conflict and low productivity." Arguably, such tensions render the implementation of a successful certification system in the Northeast highly problematic.

The question concerning certification's compatibility with meta-frames, other than ecological modernization, is intimately linked with what can be seen as an Achilles' heel of certification, that is, its inherent conservatism and potential inability to produce fundamental system transformations or transitions. If a sustainability transition is what is needed (in our case, notably, with regard to the social relations governing ethanol production in Brazil), certification may well turn out to be an instrument in favor of the status quo rather than transition. However, just as the distinction between the agrarian capitalist paradigm and the agrarian question perspective is a simplification of a more complex social situation and debate, it would be premature to condemn certification right away as an exclusively and unavoidably conservative instrument. Even Brazilian NGOs and "radical" experts have indeed advocated small-scale community certification as an instrument of progressive social change (e.g., *Repórter Brasil*, 2010c). And yet the current certification schemes are clearly developed within a globalizing ecological modernization framework. The social inequality and asymmetries of power in Northeast Brazil, together with the very high economic stakes involved in the decision making on ethanol policies in Brazil, invite caution concerning the possibilities of actually implementing socially beneficial small-scale certification schemes. While the Northeast certainly is exceptional by virtue of the extreme polarization of social relations, the lessons from the Northeast are likely to be applicable to potential and actual biofuel-producing regions in the global South characterized by similar inequalities and for which sugarcane ethanol is being seriously considered as an alternative fuel (see Chapter 12).

7. Conclusion

This chapter has argued that despite its modest share in the supply of ethanol fuel in Brazil, the Northeast should receive more attention in international debates concerning the sustainability of biofuels in general and certification in particular. Measures to minimize the social impacts of biofuel production would be particularly crucial for the development of the poor and highly disadvantaged Northeast region of Brazil. However, it is precisely in the Northeast that certification faces perhaps its greatest challenges, notably owing to the historically shaped and persistently unequal social relations. And yet reform of such power relations would be a fundamental precondition for socially equitable development and reduction of poverty in the region. The success of such restructuring hinges on the ability of the different stakeholders to overcome the polarization between the two dominant and competing worldviews. The first considers biofuels as a positive opportunity for Brazilian farmers to achieve further development within the framework of modern and globalized agriculture, whereas the other emphasizes farmers' independence and resists any attempts to integrate peasants and family farmers into capitalist relations of production. From the former perspective, certification can provide a unique opportunity to harness international pressure to increase the accountability of northeastern elite toward society at large. For those opposed to agricultural capitalism, certification represents little more than whitewashing serving solely the interests of the already powerful. Regardless of whether the integration of these contrasting perspectives turns out to be feasible, the international biofuel certification schemes would do well to pay careful attention to and learn from the history of the sociopolitical conditions in the Northeast of Brazil.

8

Implications of global ethanol expansion on Brazilian regional land use

AMANI ELOBEID, MIGUEL CARRIQUIRY, AND JACINTO F. FABIOSA
Iowa State University, Ames, USA

Abstract

A spatially disaggregated model of Brazilian agriculture is used to assess the implications of global biofuel expansion on Brazilian land use at the regional level. After establishing a baseline, two scenarios are investigated. First, an exogenous increase in the global demand for biofuel is introduced into the model, and the impact is analyzed in terms of land use change and commodity price changes, given baseline assumptions on potential land expansion. Second, the same exogenous biofuel demand shock is implemented with a different responsiveness in area expansion to price signals in Brazil, reflecting varying plausible assumptions on land availability for agricultural expansion. The motivation for this second scenario is derived from the existing uncertainties regarding land availability for agricultural expansion and the extent of future enforcement of land use policies in Brazil. We find that most of the global increase in ethanol consumption (other than in the United States) is supplied by Brazilian production expansion, in particular, from the Southeast region of the country. Interestingly, total sugarcane area expansion in Brazil is higher than the increase in overall area used for agriculture. This implies that part of the sugarcane expansion occurs in previously utilized agricultural areas (other crops and pasture) that are not replaced. This suggests that some intensification in land use results from sugarcane expansion. Halving land expansion elasticities in the second scenario results in a lower expansion of area used for agricultural activities. Intensification of beef production allows a higher proportion of the sugarcane expansion to occur in pasture areas. While tending to move upward, commodity prices remain virtually unchanged relative to those of the first scenario. Larger changes in commodity prices can be expected when the scope of intensification is exhausted.

Keywords: Brazil, partial equilibrium model, regional land use change, sugarcane ethanol

1. Introduction

The rapid increase in global biofuel production and consumption (of ethanol, in particular) has an associated derived demand for crops to produce the necessary feedstock. Per hectare yields translate feedstock needs into a corresponding demand for land. The rapid expansion in biofuel production can thus be linked to an increase in land demand for agricultural production purposes. It is in fact the land use change impact, together with the diversion of food crops into energy, that feeds most of the controversy surrounding biofuel expansion (Fabiosa et al., 2010; Fargione et al., 2008; Searchinger et al., 2008; refer to Chapter 2 for a discussion of direct and indirect land use due to biofuel expansion).

World production of fuel ethanol has increased from 39 billion L to almost 73 billion L from 2006 to 2009, according to F. O. Licht (2010). The same source expects production of the fuel to reach 83 billion L in 2010. A large proportion of the increase in ethanol consumption has been fueled by policies either mandating its use or providing financial incentives to make the fuel competitive with gasoline (see Chapter 1). The two largest producing countries, the United States and Brazil, are expected to contribute a combined 88 percent of global production in 2010 (F.O. Licht, 2010).

Brazil has been a pioneer in incorporating biofuels, particularly ethanol from sugarcane, into its fuel supply and is currently the largest exporter in the world ethanol market (see Chapter 6). As the demand for biofuels expands, Brazil is expected to continue to play a major role in meeting both domestic and global needs. Given the expected supply expansion, and the fact that Brazil is home to natural areas with a high degree of biodiversity, concern has been voiced regarding the potential of the global biofuel expansion to accelerate deforestation in the Amazon and Cerrado areas. Thus how this increased production of biofuels will affect the agricultural and biofuel sectors in Brazil, as well as how it will affect land use, is a contentious topic given the potential environmental consequences.

The area devoted to agriculture in Brazil has expanded significantly in the recent past, including the area planted to sugarcane (see Table 8.1). Table 8.1 shows that sugarcane area grew at a much higher rate than areas of other major crops. However, according to Nassar et al. (2008), most of the growth in the sugarcane crop occurred in previously utilized regions. In particular, the authors find that for the South Central region (the area with the largest sugarcane expansion), 98 percent of the sugarcane area growth in 2007 and 2008 was on land previously used for agriculture (53%) or pasture (45%). While this hints that sugarcane ethanol is not directly responsible for the clearing of new areas for agricultural activities, it also does not rule out an increase in the use of previously unused (natural or idled) land to partially replace the product from the uses displaced by sugarcane (e.g., Lapola et al., 2010). The need for additional area is only eliminated when the demand for the products from other

Table 8.1. *Area change for major land-using agricultural activities in Brazil* *(1,000 ha)*

	1999–2000	2008–2009	Change	% Change
Sugarcane	4,880	8,423	3,544	73
Major crops[a] (excluding sugarcane)	36,594	46,290	9,697	26
Major crops (including sugarcane)	41,473	54,714	13,240	32
Pasture	195,025	203,873	8,848	5
Total	236,498	258,587	22,089	9

[a] Major crops include corn (first and second crops), soybeans, rice, cotton, dry beans (first and second crops), wheat, barley, and sugarcane.

Sources: Prepared based on data from the Brazilian Institute of Geography and Statistics (http://www.ibge.gov.br/home/) and the Brazilian Ministry of Agriculture's National Supply Company (http://www.conab.gov.br/).

land-using activities declines or when crop yields per hectare increase by a sufficient amount to compensate for the area lost to sugarcane.

In terms of regional distribution of sugarcane, the South Central region (comprising the states of São Paulo, Minas Gerais, Paraná, Mato Grosso, Mato Grosso do Sul, and Goias) accounted for about 82 percent of the total area in 2008 and for 96 percent of the expansion between 2005 and 2008 (Nassar et al., 2008). While the Northeast region's area seems to be relatively stable over time (see Chapter 7), these authors indicate that states in the northeast Cerrado region (e.g., Maranhão, Tocantins, and Piauí) are a promising area for the expansion of sugarcane.

In this study, we analyze the regional land use changes in Brazil that would result from an increase in the consumption of ethanol beyond the levels projected in a business-as-usual scenario. Special attention is paid to the regional expansion of sugarcane area and to the additional area that needs to be incorporated into agricultural uses to accommodate that expansion. The impact of the expansion on the prices of major commodities is also estimated. This analysis is conducted under two different scenarios. In the first scenario, we assume that the enforcement of the land use reserve remains at the levels observed in the recent past and that abundant additional land can be readily incorporated into production.[1] In the second scenario, we assume a less abundant supply of land for agricultural expansion. This could be the result, for example, of enhanced enforcement of the land reserve requirements. In this second scenario, the supply of agricultural commodities in Brazil becomes more inelastic (compared to the previous scenario), resulting in area expansion in other regions of the world, coupled with higher prices. A different pattern of substitution can also be expected within the country, as different activities can react differently to the limitations to land

---

[1] In current Brazilian law, producers must keep in reserve a portion of their land (i.e., in its natural form). This proportion varies regionally from 20% in the established regions (e.g., South) to 80% in the Amazon area. In São Paulo State, it is 20% (see also Chapter 6).

expansion. A model of world agriculture that is able to project land use, production, consumption, and trade as well as commodity prices is used for the analysis.

Section 2 provides a description of the models used in the study, along with additional details relative to the regional Brazil and world ethanol components. Section 3 describes the scenarios to be analyzed. Results from the models are presented and discussed in Section 4. Finally, Section 5 offers some concluding remarks.

2. Model description

2.1. Overview of the modeling system

The international FAPRI model is a system of econometric, multimarket, nonspatial, partial equilibrium models.[2] It covers all major temperate crops, ethanol, sugar, biodiesel, dairy, and livestock products in all the major producing and consuming countries (see Figure 8.1). The model is run in yearly time steps. To name a few applications, this modeling system has been used extensively to create market outlooks, to analyze the impacts of technical change, and to provide policy analysis, including land use change calculations, for the U.S. Environmental Protection Agency's rulemaking under the Renewable Fuel Standard (see, e.g., Fabiosa et al., 2007, 2010; FAPRI, 2010; Hayes et al., 2009; Tokgoz et al., 2008). Additional validation through external reviews and internal updates is periodically performed.

Interactions among markets are reflected through extensive linkages that capture derived demands for feed in the livestock sector, feedstock in biofuel production, substitution possibilities between close substitutes, and competition for land. The modeling of these biological, technical, and economic relations is based on accepted relationships in agricultural production and markets and on the analysis of historical data.

The model finds a set of prices for each commodity such that supply equals demand for all commodities and countries. Through the linkages, changes in one commodity or country affect the markets for other commodities or countries. In general terms, agricultural production results from the area allocated to the different crops multiplied by the crop's yield. The area allocated to the different crops depends on crops' relative expected returns. This captures the competition for land between these activities. Beginning stocks complete the domestic supply quantities available. The domestic demand specification depends on the commodity and can include food uses, feed uses, industrial uses, and ending stocks.

2.2. The regional Brazil model

Brazil encompasses widely varying ecosystems, ranging from grassland and crops in the South to tropical forests in the North and semiarid areas in the Northeast. The different regions present large disparities in terms of infrastructure and natural

---

[2] FAPRI is the Food and Agricultural Policy Research Institute at Iowa State University. A more detailed description of the FAPRI modeling system, including data and elasticities, is provided in FAPRI (2011).

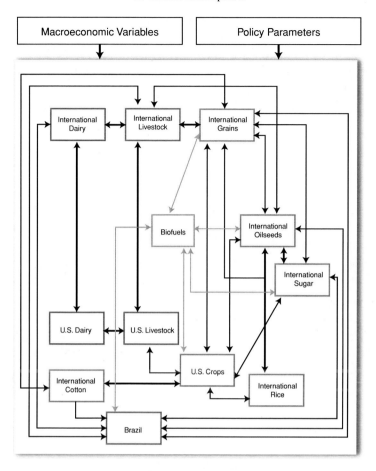

Figure 8.1. FAPRI model interactions. The model interactions represent trade, prices, and physical flows. For the purposes of this study, all models in the system were run, with the exception of international rice.

resources available to increase agricultural production. Thus, while rapid expansion of production of some commodities may be achieved only by displacing other agricultural activities in land-constrained regions, increases in area used by all activities may be observed in other parts of the country. This points to distinct dynamics in the competition for land across space. Environmental (both local and global), social, and economic impacts hinge critically on the nature of these land use changes. Therefore it is becoming increasingly important to recognize the spatial dimension of the agricultural expansion. Given the emerging importance of Brazil, both in terms of its capabilities to expand area (and production) in response to demand changes and its potential for greenhouse gas (GHG) emissions from land clearing, a regional model of Brazilian agriculture was developed.[3] This model is fully integrated as a part of the FAPRI modeling system.

---

[3] The Brazilian model was developed by the Center for Agricultural and Rural Development at Iowa State University in collaboration with the Institute for International Trade Negotiations, Brazil.

The model of Brazilian agricultural production incorporates major crops, biofuels, and livestock interacting and competing for agricultural resources, in particular, land. Outputs from the model include projections of supply and utilization variables and the amount of land allocated to the activities considered. On the crop side, corn (first and second crops), the soybean complex (including soybean meal, soybean oil, and biodiesel), the sugarcane complex (including sugar and ethanol), rice, cotton, and dry beans (multiple cropping, depending on the region) are considered. The modeled animal products are beef, pork, poultry, and dairy. In terms of land allocation, the area used by a given activity depends on its expected real returns in comparison to expected returns of activities that compete for the resource. Land used for pasture is explicitly modeled. Since not all of the regions considered are equally suited for different activities, the competition for land is contingent on the location. As such, not all activities compete with each other with the same intensity in all regions. Additionally, the model also allows for production costs, yields, and prices to vary by region.

Through the use of spatially disaggregated information on historical production activities and land availability, the model is able to determine the relative profitability of different activities at the local level, which will drive regional supply curves for relevant commodities and their associated land use. For this modeling effort, Brazil is divided into six regions: South, Southeast, Center West Cerrado, Amazon Biome, Northeast, and North-Northeast Cerrado. Figure 8.2 presents the regional disaggregation of Brazil, including the states that make up each region. The model is able to capture the regional differences in terms of capabilities and consequences of the expansion. In this way, the impacts of land use changes derived from increasing demand for agricultural products can be more precisely established.

Two different procedures are used to project agricultural area and allocate it to land-using activities. For crops deemed not to directly compete for land resources during the main growing season, behavioral equations that project agricultural area are used. These equations are mainly driven by real relative returns of the different activities. Wheat, barley, the second crop of corn, and the second crop of dry beans fall into this category. The area allocated to a second group of activities, competing for land resources in time and space, is modeled using a two-step approach. The total area utilized for agricultural activities is determined first. A second step allocates this area to the competing land uses. Corn, soybeans, rice, cotton, dry beans, sugarcane, and pasture are modeled through this procedure.

The first step, the calculation of the total area to be used for agriculture in each region, is dependent on expected returns to agriculture and potential land availability, as follows:

$$A_{jt}^{ag} = A_j^T m_j(\bar{r}_{jt}), \tag{8.1}$$

where $\bar{r}_{jt}$ denotes expected returns to land use (crops and pasture) in region j and year t and $m_j(\bar{r}_{jt})$ is the share of the potential agricultural land (A_j^T) that is used in

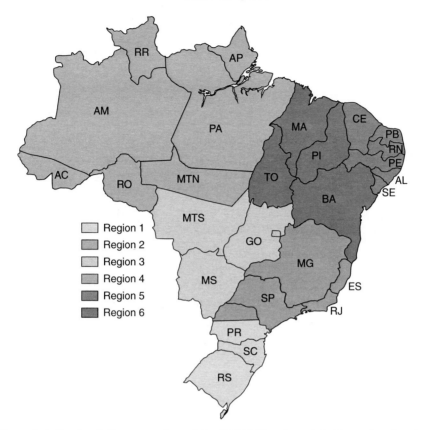

Figure 8.2. Regional disaggregation of the CARD Brazil model. (See color plate.)
Region 1: South
Rio Grande Do Sul; Santa Catarina; Paraná
Region 2: Southeast
São Paulo; Rio de Janeiro; **Espírito** Santo; Minas Gerais
Region 3: Center West Cerrado
Mato Grosso Do Sul; Goias; Distrito Federal; Mato Grosso
Region 4: Amazon Biome
Rondonia; Amazonas; Pará; Roraima; Amapá; Acre; Mato Grosso
Region 5: Northeast
Ceará; Paraíba; Rio Grande Do Norte; Pernambuco; Alagoas; Sergipe
Region 6: North-Northeast Cerrado
Tocantins; Bahia; Maranhão; Piauí

that region and year. Expected returns to agriculture are projected as (area) weighted average expected returns for the different activities covered, using

$$\bar{r}_{jt} = \bar{r}_{jt-1} * \sum_{i=1}^{I} \left(\frac{\tilde{A}_{ijt}}{A_{jt}^{ag}} \left(1 + \frac{r_{ijt} - r_{ijt-1}}{r_{ijt-1}} \right) \right), \tag{8.2}$$

where $\tilde{A}_{ijt}$ and r_{ijt} denote the area allocated and expected returns to activity $i = 1,$ $2, \ldots, I,$ in region j and year t, respectively. A linear allocation method proposed by

Table 8.2. *Regional land use elasticities and own-price elasticities for activities in Brazil model*

Region	$\varepsilon_{r_j}^{Ag,j}$	Corn, first crop	Soybeans	Cotton	Rice	Dry beans, first crop	Sugarcane	Pasture
South	0.06	0.18	0.43	0.21	0.15	0.09	0.40	0.03
South East	0.07	0.20	0.43	0.21	0.12	0.10	0.40	0.05
Center West	0.18	0.20	0.48	0.25	0.13	0.10	0.43	0.11
North	0.25	0.20	0.45	0.25	0.15	0.09	0.20	0.24
Northeast Coast	0.01	0.22	$-^a$	0.20	0.13	0.10	0.39	0.01
Northeast Cerrado	0.10	0.19	0.44	0.22	0.13	0.10	0.40	0.07
Brazil	0.13							

^{a} Soybeans are not planted in this region.

Holt (1999) is used to share the area out to the different activities. The share of the total area allocated to a given activity (v_{ijt}) is determined by

$$v_{ijt} = b_{ij} + \sum_{i=1}^{I} s_{ij} * r_{ijt}, \qquad (8.3)$$

where s_{ij} are coefficients and $\sum_{i=1}^{I} v_{ijt} = 1$ for all j and t. Therefore the area allocated to a crop is given by

$$\tilde{A}_{ijt} = A_{jt}^{ag} * v_{ijt}. \qquad (8.4)$$

In this framework, the own-price elasticity for the area allocated to a crop can be decomposed into a *scale effect* and a *competition effect* as $\varepsilon_{ij} = \varepsilon_{ij}^{scale} + \varepsilon_{ij}^{comp}$. The first term captures the additional area for a crop given an expansion in total area in response to that crop's returns. The second term governs the area in which the crop competes away from other activities as its expected returns increase. It is easy to show that the scale effect is $\varepsilon_{ij}^{scale} = \varepsilon_{r_j}^{Ag,j} * \varepsilon_{r_{ij}}^{r_j}$, where $\varepsilon_{r_j}^{Ag,j}$ is the elasticity of agricultural area to average expected returns to agriculture and $\varepsilon_{r_{ij}}^{r_j}$ denotes the elasticity of expected agricultural returns to the returns of activity i. The subscript j denotes the region.

Table 8.2 presents the elasticities used in the model. Clearly the Center West Cerrado and Amazon areas are the regions (given the land availability) that will present the highest response to changes in agricultural returns. Long-established regions and regions with land limitations have lower area elasticities. Also, soybeans and sugarcane are the most return-responsive crops in the model.

The supply side of the livestock sector is also regionalized in the model. The products modeled are pork, beef, poultry, and dairy. While poultry production is modeled directly through a behavioral equation depending on regional prices and costs of production, the stocks of animals are tracked over time, and production levels are consistent with the evolution of these stocks. Given stocks of cows and sows, the numbers of calves and piglets are obtained (through projected birthrates). Adult animals not part of the breeding herd are allocated to an "other" category. Meat production numbers are obtained by multiplying the projected number of animals slaughtered in each category by a slaughter weight. The numbers of slaughtered and dead animals are used to calculate the beginning stocks for the following year.

It is worth noting that the model allows for feedback between the pasture area and the size of the cattle herd. This is an important feature as beef production is by far the largest user of pasture. The link is captured by modeling the stocking rate directly, which depends on the profitability of beef production. This profitability will in turn affect the amount of area devoted to pasture through the land allocation mechanism described earlier.

2.3. The ethanol model

While its structure is country specific, in its basic form, the international ethanol model is based on behavioral equations for production, consumption, stocks, and trade. Of the eight countries covered, complete models are set up for the United States, Brazil, China, the European Union, and India. Net trade equations are established for Japan, South Korea, and an aggregate called "rest of the world." A representative ethanol price for the world (Brazilian anhydrous ethanol price) is solved endogenously to equate excess supply and excess demand for all the countries. For most countries, the domestic price is determined through a price transmission from the world price, adjusted by exchange rate and relevant policies. An exception is the United States, which is nearly insulated from the world market, given its import tariffs on nonpreferential ethanol imports.[4] The U.S. model solves endogenously for the ethanol price that clears the domestic market (unless international prices are low enough relative to the domestic price).[5]

The derived demand for ethanol feedstock is country specific. Whereas sugarcane is the main feedstock in Brazil, corn and wheat are responsible for most of the ethanol production in the United States and the European Union, respectively. Brazil is the only consistent net exporter of ethanol in the model. The area allocated to sugarcane in the country depends on the expected returns to this crop, relative to other potential activities competing for land. Expected returns to sugarcane follow from a composite of the expected prices of sugar and ethanol and the sugar content of the cane. The fraction of the total recoverable sugar (the feedstock for sugar and ethanol) used to

---

[4] For more details on the U.S. ethanol model, see FAPRI (2008).

[5] The U.S. ethanol model is embedded in the U.S. crops model (see Figure 8.1) and has a more detailed model structure than what is described here. The U.S. crops model was developed and is maintained by FAPRI at the University of Missouri, Columbia. For more details, see FAPRI (2004).

produce ethanol depends on the price of ethanol relative to that of sugar. The remainder is used for sugar production.

On the domestic demand side (for transport), ethanol is consumed in anhydrous and hydrous forms. The anhydrous form is consumed in mandatory blends with gasoline (25% ethanol), by gasoline cars. Hydrous ethanol is mainly used by flexible-fuel vehicles (FFVs) but also by gasohol cars. FFV owners can choose between ethanol and gasoline (blended), and their choice is quite sensitive to the relative prices of these two fuels (see also Chapter 6).

3. Scenarios

3.1. Baseline

To isolate the impacts of a specific change being analyzed, a reference trajectory or baseline needs to be established for all the variables of interest. This baseline reflects continuity of current policies, in a business-as-usual environment. As such, the baseline already incorporates a significant global expansion of the ethanol sector and growth of agricultural areas in most regions of Brazil, including an expansion of sugarcane area. Scenario analysis allows us to study how changes in a single or a subset of factors affect the market outcome and the impacts on the variables of interest. The new equilibrium is then compared to the benchmark or baseline trajectory.

3.2. Scenario 1

For this scenario, we shock the demand for ethanol in each country with a 25 percent exogenous (and permanent) expansion. After introducing the shock, all the markets are allowed to react to the expanded ethanol demand. The initial impact of the shock will be an increase in the price of ethanol, which will discipline the demand expansion and lead to enhanced ethanol supplies. The impact of the derived additional demand for ethanol feedstocks as well as the increased supply of ethanol by-products will be then transmitted to the markets for other commodities and countries. As a result, we expect additional land to be used for agricultural production as well as higher crop prices as the competition for area intensifies. Because Brazil is the largest world ethanol exporter and has a demonstrated potential to expand agricultural production, a large proportion of the adjustment is expected to occur in that country. This ability to expand agricultural production is expected to moderate the price increase brought about by the expanded ethanol demand.

3.3. Scenario 2

For the second scenario, we combine the exogenous expansion in global ethanol demand with a limit of land use expansion in Brazil. For example, the Brazilian

Table 8.3. *Change in ethanol production, consumption, and trade in 2022 for scenario 1*

Country	Production		Consumption		Net exports[6]	
	Million liters	%	Million liters	%	Million liters	%
Brazil	13,948	21.2	8,281	20.8	5,670	21.8
Canada	2	0.1	1,189	24.7	−1,187	33.7
China	100	3.8	712	22.7	−611	123.2
European Union	156	2.12	2,306	23.4	−2,151	80.8
India	179	7.2	650	21.0	−470	71.2
United States	14,103	18.5	14,215	15.9	−135	−1.0
Japan	–	–	489	22.4	−489	22.4
South Korea	–	–	325	22.9	−325	22.9
Rest of the world	–	–	302	24.0	−302	24.0
World[7]	28,488	18.3	28,469	18.3	5,670	21.8

government may decide to impose tighter enforcement of regulations, limiting the area expansion in the country. This additional factor has the impact of limiting the country's ability to respond by increasing agricultural area. The motivation for this scenario is the growing pressure being exerted by the international community, environmental organizations, and the Brazilian government to curb land use conversion in general and deforestation in particular. This scenario is implemented by halving the area expansion elasticities of the different regions. Thus the same increase in returns to agricultural production will result in a lower expansion in output. In other words, the same supply expansion will need higher price changes to materialize.

While restrictions in area expansion through, for example, tighter regulations are expected to reduce land use change, and therefore deforestation, in Brazil, a partially compensating change can be expected in other regions (leakage), coupled with larger commodity price changes. In this case, curbing the increase in GHG emissions comes at the cost of higher food prices.

4. Results

4.1. Scenario 1

The increase in global ethanol use affects not only the production but also the trade of the biofuel. The changes, which are country specific, are presented in Table 8.3 for year 2022, the last year of modeling, which was selected to coincide with the last year

---

[6] A positive number denotes an increase in net exports (reduction in net imports). A negative number represents an increase in net imports (reduction in net exports).

[7] World production and consumption changes differ by changes in ending stocks.

Table 8.4. *Change in the prices and areas of selected commodities in 2022 for scenario 1*

	Price change %	Area change	
		1,000 ha	(%)
Ethanol	35.79	–	–
Sugar	4.27	–	–
Sugarcane	–	1,384	4.74
Corn	2.71	1,606	1.00
Soybeans	0.61	−362	−0.32
Wheat	1.06	−99	−0.04
Sorghum	1.43	55	0.13
Barley	1.37	−5	−0.01
Other crops[a]	–	−94	−0.11
Total	–	2,485	0.35

[a] Other crops include rapeseed, sunflower, peanuts, and sugar beet.

of volumes established in the Renewable Fuels Standards of the United States. As mentioned before, the increase in ethanol prices in response to the additional demand disciplines the global utilization increase to be below the size of the shock (25%). At the new equilibrium, global consumption increased by 18.3 percent, or 28,469 million L, relative to the baseline. With the exception of Brazil, all countries listed in Table 8.3 are net importers of ethanol.

While the consumption in all countries increased by a percentage similar to that in the shock, production and trade changes varied by country. For the case of the United States, the increase in consumption is mostly supplied by a commensurate expansion in domestic production, with a relatively minor change in trade. It is also worth noting that the demand for high blends, such as E-85, is fairly elastic, as FFV drivers can revert to gasoline whenever the price of ethanol increases relative to that of the fossil fuel. It is this fact that reduces consumption more in the United States relative to other countries presented in Table 8.3 (when compared to the initial increase in ethanol demand). Given the large market penetration of FFVs in Brazil, the demand for ethanol is also relatively elastic in that country. Table 8.3 also shows that a high proportion of the additional demand in countries other than the United States is supplied by increased exports from Brazil. An important implication is that the expansion of ethanol production based on grains is limited to the additional fuel consumed in the United States and, to a lesser extent, to expanded production in China and the European Union. This muted grain-based ethanol production expansion will dampen the effects in the market for grain feedstocks, mostly corn and wheat. Table 8.4 shows the percentage change of the price of selected commodities, and the global change in area harvested of these commodities.

The prices of all the commodities presented here increase in response to the increase in global ethanol production and consumption. The shock is introduced in the model as an exogenous increase in demand. Thus the new equilibrium price of ethanol for the scenario is higher than in the baseline. The prices of feedstocks for ethanol production, such as corn, also increase, reflecting the enhanced derived demand for these products. The increase in the prices of most of the other commodities is due to their reduced supply, as additional land is claimed by ethanol feedstocks.

As previously mentioned, Table 8.3 shows that most of the increased consumption (in countries other than the United States) is met by the expansion of Brazilian production and exports. The feedstock needed for the additional ethanol production in Brazil is obtained through an increase in the area devoted to sugarcane and an increase in the proportion of the recoverable sugars in the sugarcane used for ethanol, at the expense of sugar. This latter source reflects a decline in the production of sugar and an increase in the price of the sweetener (see Table 8.4). The regional distribution of the increase in sugarcane area is presented in Table 8.5.

The estimated country-level increase in sugarcane area as a result of the surge in ethanol demand is about 1.4 million ha, an 11.2 percent increase from baseline levels. As expected, most of the expansion is projected to occur in the Southeast, the region with the largest sugarcane area and the highest growth rate in the recent past. This region is followed by the Northeast Coast and the Center West, a region in which the ethanol industry is currently expanding.

While the increase in ethanol consumption leads to the expansion of sugarcane area as well as areas of other crops (especially the second corn crop), total agricultural area expansion is lower than the combined increase in all crops. The area planted to crops increases by 1.6 million ha. However, at 802,000 ha, the expansion in the land used for agriculture is lower. This implies that the model is projecting some of the crop expansion to occur in areas already in use for agriculture. In particular, some of the crop area expands over pasture, partially offsetting the demand for additional land and the pressure on natural landscapes. The increase of cropped area into pasture is accommodated by an increase in the intensity with which pastures are used, as evidenced by higher stocking rates (stock of cattle divided by pasture area), as shown in Table 8.6. The largest levels of pasture use intensification can be observed in the regions with difficulties in incorporating additional land and facing the most pressure for sugarcane expansion (e.g., Southeast and Northeast Coastal). While additional sugarcane area is expected in the Center West, the availability of land for expansion dampens the need for intensification in pasture use. The rest of the difference between the increase in total agricultural area and the increase in total crop area is accounted for by an increase in the area that is double-cropped (see Table 8.5). Thus intensification in land use reduces the need for the expansion of agriculture into previously unused areas.

An important portion of the expansion in crop area (other than sugarcane) can be attributed to the need to partially replace grains used to produce ethanol. As an

Table 8.5. *Regional changes in the area used for agriculture in Brazil in 2022 for scenario 1*

Region (1,000 ha)	Sugarcane (1)	Other first crops[a] (2)	Second crops[b] (3)	Area planted (4) = (1) + (2) + (3)	Pasture (5)	Area used (6) = (4) + (5) − (3)
South	74.7	−16.3	106.7	165.1	5.8	64.2
Southeast	991.2	−236.5	13.6	768.3	−377.3	377.5
Center West	115.9	104.8	102.7	323.4	−94.7	126.0
North	10.0	57.9	3.1	71.0	66.7	134.5
Northeast Coast	143.2	36.8	0.0	180.0	−127.1	52.9
Northeast Cerrado	17.3	53.1	12.5	82.9	−23.8	46.6
Brazil	1,352.3	−0.1	238.6	1,590.8	−550.5	801.7

[a] Includes corn, soybeans, cotton, rice, and dry beans.
[b] Includes the second crops of corn and dry beans, wheat, and barley. As winter crops, the latter two crops are assumed to be mostly double-cropped with summer crops.

Table 8.6. *Change in the stocking rate of pastures (stock of cattle divided by pasture area) by region in 2022 for scenario 1*

Region	Change in stocking rate
South	0.069
Southeast	0.702
Center West	0.269
North	0.234
Northeast Coast	0.942
Northeast Cerrado	0.196
Brazil	0.356

example, an increase in corn-based ethanol production in the United States will result in a reduction in exports of about 7.5 million metric tons (11%). Ceteris paribus, the generated excess demand for the rest of the world will push corn prices up and increase crop area in Brazil.

4.2. Scenario 2

We turn our attention now to the implications of restricting the ability of producers in Brazil to increase the area under sugarcane production. Given the additional constraints to land expansion in Brazil, we would expect that larger price increases would be needed to bring about a sufficient supply of agricultural products and to increase agricultural area in Brazil and in other countries to compensate for the diminished supply expansion in Brazil. The results indicate, however, that the restriction in land expansion has a limited impact on prices and crop areas, given the size of the demand shock and the scope for intensification in production of the livestock sector and of double-cropping.

The equilibrium changes in production, consumption, and trade of ethanol as a result of the introduced shock to the system are virtually unchanged from those observed in the first scenario and thus are not repeated here. This is because additional ethanol supplies were obtained with a marginal price change in the model, limiting the price transmission to other commodities. Again, most of the additional demand is met through expanded exports by Brazil. Most of the consumption expansion in the United States is supplied through domestic sources.

Given the constrained ability of Brazilian producers to respond to price changes as land use restrictions are more tightly enforced in this scenario, prices for ethanol and its feedstocks were expected to increase more than in scenario 1 (see Table 8.7). However, as mentioned, the price changes are only marginally different from those observed in the first scenario. Additionally, the total area devoted to agriculture does

Table 8.7. *Change in the prices and areas of selected commodities in 2022 for scenario 2*

	Price change %	Area change	
		1,000 ha	(%)
Ethanol	35.85	–	–
Sugar	4.34	–	–
Sugarcane	–	1,380	4.72
Corn	2.72	1,604	0.99
Soybeans	0.61	−363	−0.32
Wheat	1.07	−99	−0.04
Sorghum	1.44	55	0.13
Barley	1.37	−5	−0.01
Other crops[a]	–	−95	−0.11
Total	–	2,478	0.35

[a] Includes rapeseed, sunflower, peanuts, and sugar beet.

not expand as much as before. The reduced ability to expand area in Brazil does not significantly constrain the country's ability to increase ethanol supply because of the same intensification in land use observed in scenario 1.

Driven by the assumption of a lower area expansion elasticity, the area used for agriculture in Brazil increases by a smaller amount compared to the first scenario (see Table 8.8). While the total area planted with crops (column 4 in Tables 8.5 and 8.8) increases by a similar amount relative to scenario 1, the area used for agricultural activities (column 6) is about 15 percent lower in the second scenario. In terms of total expansion, we find that the crop area growth (including sugarcane) occurs to a higher extent in pasture area, increasing the intensification of land use and reducing the need to incorporate additional area into production. Thus deforestation is reduced relative to scenario 1, limiting negative environmental impacts, including GHG emissions.

The relatively small impact of limiting producers' ability to expand into new areas through a policy such as tightening the enforcement of land reserve restrictions is crucially dependent on the size of the demand shock introduced and the room for intensification embedded in the established baseline. As the scope for land use intensification is exhausted, additional biofuel quantities can be expected to be produced only by incorporating new land into production. In this situation, limitations on land expansion would have larger consequences for commodity prices.

5. Conclusions

Global biofuel production and consumption have an associated derived demand for crops to produce the necessary feedstock and corresponding land use requirements. A spatially disaggregated model of Brazilian agriculture, part of the FAPRI modeling

Table 8.8. *Regional changes in the area used for agriculture in Brazil in 2022 for scenario 2*

Region (1,000 ha)	Sugarcane (1)	Other first crops[a] (2)	Second crops[b] (3)	Area planted (4) = (1) + (2) + (3)	Pasture (5)	Area used (6) = (4) + (5) − (3)
South	74.6	−25.9	107.4	156.1	−5.4	43.3
Southeast	986.8	−242.1	13.7	758.4	−429.8	314.9
Center West	117.7	114.7	103.2	335.6	−56.6	175.9
North	9.8	54.2	3.1	67.1	15.3	79.3
Northeast Coast	141.3	33.2	0.0	174.5	−143.1	31.4
Northeast Cerrado	17.3	52.3	12.6	82.2	−34.7	34.9
Brazil	1,347.5	−13.6	240.1	1,574.0	−654.2	679.7

[a] Includes corn, soybeans, cotton, rice, and dry beans.
[b] Includes the second crops of corn and dry beans, wheat, and barley. As winter crops, the latter two crops are assumed to be mostly double-cropped with summer crops.

system of world agriculture, is used to assess the implications of global biofuel expansion on Brazilian land use at the regional level.

We find that most of the expansion in global ethanol consumption outside the United States is met by Brazilian ethanol production, which leads to an increase in the area devoted to sugarcane. However, a large proportion of the sugarcane area expansion occurs in areas already in agricultural use. For example, for the Southeast region, about 62 percent of the expansion of sugarcane area is accommodated by a decline in the areas of pasture and other crops. This proportion increases to 68 percent when land expansion limitations are introduced (scenario 2). In the Northeast region, virtually all the sugarcane area expansion comes at the expense of pasture. The trade-offs between crops and pastures are not as apparent in regions with larger reserves of available land such as the Center West and the North.

The results suggest that reducing the overall responsiveness of Brazilian agriculture may limit the land use changes brought about by biofuel expansion, which would in turn reduce its environmental impacts in terms of land expansion. The impacts on food prices are limited here because of the ability of local producers to increase the intensity of land use in both crop and livestock production. For crops, the intensification of land use is achieved in this case by increasing the prevalence of double-cropping and by raising crop yields. For livestock, increasing the number of heads of cattle per hectare of pasture (stocking rate) frees up area that can be used for crops. However, both these land use intensification mechanisms have their limits. Once intensification possibilities are exhausted, larger quantities of land will need to be incorporated into production, and higher commodity prices per unit of additional biofuel demand will need to occur as biofuel production expands.

Part Three
Asia

9

Biofuel expansion in Southeast Asia: Biodiversity impacts and policy guidelines

JANICE S. H. LEE, JOHN GARCIA-ULLOA, AND
LIAN PIN KOH
ETH Zürich, Zürich, Switzerland

Abstract

Over the last few decades, oil palm (*Elaeis guineensis*) agriculture in Southeast Asia has created new opportunities for poverty alleviation and economic development. Conversely, this crop has also become a key driver of land use change in the region, with potentially dire consequences for biodiversity, human livelihoods, and the global climate. Studies suggest that oil palm growers could marginally increase the species richness of butterflies and birds on their plantations by changing management practices or by preserving remnant forest patches within their estates. However, the magnitude of these biodiversity enhancements is low relative to the biodiversity of undisturbed tropical forests. This suggests that little can be done to make oil palm plantations more hospitable to biodiversity. Unless future oil palm expansion is diverted to degraded lands, such as preexisting croplands or anthropogenic grasslands, rising global biofuel demand is likely to exacerbate the high rates of forest conversion and threats of extinction to species in major oil palm–producing countries such as Indonesia and Malaysia. This chapter reviews the biodiversity impacts of oil palm expansion in Southeast Asia and provides a number of policy recommendations that can help mitigate the negative biodiversity impacts of palm oil biodiesel production.

Keywords: biodiversity, oil palm, Southeast Asia

1. Introduction

Approximately 80 percent of total world energy supply is derived from fossil fuels such as oil, natural gas, and coal. Fossil fuels are finite sources of energy and are estimated to last anywhere from 41 to ~700 years, depending on production and consumption rates (Goldemberg, 2007; Goldemberg and Johansson, 2004). Growing demand for energy from industrialized nations, such as the United States, as well as from emerging economies, such as China and India, will continue to place tremendous pressures on world petroleum supplies in the next few decades (Worldwatch Institute, 2007). As such, many countries are seeking to diversify their energy portfolios. Growing concerns over anthropogenic climate change have also driven countries to search for alternatives to fossil fuels that can help lower greenhouse gas (GHG) emissions and slow the pace of global warming (Koh and Ghazoul, 2008). These pressing global energy and environmental challenges have at least partly driven the recent worldwide interest in biofuels (see Chapter 1).

In the Southeast Asian context, at least four countries are currently promoting biofuel policies: Indonesia, Malaysia, Thailand, and the Philippines. Even though different policy concerns have influenced each of these countries to promote biofuel production, it seems that energy security, trade balance, and economic development are shared by most countries (Table 9.1). Social issues such as rural development and agricultural employment are only significant drivers of biofuel expansion in Indonesia and Thailand. What is striking is that environmental concerns have played only a minimal role.

A number of different biodiesel and bioethanol production practices are pursued in the four countries, with feedstock availability being the main limiting factor of which biofuel option is finally adopted in each country (Table 9.2). Palm oil biodiesel is the main biofuel practice pursued in Indonesia and Malaysia. This is not surprising, given that in 2008, the two countries accounted for 83.9 percent of global production and 82.0 percent of palm oil exports (FAO, 2011).[1] Additionally, palm oil produced in the tropics has a much higher oil yield (~5,000 L/ha, compared to ~1,200 L/ha for rapeseed), and hence it entails lower production costs (Worldwatch Institute, 2007). Indonesia recently announced plans to double oil palm production by 2020 (Koh and Ghazoul, 2010). This is a worrying sign for many tropical biologists because palm oil production is located within the tropics, where the majority of the world's remaining and most imperiled biodiversity is located (Conservation International, 2010). Furthermore, high proportions of still forested lands in these regions are suitable for palm oil biofuel production. A recent study estimated that an increase in global biodiesel production capacity to meet future biodiesel demands (an estimated 277 million metric tons per year by 2050) may lead to potential habitat losses of between 0.4 million and 114.2 million ha in the tropics (Koh, 2007). Without proper mitigation guidelines, the future expansion of biodiesel feedstock production in the tropics will likely threaten their native biodiversity (Koh and Ghazoul, 2008; Mittermeier et al., 2004).

Some researchers argue that next-generation biofuels, produced from nonfood feedstocks such as agricultural wastes, can fulfill many of the promises of renewable fuels without many of the environmental ills (Shi et al., 2009). These second- and

---

[1] It is worth mentioning that in 2008, palm oil was the sixth most widely traded agricultural commodity in monetary terms, at USD 29.3 billion, with the biggest importers being China, India, Pakistan, Germany, and the Netherlands (FAO, 2011).

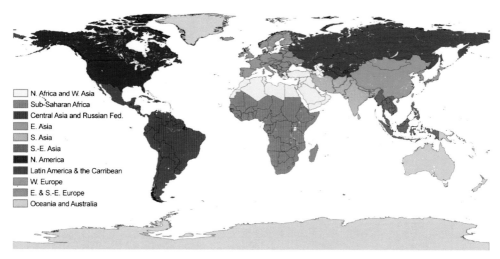

Plate 2.1. World regions used in this study.

N. Africa and W. Asia
Sub-Saharan Africa
Central Asia and Russian Fed.
E. Asia
S. Asia
S.-E. Asia
N. America
Latin America & the Carribean
W. Europe
E. & S.-E. Europe
Oceania and Australia

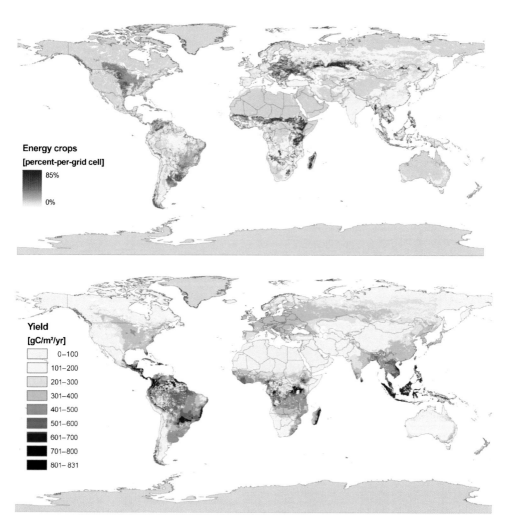

Energy crops
[percent-per-grid cell]
85%
0%

Yield
[gC/m²/yr]
0–100
101–200
201–300
301–400
401–500
501–600
601–700
701–800
801–831

Plate 2.5. (a) Location and (b) yield of dedicated energy crops in 2050 in the baseline scenario.

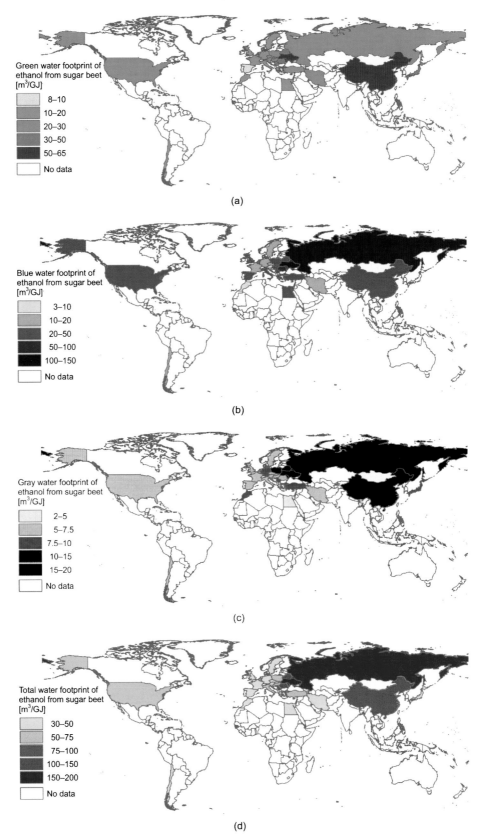

Plate 4.1. Weighted average (a) green, (b) blue, (c) gray, and (d) total WFs of ethanol from sugar beet (m³/GJ) for five regions.

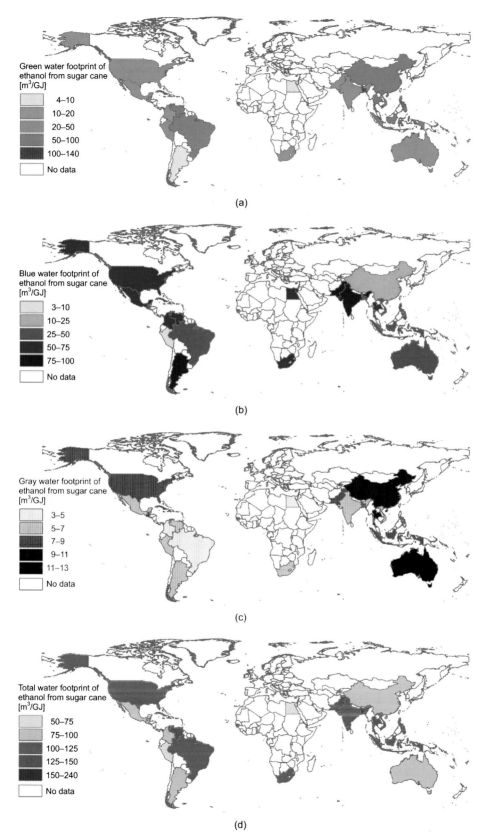

Plate 4.2. Weighted average (a) green, (b) blue, (c) gray, and (d) total WFs of ethanol from sugarcane for four regions.

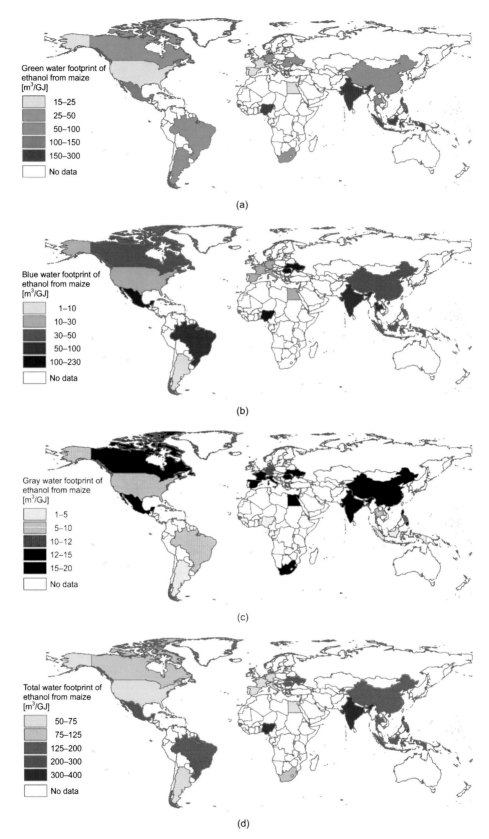

Plate 4.3. Global weighted average (a) green, (b) blue, (c) gray, and (d) total WFs of ethanol from maize for five regions.

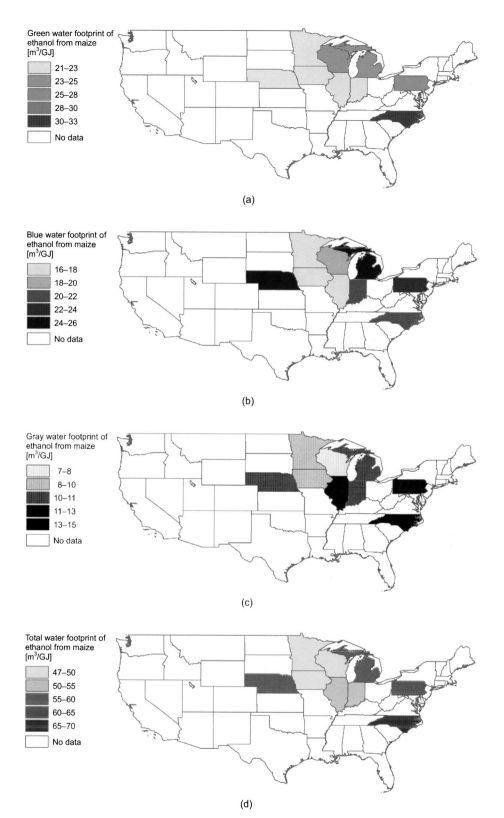

Plate 4.4. (a) Green, (b) blue, (c) gray, and (d) total WFs of ethanol from maize in the United States for 10 states.

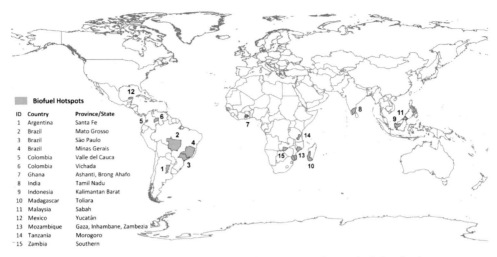

Plate 5.1. A preliminary map of global biofuel hotspots in tropical developing countries at subnational level.

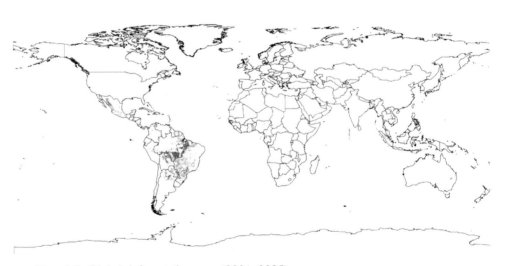

Plate 5.2. Global deforestation map (2001–2005).

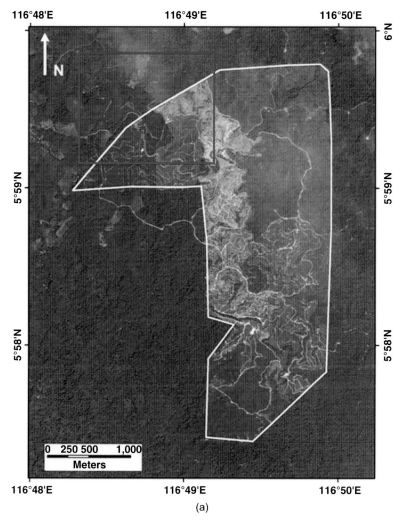

Plate 5.3(a). Deforestation in Sabah, Malaysia, represented by satellite images from three different sensors with different spatial resolutions: image of DigitalGlobe from Google Earth, August 2007.

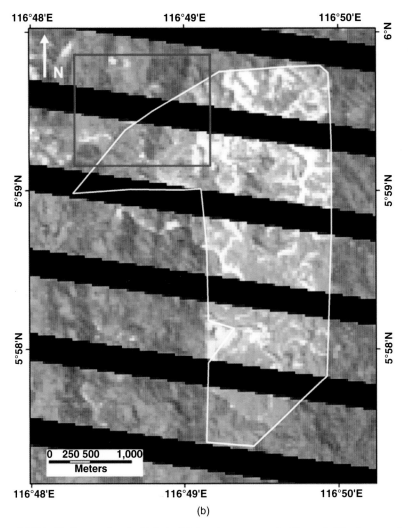

Plate 5.3(b). Deforestation in Sabah, Malaysia, represented by satellite images from three different sensors with different spatial resolutions: color composite image of Landsat-7 ETM+, July 2008.

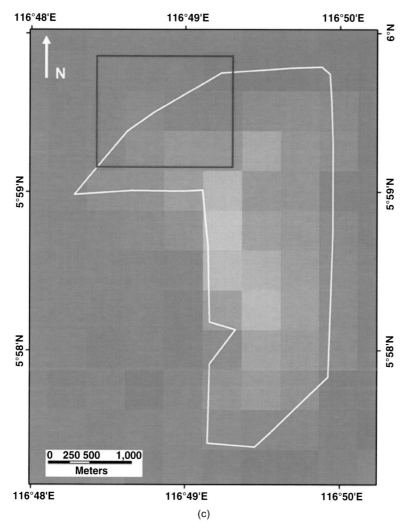

Plate 5.3(c). Deforestation in Sabah, Malaysia, represented by satellite images from three different sensors with different spatial resolutions: color composite image of MODIS MOD09, August 2007.

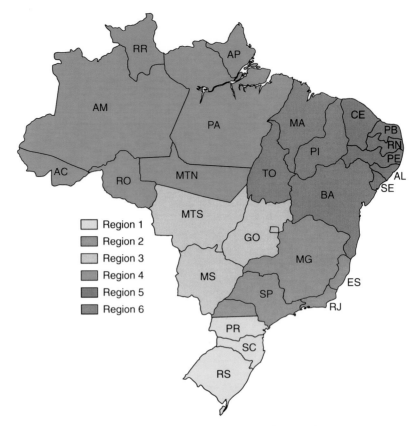

Plate 8.2. Regional disaggregation of the CARD Brazil model.
Region 1: South
Rio Grande Do Sul; Santa Catarina; Paraná
Region 2: Southeast
São Paulo; Rio de Janeiro; **Espírito** Santo; Minas Gerais
Region 3: Center West Cerrado
Mato Grosso Do Sul; Goias; Distrito Federal; Mato Grosso
Region 4: Amazon Biome
Rondonia; Amazonas; Pará; Roraima; Amapá; Acre; Mato Grosso
Region 5: Northeast
Ceará; Paraíba; Rio Grande Do Norte; Pernambuco; Alagoas; Sergipe
Region 6: North-Northeast Cerrado
Tocantins; Bahia; Maranhão; Piauí

Plate 9.1. Mature oil palm tree on a plantation in Indonesia.[2] Photo by Janice S. H. Lee.

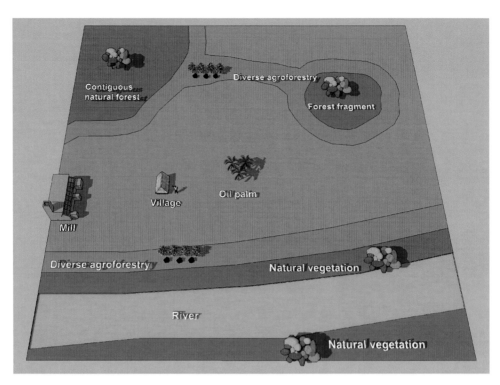

Plate 9.2. Illustration of a designer landscape for sustainable oil palm production. *Source:* Koh et al. (2009b).

Plate 10.2. Jatropha plantation on an unutilized hillside in Xinping County, Yunnan Province, in 2008.

Plate 10.3. Jatropha seedling preparation in Xinping County, Yunnan Province.

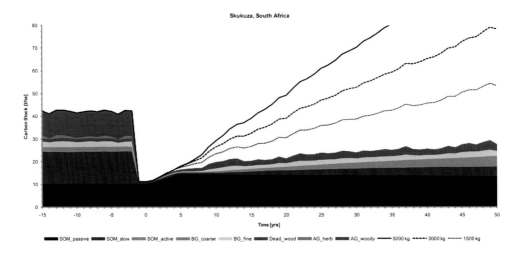

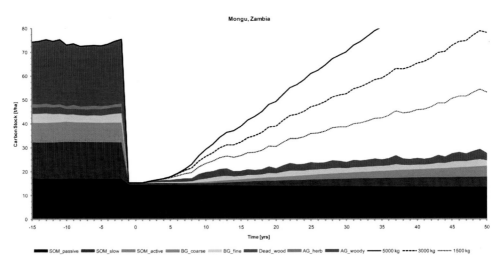

Plate 13.1. The net carbon balance of jatropha biodiesel production over time at the selected sites in semiarid savanna (Skukuza) and miombo woodland (Mongu).The black lines indicate the net carbon balance for different yield scenarios (kg/yr).

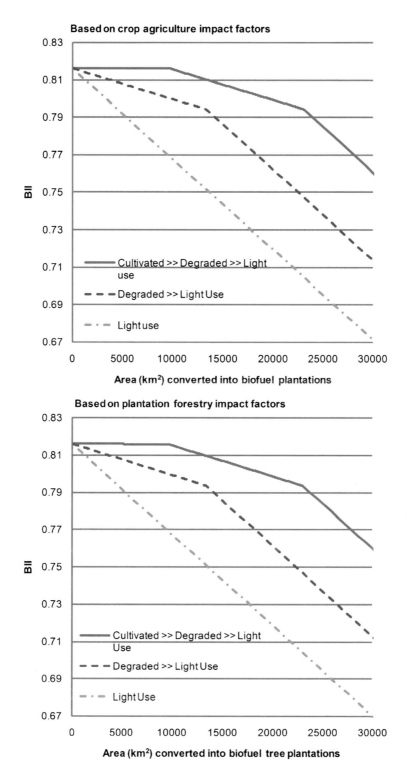

Plate 13.2. The influence of type of land allocated to biofuel on the BII impacts for annual and tree biofuel crops for Eastern Cape South Africa. *Source*: von Maltitz et al. (2010).

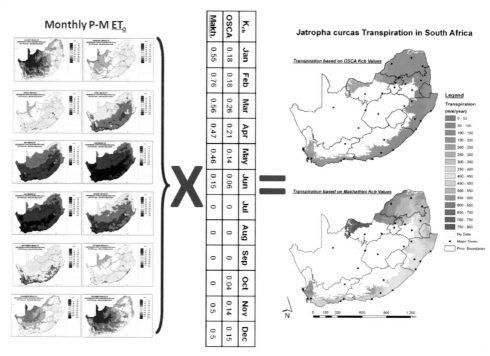

Plate 13.4. Illustration of the method used to estimate spatial water use patterns of *Jatropha curcas* in South Africa. *Source*: Gush and Hallowes (2007).

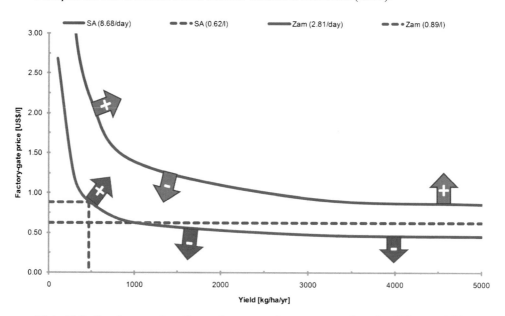

Plate 13.7. Break-even plot of complementary factory-gate prices for different yields that support the legal minimum wage payment in South Africa and Zambia. Areas above the curves indicate net profits, and areas below represent net losses. Values in parentheses for the respective countries represent minimum legal wage and basic fuel price in USD/d and USD/L, respectively.

Table 9.1. *The main drivers of biofuel production in Southeast Asia*

Country	Economy				Social		Environment	
	Energy security	Trade balance	Price of petroleum	Economic development	Increase agricultural employment	Rural development	Climate change	Air pollution
Indonesia	✓			✓	✓	✓		
Malaysia		✓	✓	✓			✓	
Philippines	✓	✓		✓				✓
Thailand	✓	✓		✓	✓	✓		

Sources: Braimoh et al. (2010); Zhou and Thomson (2009).

Table 9.2. *Biofuels and feedstocks for the main producing nations of Southeast Asia*

Country	Main option	Feedstock	Secondary option	Feedstock
Malaysia	Biodiesel	Palm oil		
Indonesia	Biodiesel	Palm oil (primarily), jatropha (secondarily)	Bioethanol	Cassava and sugarcane
Philippines	Bioethanol and biodiesel	Sugarcane (for bioethanol) and coconut oil (for biodiesel)		
Thailand	Bioethanol	Sugarcane and cassava	Biodiesel	Palm oil and jatropha

Sources: Zhou and Thomson (2009); Braimoh et al. (2010).

third-generation biofuels are currently too costly to be produced on a commercial scale in most Southeast Asian countries. Nevertheless, they may become more readily available and affordable in the future through technological breakthroughs, contingent on strong governmental support and a string of local government and international subsidies and initiatives (Doornbosch and Steenblik, 2007). Even so, next-generation biofuels may not be completely free of environmental trade-offs. A recent study demonstrated how a global aggressive cellulosic biofuel program could result in major losses of biodiversity across many parts of the tropics, both directly, by replacing native habitats, and indirectly, by displacing other agricultural land uses onto native habitats (Melillo et al., 2009). For at least the next decade, first-generation biofuels will still be in demand (OECD-FAO, 2008). In the tropics, where myriad anthropogenic factors are already driving intense land use conflicts, adding biofuels as another demand on the land will make the preservation of the remaining natural habitats an even greater challenge.

In this chapter, we discuss the impacts of palm oil biodiesel expansion on biodiversity in Southeast Asia. Initially, Section 2 discusses how oil palm plantations affect biodiversity in the region. Consequently, we identify the major drivers that are responsible for this biodiversity loss. In Section 3, we turn our attention to some policy guidelines that can help mitigate negative biodiversity impacts from palm oil biodiesel production. Even though our proposals are drawn from lessons learned in Southeast Asia, the same principles could also be applied in countries and regions that are currently promoting the use of palm oil for biodiesel production (e.g., West Africa, Brazil, Colombia, and Central America).

2. Oil palm expansion and tropical biodiversity

2.1. Biodiversity loss

Oil palm plantations contain less than half as many vertebrate species as primary forests (Fitzherbert et al., 2008) (see Table 9.3). Forest bird species declined by 73 to

Table 9.3. *Comparison of the number of species of invertebrate and vertebrate groups between natural forests and oil palm plantations in Southeast Asia[a]*

Study	Country	Taxonomic group	No. of species		
			Forest	Plantation	Shared
Chang et al. (1997)	Malaysia	Mosquitoes	6	6	6
Chung et al. (2000)	Malaysia	Subterranean beetles	306	64	–
Chung et al. (2000)	Malaysia	Arboreal beetles	174	40	–
Chung et al. (2000)	Malaysia	Ground beetles	557	75	–
Bruhl (2001)	Malaysia	Ants (site 1)	20	11	6
Bruhl (2001)	Malaysia	Ants (site 2)	8	15	6
Bruhl (2001)	Malaysia	Ants (site 3)	4	8	1
Liow et al. (2001)	Malaysia	Bees	8	17	–
Benedick (2005)	Malaysia	Butterflies	26	12	1
Hassall et al. (2006)	Malaysia	Terrestrial isopods	8	4	0
Chey (2006)	Malaysia	Moths (site 1)	75	85	28
Chey (2006)	Malaysia	Moths (site 2)	133	73	28
Chey (2006)	Malaysia	Moths (site 3)	78	90	11
Koh and Wilcove (2008)	Malaysia	Butterflies	63	–	12
Danielsen and Heegaard (1995)	Indonesia	Birds	67	17	3
Danielsen and Heegaard (1995)	Indonesia	Bats	8	1	1
Scott et al. (2004)	Indonesia	Small mammals	5	3	2
Aratrakorn et al. (2006)	Thailand	Birds	108	41	21
Peh et al. (2005, 2006)	Malaysia	Birds	152	–	36
Maddox et al. (2007)	Indonesia	Medium/large mammals	38	4	4

[a] *Source:* Danielsen et al. (2009).

77 percent (Koh and Wilcove, 2008), and only 10 percent of mammal species were detected in oil palm plantations when compared to undisturbed forest in the region (Maddox et al., 2007). Endangered species such as the Sumatran tiger (*Panthera tigris sumatrae*), tapirs (*Tapirus indicus*), and clouded leopards (*Neofelis nebulosa*) were never recorded on oil palm plantations, and most mammals even preferred marginal and heavily degraded landscapes, such as shrublands, to oil palm (Maddox et al., 2007). Mammals that do occur on oil palm plantations tend to be of low conservation value and are dominated by a few generalist species such as the wild pig (*Sus scrofa*), bearded pig (*Sus barbatus*), leopard cat (*Prionailurus bengalensis*), and common palm civet (*Paradoxurus hermaphroditus*) (Maddox et al., 2007).

Conversely, invertebrate taxa showed greater variation in their sensitivity to the oil palm habitat (Fitzherbert et al., 2008). For example, the conversion of forests to oil palm caused forest butterfly species to decline by 79 to 83 percent (Koh and Wilcove, 2008), whereas ants, moths, and bees showed a higher total species richness in oil palm plantations than forests (Danielsen et al., 2009). Nevertheless, studies consistently show a dominance of nonforest invertebrate species in oil palm plantations (Danielsen et al., 2009). Comparing across both vertebrate and invertebrate taxa, a mean of only 15 percent of species recorded in primary forest could be found on oil palm plantations (Fitzherbert et al., 2008). Not surprisingly, plant diversity within oil palm plantations was impoverished when compared to forests due to regular maintenance and replanting (every 25–30 years) of oil palm fields (Danielsen et al., 2009; Fitzherbert et al., 2008). In the following subsections, we elaborate on the key drivers of biodiversity loss in the context of biofuel expansion: habitat loss, pollution, and interaction with other frontier-opening activities.

2.2. Drivers of biodiversity loss

2.2.1. Habitat loss

Currently biodiversity in Southeast Asia is under the greatest threat of biofuel expansion, mainly from oil palm, sugarcane, and soybean expansion (Koh, 2007; Koh and Wilcove, 2007, 2008, 2009; Wilcove and Koh, 2010). Conversion of natural habitats into biofuel monocultures, by definition, implies a drastic loss in biodiversity and change in the composition of species communities in the area. Compared to a primary forest, biofuel plantations are structurally less complex, more transient, and often highly fragmented (Fitzherbert et al., 2008).

Using land use data compiled by the Food and Agriculture Organization of the United Nations (FAO), Koh and Wilcove (2008) show that between 1990 and 2005, 55 to 59 percent of oil palm expansion in Malaysia and at least 56 percent of that in Indonesia had occurred at the expense of forests. Unless future expansion of oil palm can be diverted to nonforested lands, such as preexisting croplands or anthropogenic grasslands, rising global biofuels demand likely will exacerbate the high rates of forest conversion and threats of extinction to species in major oil palm–producing countries (Koh, 2009; Koh et al., 2009a; Koh and Wilcove, 2007, 2008, 2009). Of course, the expansion of the biofuel industry is not the only cause of habitat loss in these areas. Other causes include large-scale commercial logging, pulp and paper industries, cattle ranching, shifting cultivation, mining, urban development, and agricultural expansion of other crops (Angelsen and Kaimowitz, 1999). However, growing global demand for palm oil will likely exacerbate deforestation across the tropics over the next decade (IATP, 2008).

2.2.2. Environmental pollution

Apart from habitat and biodiversity loss, biofuel industries that engage in poor production practices could also cause indirect ecological damage through environmental pollution and degradation. Inappropriate management practices such as intensive application of fertilizers or pesticides as well as using fires for land clearing could lead to environmental problems such as soil degradation and water and air pollution, which in turn could result in long-term ecological impacts in the tropics. For example, mature oil palm plantations in Malaysia have a soil erosion rate of approximately

7.7–14 metric tons/ha/yr (Hartemink, 2006). Soil erosion in oil palm plantations can be even more serious in the early years, when a complete palm canopy is not yet established, which is why maintaining a legume crop cover is important to protect against soil erosion (Corley and Tinker, 2003). Surface runoff as a result of soil erosion brings organic matter and agrochemicals into aquatic systems, which can lead to deterioration of aquatic habitats and affect the biodiversity downstream.

Additionally, by-products of the industrial processing of oil palm fruits into crude palm oil are highly polluting and are a significant source of pollution if released into the environment without proper treatment. Palm oil mill effluent is rich in organic matter and contributes to eutrophication and depletion of dissolved oxygen levels in aquatic systems if left untreated (Donald, 2004). Despite the existence of present technologies to treat mill effluents, accidental leakages and discharge from small mills occasionally occur, leading to adverse impacts on aquatic ecosystems (Sheil et al., 2009).

Burning is a common crop management practice for clearing natural vegetation for oil palm expansion in parts of Southeast Asia and in the tropics in general (Casson, 2003; Sheil et al., 2009). Fires often are the quickest and cheapest method to clear land (Guyon and Simorangkir, 2002). Oil palm expansion was partially responsible for the devastating 1997–1998 forest fires in Indonesia. Satellite imagery showed that many fires were started by oil palm producers for clearing land for oil palm cultivation (Dennis et al., 2005). The dry conditions brought about by the El Niño phenomenon exacerbated the fires, which burned 11.6 million ha of land, more than half of which were montane, lowland, and peat forests (Tacconi, 2003).

2.2.3. Interaction with other frontier-opening activities

The development of biofuel plantations is associated with other drivers of habitat loss and degradation such as logging and the establishment of roads and waterways. These infrastructural developments increase the accessibility of natural resources for further exploitation. Some claim that oil palm plantations often associate with logging companies because profits from the sale of timber can help offset part of the establishment costs of an oil palm plantation (Casson, 2000). In some cases, applications for licenses to establish oil palm estates are merely a pretext for some companies to clear-cut forests (Casson, 2000). This may explain why out of 5.3 million ha of land allocated to oil palm development in Kalimantan, only fewer than 1 million ha have actually been planted (Casson et al., 2007).

3. Reconciling biofuel expansion with biodiversity conservation

Reconciling biofuel expansion with biodiversity conservation is not a straightforward process due to the links between the biofuel industry and both the agricultural and energy sectors. A careful assessment of land use allocation options and major restructuring of the agricultural management system may be required for biofuel expansion to proceed with little or no environmental costs. Additionally, the development of energy-efficient transportation systems and advancement of second- and third-generation biofuels will help alleviate demand for conventional biofuel feedstocks. However, these actions will require a considerable amount of time, a great many resources, and long-term commitment from society. From a biodiversity perspective, there is an added urgency also to work on immediate solutions to minimize the loss of threatened biodiversity to biofuel expansion.

3.1. Degraded lands

Clearly the obvious solution is to avoid planting biofuel feedstocks on native natural habitats (IATP, 2008). The replacement of biodiverse habitats with monoculture plantations is a serious threat to biodiversity. Moreover, as many of the regions being targeted for biofuel production contain high concentrations of flora and fauna that cannot be found elsewhere (i.e., that are endemic to these regions), the loss of these habitats would result in the global extinction of numerous species (Myers et al., 2000). The destruction of critical ecosystems for biofuel production often negates any environmental benefits accrued from the use of biofuels (Gibbs et al., 2008).

Some researchers have argued for the use of so-called degraded lands for biofuel cultivation. However, this proposal is not as straightforward as it seems due mainly to the difficulties in determining what constitutes degraded lands. For example, if degraded lands were to include secondary logged forests, which still preserve a significant portion of primary forest biodiversity, then biodiversity losses would continue as such forests are converted to oil palm plantations (Barlow et al., 2007; Dunn, 2004; Koh and Wilcove, 2008). In many cases, secondary vegetation habitats are utilized by high-conservation-value (HCV) species such as the Sumatran tiger; therefore it is foolhardy to define degraded lands simply on the basis of vegetation structure (Maddox et al., 2007). Even if degraded lands can be satisfactorily defined, there remain other challenges to developing oil palm on these lands. For example, to convert *alang-alang* grasslands into oil palm plantations, copious amounts of fertilizers and pesticides need to be applied (Fairhurst and McLaughlin, 2009). Many degraded lands remain undeveloped for numerous reasons, not least of which is unclear land tenure and ownership status, which poses an added investment risk to oil palm developers (Cotula et al., 2008). Furthermore, degraded lands may not be completely worthless. Oil palm developers would have to compete for these lands with other activities such as forest restoration, cattle ranching, human settlement, and urbanization. Hence strategies to expand biofuel production into degraded lands must be approached with caution.

3.2. Payment for ecosystem services

Apart from their biodiversity values, it is imperative to recognize the ecosystem services natural habitats provide, including genetic diversity, carbon sequestration, water cycling and purification, climate regulation, and provision of nontimber products (Constanza et al., 1997). The establishment and enforcement of protected areas remains a top strategic priority for protecting biodiversity, but these legislative tools could be supplemented with innovative schemes such as payment for ecosystem services or reducing emissions from deforestation and degradation (REDD), which create financial incentives to divert agricultural expansion away from forests and onto preexisting croplands or degraded lands. The question follows, then, is whether such financial incentives are sufficient to counter strong market forces that favor habitat conversion. Recent REDD scheme partnerships between nongovernmental organizations and private companies (Fischer, 2009) are positive steps toward greater collaboration and engagement of various stakeholders in conserving tropical forests. However, few studies have compared the feasibility of such schemes against current market prices for biofuel feedstocks. Butler et al. (2009) compared the profitability of converting forests into oil palm plantations against conserving forests for a REDD scheme. Under current voluntary carbon markets, it is more profitable to convert forest into oil palm (yielding net present values of USD 3,835–9,630) than to preserve it for carbon credits (USD 614–994). However, should REDD become a legitimate emissions reduction activity under the second commitment period of the Kyoto Protocol (2013–2017), carbon credits traded in Kyoto-compliance markets would have a fighting chance to compete with oil palm agriculture or other similarly profitable human activity as an economically attractive land use option.

Even then, the extent to which REDD can contribute toward mitigating global climate change will depend on its adoption by society (Ghazoul et al., 2010a, 2010b). This, in turn, will depend on REDD appropriately compensating the full range of economic, social, and political opportunity costs, which remain largely unknown. There is therefore an urgent need for a comprehensive trade-off analysis of REDD to identify (1) direct economic costs due to forgone activities, (2) indirect societal costs associated with loss of employment and value-adding downstream industries, (3) less tangible political and socioeconomic issues relating to variations in investment in national and local development, and (4) new opportunities and benefits afforded by intact forests under REDD protection. The authors argue that only until we have understood the complex issues affecting society's response to REDD can we assess its full potential.

3.3. Improve management practices

Koh (2008a) demonstrates that certain local vegetation characteristics, such as percentage ground cover of weeds, epiphyte prevalence, and presence of leguminous crops, can help enhance native bird and butterfly species richness on oil palm plantations (Figure 9.1). At a landscape level, the percentage of natural forest cover explained 1.2 to 12.9 percent of variation in butterfly species richness and 0.6 to 53.3 percent of variation in bird species richness. Adoption of such measures as part of plantation

Figure 9.1. Mature oil palm tree on a plantation in Indonesia.[2] Photo by Janice
S. H. Lee. (See color plate.)

management practices could make oil palm plantations more hospitable for native
biodiversity. However, admittedly, the magnitude of such biodiversity enhancement
is nowhere near that of a pristine tropical rainforest.

Many oil palm plantations have already adopted integrated pest management practices that favor the use of nonchemical pest control methods, such as the establishment

---

[2] In this case, the company management followed recommendations of biodiversity-friendly practices and allowed
epiphytes to grow around the oil palm trunk. These epiphytes may serve as nesting and foraging grounds for several
bird species, which in turn provide a useful pest control service for the oil palm.

of beneficial plants (e.g., *Euphorbia heterophylla*), to attract insect predators and par-asitoids of oil palm pests (e.g., the wasp, *Dolichogenidea metesae*) (Basri et al., 1995; Corley and Tinker, 2003). In a recent study, Koh (2008b) conducted a bird-exclusion experiment on oil palm plantations in Sabah, Malaysia, to demonstrate the pest control service provided by insectivorous birds for oil palms. Other means of mitigating the impacts on biodiversity loss within the oil palm landscape include the establishment of riparian buffer zones to mitigate against aquatic pollution, preservation of HCV forests and habitat corridors to increase connectivity among forest remnants, and the formation of wildlife buffer zones to ameliorate the deleterious effects of forest edges between forests and the surrounding oil palm matrix (Fitzherbert et al., 2008; Maddox et al., 2007).

3.4. Certification schemes

The recognition of the need to encourage the adoption of environmentally friendly management practices in the production of palm oil for biodiesel has led to the cre-ation of multistakeholder organizations such as the Roundtable of Sustainable Biofuels and the Roundtable of Sustainable Palm Oil (RSPO). These organizations typically engage a diverse range of biofuel-sector stakeholders, including governments, non-governmental organizations, producers, consumers, and suppliers, to create, verify, and certify performance standards for sustainable production of biofuel feedstocks and biofuels (UNEP, 2009) (see Chapter 7). Within these organizations, conservation groups have a platform to engage producers and inform them of suitable new areas for biofuel expansion that will lead to the least ecological damage. Independent Envi-ronmental Impact Assessments of future biofuel expansions and Life-Cycle Analyses of biofuel products provide greater transparency in the costs of production of biofu-els and reassure consumers that biofuels purchased are sustainably produced (UNEP, 2009). However, critics of biofuel certification schemes argue that market-based prod-uct certification often covers only a fraction of the market size (Laurance et al., 2010b; Liu et al., 2004; Sto et al., 2005) and may be misleading as some production appears to be sustainable but in actual fact is not (Doornbosch and Steenblik, 2007; Laurance et al., 2010b). Most important, it has no control over the extent of indirect land use change resulting from displacement of other land use activities by biofuel production (Doornbosch and Steenblik, 2007). However, most conservationists would agree that these certification schemes are important steps in the right direction. For example, the RSPO certification criteria make clear provisions that farming prac-tices that might be harmful to biodiversity are identified and mitigated (Principle 5), while new plantings (since November 2005) are not to replace primary forest or any area required to maintain or enhance one or more HCVs (Criterion 7.3) (RSPO, 2007).

3.5. Designer landscapes

From an ecological perspective, two concepts have been proposed to minimize the adverse impacts of agricultural expansion on biodiversity, land sparing, and wildlife-friendly farming. The former seeks to minimize land area required for farming by land intensification through maximizing yields, whereas the latter aims to enhance biodiversity within a low-intensity agricultural landscape (Fischer et al., 2008). In the context of oil palm agriculture, the land-sparing approach could boost global palm oil production by almost 50 percent without any expansion of cropland (Koh et al., 2009b). This would be equivalent to 26 million ha of land spared from conversion to oil palm. Conversely, proponents of wildlife-friendly farming argue that land sparing is counterproductive to the livelihoods of many rural communities. Instead, biodiversity conservation and rural development objectives would be better served by encouraging low-intensity and diverse cropping practices such as agroforestry.

Obviously, neither approach is without environmental or socioeconomic trade-offs. Therefore Koh et al. (2009b) propose a combination of both approaches by way of carefully designing landscapes based on optimal requirements for sustaining biodiversity, economic, and livelihood needs. Agroforestry (wildlife-friendly farming) zones around HCV areas can be used as corridors to connect surrounding fragments of HCV forests, act as buffer zones to mitigate human encroachment into HCVs, and reduce edge and matrix effects from the intensively cultivated biofuel feedstock landscape (land sparing). Such an approach offers a possibility for a more environmentally sustainable and socially responsible pathway for future biofuel expansions. Engagement with local stakeholders and support from local authorities are particularly important in developing nations like Indonesia, where rural development and improvement of people's lives are urgent priorities.

3.6. Understanding trade-offs of alternative development options

The global human population is expected to reach 9.1 billion people by 2050. These people will demand more water, food, and fuel. For example, the Indonesian government recently announced plans to double its production of oil palm by 2020. At the same time, we also need to preserve our remaining forest and biodiversity, reduce GHG emissions, and sustain economic growth and development. Faced with these Herculean and often conflicting priorities, decision makers need to be able to assess the consequences and trade-offs of alternative development options. Koh and Ghazoul (2010) analyzed the outcomes of prioritizing oil palm production, food production, forest preservation, or carbon conservation based on a spatially explicit scenario analysis. Every single-priority scenario had substantial trade-offs associated with other priorities. An optimal solution was a hybrid approach in which expansion targets

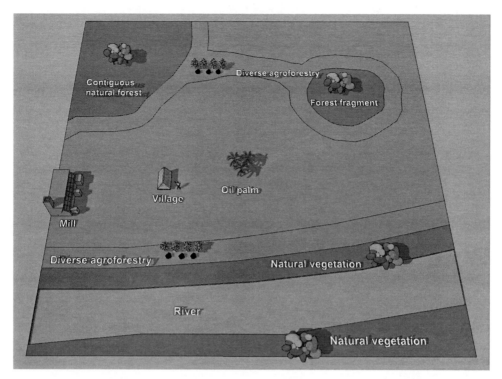

Figure 9.2. Illustration of a designer landscape for sustainable oil palm production. *Source:* Koh et al. (2009b). (See color plate.)

degraded and agricultural lands that are most productive for oil palm, least suitable for food cultivation, and contain the lowest carbon stocks. This approach avoided any loss in forest or biodiversity and substantially ameliorated impacts of expansion on carbon stocks and food production capacity. Their results suggest that if properly planned, the trade-offs of oil palm expansion can be largely avoided and the other demands of society can at least partially be met.

4. Conclusions

The direct conversion of natural habitats into agricultural landscapes for biofuel feed-stocks is the biggest threat arising from biofuel expansion in Southeast Asia. Palm oil plantations are depauperate in biodiversity compared to the natural habitats they replace. Biofuel feedstock production could also result in various kinds of environmental pollution. Proposals to reconcile biodiversity conservation with oil palm expansion include the use of degraded lands for future expansions, adoption of payments for ecosystem services and sustainable practices certification schemes, and the development of carefully designed landscapes for oil palm expansion. But perhaps most important, policy and decision makers need to have a clear understanding of the consequences and trade-offs of pursuing ambitious biofuel policies in their countries, without which these policies may do more harm than good to the environment. Hence emphasis on multistakeholder collaboration to develop more sustainable policies for expanding the biofuel industry is the best immediate remediation for ensuring the protection of the remaining natural habitats and biodiversity across the tropics.

10

Jatropha production for biodiesel in Yunnan, China: Implications for sustainability at the village level

DAISUKE SANO, JANE ROMERO, AND MARK ELDER
Institute for Global Environmental Strategies, Hayama, Japan

Abstract

This chapter examines the implications for sustainability of jatropha biodiesel production in China based on the field survey conducted in Yunnan Province in 2008. Jatropha was identified as a potential biodiesel feedstock in the southwestern region of China. Biofuel production aims to increase the national renewable energy supply and local communities' economies in conjunction with afforestation efforts, while avoiding food–fuel conflicts. Results of the survey indicate that a food–fuel conflict regarding farmland seemed to be avoided while the policy of "no fuel crops on farmland" was in place and while additional land from unutilized forestry land was provided. However, there is concern that some influence on food production could still occur in the future. This is because jatropha is a very labor-intensive crop that could be negatively affected by potential labor shortages due to increasing labor migration out of rural areas. Jatropha production appears profitable only in the short term because of the large demand for seeds and seedlings, but its long-term economic viability as a biodiesel feedstock remains largely uncertain. To cope with uncertainties, risk-averse farmers in our survey adopted a wait-and-see attitude despite the government-led promotion measures. In this sense, economic viability may be more of a limiting factor than resource availability. A full-scale sustainability assessment should be conducted when jatropha trees mature in the future.

Keywords: China, economic viability, fuel–food conflict, jatropha, labor

1. Introduction

Since China enacted the Renewable Energy Law in 2005 in response to its rapid increase in energy use, various types of renewable energy have been strongly promoted, including biofuels. China currently gets only 8 percent of its primary energy from renewable energy, and its target is set at 10 and 15 percent by 2010 and 2020, respectively (NDRC, 2007). This law promotes a wide range of renewable energy sources aiming at a broad set of objectives, including diversifying energy supplies, safeguarding energy security, protecting the environment, and realizing the sustainable development of economy and society.

China's renewable energy policies stress the large-scale provision of electricity nationwide in the midst of rapid industrialization. The great majority of investment in renewable energy is made for wind power (70%), followed by other renewables (17%) and solar (8%), with biofuels contributing only 3.6 percent (Pew Charitable Trusts, 2010). Although biofuels' contribution to renewable energy is relatively small, biofuels are primarily expected to improve energy diversification and rural development through expanded agricultural activities.

While China became the third largest bioethanol producer in the world at 1.33 million metric tons in 2007 (FAO, 2008b; Huang et al., 2008), biodiesel production still remains small at 300,000 metric tons (in 2007), produced mainly from recycled waste oil. There are about a dozen operating plants using waste oil as feedstock and 20 planned plants that will operate not only on waste oil but also on other biodiesel feedstocks such as jatropha. The production capacity per plant is relatively small due to the insufficient supply of feedstock. This is a result of difficulties in feedstock collection and of the lack of an established feedstock market. Furthermore, unlike bioethanol, there is no blending mandate for biodiesel. Since China is a net importer of vegetable oil (Huang et al., 2008), it is seeking alternative feedstocks for larger-scale biodiesel production from nonfood sources (ICRAF China, 2007).

Jatropha has been identified as one of the most promising feedstocks for biodiesel production in the country. In 2007, the State Forestry Administration began the promotion of jatropha production in collaboration with the state-owned petroleum companies as well as food processing companies such as the China National Cereals, Oils, and Foodstuffs Corporation and a few other companies located in the southwestern region of China. Foreign investment from the United Kingdom, the United States, and Germany is also flowing into these provinces. In Yunnan Province, jatropha production expanded to approximately 50,000 ha in 2008 out of the future target of 667,000 ha. Jatropha production is supervised by the Forestry Department as part of its afforestation efforts, while it is seen as an additional income opportunity for farmers.

While various potential benefits from biofuel at the national level (e.g., diversified–decentralized sources of energy and rural–industrial development) have been identified in the academic literature (e.g., Gan and Yu, 2007; Zhong et al., 2010), little is known about the reality on the ground surrounding the introduction of jatropha production at the village level. This chapter reviews the basic conditions of jatropha production in villages in Yunnan, the largest jatropha-producing province in the southwest of China, and attempts to analyze the implications for sustainability both at the village and the household level based on collected survey data.

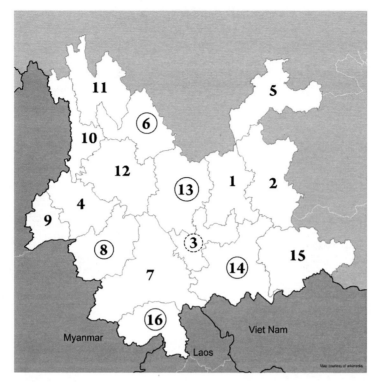

Figure 10.1. Location of field survey in Yunnan Province, China.[1] *Source:* modified map originally from Wikipedia (http://en.wikipedia.org/wiki/Yunnan).

2. Data sources

Two types of data were collected in Yunnan Province in December 2008. Village-level data were obtained from the local authorities, and household-level data were collected through structured household surveys using questionnaires. Household data in the province were collected from 85 households, including 17 non-jatropha-producing households, from 18 villages in five cities and prefectures across the province (numbers 6, 8, 13, 14, and 16 in Figure 10.1). The total area used for jatropha production in the 18 villages was around 9,800 ha and that recorded in the household survey was 2,570 ha. This corresponded to approximately 20 percent and 5 percent of the total 50,000 ha of jatropha area in Yunnan Province in 2008, respectively. An additional informal survey was conducted in Xinping County to supplement the survey (number 3 in Figure 10.1).

---

[1] Circled city/prefecture: Lijiang City (6), Lincang City (8), Chuxiong Prefecture (13), Honghe Prefecture (14), and Wenshan Prefecture (16).

Table 10.1. *Selected economic indicators for the surveyed villages for 2008*

Village	Average annual income in households (USD/Capita)[a]	No. of households	Share of agricultural population (%)	Share of households owning TVs (%)
1	144	266	99	80
2	158	406	98	74
3	161	654	99	46
4	175	443	100	45
5	175	552	96	65
6	175	678	95	100
7	184	231	99	85
8	204	410	99	73
9	234	60	100	67
10	244	557	100	97
11	248	21	6	100
12	272	1,460	98	86
13	277	1,222	23	90
14	289	1,380	100	99
15	326	518	100	87
16	327	964	88	84
17	365	165	53	100
18	467	474	96	100
Yunnan average	385[b]	–	82[c]	84[d]
National average	604[b]	–	60[c]	94[d]

[a] Exchange rate as of December 2008 (USD 0.146/RMB; http://www.x-rates.com/).
[b] Data from 2007 (NBS, 2008).
[c] Data from 2005 (NBS, 2007).
[d] Data from 2007 (color TV) (NBS, 2008).

3. Overview of the jatropha-producing villages and households in Yunnan

3.1. Economic status of the surveyed villages

The average income per capita in most villages surveyed was below the average in Yunnan Province, which was also below the national average. Agriculture was the major economic activity, except for a few villages (Table 10.1). Although all of these villages already had access to electricity, a lower rate of ownership of color TVs among lower-income villages was observed (Table 10.1). Three villages out of 18 had no access to gas or biogas, implying that households need to spend a certain amount of time to collect firewood for cooking or heating. These facts indicate that many of the households in these villages could benefit from potential extra income to further improve their standard of living using their available time for biofuel-related activities.

3.2. Land use for jatropha production as a new cash crop

Jatropha production in Yunnan takes place on unutilized hillsides, and usually, no land is cleared for planting jatropha seedlings. Shrubs and grasses remain between

Figure 10.2. Jatropha plantation on an unutilized hillside in Xinping County, Yunnan Province, in 2008. (See color plate.)

planted young jatropha trees, and thus the production is very extensive (Figure 10.2), unlike on typical large-scale farms or plantations. Most of the jatropha is planted in unutilized forest or meadowland in most of these villages, with a few exceptions (villages 6, 9, 11, and 17) (Table 10.2).

Jatropha production is an additional activity for farmers on top of their subsistence farming. According to the interviews conducted in Xinping County, an average household has access to 0.3–0.4 ha of communal arable land and 20 ha of unutilized hillside where they can grow main crops and tree crops (e.g., jatropha and walnuts). Since all of the land is owned by the government for communal use, the decision on land use is made by the government, with the exception of a small patch of land on which each household grows its own rice and vegetables for home consumption. Forests or meadows in the surveyed villages are commonly managed by the households or villages as a part of a communal area (Table 10.2).

Jatropha is a new cash crop option suitable for Yunnan but is not the only one. With the conditions of mild climate (annual average temperature at 20°C), adequate rainfall, and reasonable soil conditions, farmers already have choices of cash crops such as sugarcane,[2] bamboo, and tobacco. Yunnan has more favorable natural conditions for jatropha production (e.g., temperature, soil moisture, soil quality, and land slope) than other provinces in southwestern China (Wu et al., 2010). Jatropha trees usually

---

[2] Farmers rotate sugarcane and paddy rice in the same plot. Typically, sugarcane is harvested once a year and rice is harvested twice a year.

Table 10.2. *Land used for jatropha production in surveyed villages*

Village	Village area (ha)	Forest/meadow area (ha)	Forest/meadow managed by (%)				Jatropha area in (ha)	Ratio of unutilized forest/meadow land for jatropha production (%)
			Village	Company	Households	Other		
1	6,400	6,196	26	74			3,933	66
2	1,168	884	24	60	16		693	96
3	913	533		44	56		336	100
4	1,733	487	1		99		240	86
5	1,705	515	11		89		151	77
6	1,780	1,000	100				1	0[a]
7	2,864	2,800	29		71		107	97
8	4,573	1,058	7		91	2	120	75
9	153	133	30		70		40	0[a]
10	1,288	1,062	100				391	100
11	267	140	52	48			59	0[a]
12	5,981	5,600	50	5	45		573	93
13	6,047	4,667		1	99		1,333	95
14	1,451	1,200	9		91		373	98
15	2,567	1,986			100		305	99
16	3,418	2,967	33		67		670	99
17	667	20			100		4	0[a]
18	3,153	1,067	100				487	96

[a] Jatropha is planted on unutilized farmland or on land with other unspecified purposes.
Source: survey data collected by the authors.

bear fruit in their third year, and thus newly introduced jatropha trees were not yet mature when we conducted our survey. The decision to harvest was determined by the farmers, depending on the price of jatropha fruits. If the price was low, farmers did not harvest the jatropha fruits. The price of jatropha fruit was reported at 4 RMB/kg (fresh/wet-base, 60 cents/kg) on average in Xinping County in 2008, and it was considered reasonable but not attractive.

3.3. Mode of jatropha business

There are three modes of jatropha production in Yunnan: (1) market based by individual farmers, (2) led by the government, and (3) induced by the state-owned petroleum company (PetroChina), which receives subsidies from the government.[3] The first mode of production is rare, but some interested farmers buy seedlings and plant them in the unutilized hillside they can access. For the second case, free seedlings are provided by the Forestry Department, and jatropha is typically grown without fertilizers or any other agricultural chemicals. For the third case, free seedlings and land rent[4] are provided by the petroleum company.

Jatropha was often planted through the initiative of PetroChina but was managed by households in the surveyed villages, except for a few cases, in which jatropha was entirely planted and managed by households (villages 4, 6, 8, 14, and 17) on a smaller scale (Table 10.3). In Xinping County, PetroChina planted jatropha on approximately 2,000 ha (in 2007) and had plans to increase it to approximately 6,700 ha. So far, jatropha has been planted solely by PetroChina, and no farmers are involved.

3.4. Characteristics of jatropha-producing households

According to the survey results, the majority of households (55 out of 66) had jatropha plots smaller than 10 ha, and the rest (11 households) had plots larger than 53.3 ha (Table 10.4). On the basis of these results, households were classified for the analysis into the categories "nonjatropha producers," "small-scale producers" (less than 10 ha), and "large-scale producers" (larger than 53.3 ha). The households without jatropha plots (17 households) tended to have older workers, less education, and a larger share of communal land (Table 10.4). This implies that their economic activities were connected with the communal land and that there was little room to consider jatropha production as an additional activity. The largest portion of the areas planted with jatropha (over 90%) was managed by households commissioned by PetroChina, with an average of over 200 ha per household. However, most of the households

---

[3] 200 RMB/mu (450 USD/ha) to the company, one time only.
[4] PetroChina pays contracted farmers approximately 11 USD/ha/yr (three-year contract).

Table 10.3. *Jatropha production management in surveyed villages*

	Jatropha planted by (%)		Jatropha land managed by (%)			
Village	Company	Households	Village	Company	Households	Others
1	91	9	92	0	8	0
2	63	37	15	77	8	0
3	100	0	0	100	0	0
4	0	100	0	0	100	0
5	99	1	67	0	33	0
6	0	100	0	0	100	0
7	_[a]	_[a]	50	0	50	0
8	0	100	0	0	100	0
9	100	0	0	0	100	0
10	100	0	100	0	0	0
11	100	0	0	0	100	0
12	36	64	47	0	53	0
13	90	10	0	0	100	0
14	0	100	0	0	100	0
15	100	0	0	0	99	1
16	100	0	30	7	60	4
17	0	100	0	0	100	0
18	3	97	100	0	0	0

[a] No data.

Table 10.4. *Jatropha plot size distribution and characteristics of surveyed households*

		Household average of					
Group[a]	Jatropha area (ha)	No. of households (persons)	Labor age[b] (years)	Schooling (years)	Agricultural labor (persons)	Total arable land (mu)	Communal land share in total arable land (%)
1	0	17	42.0	4.1	3.5	12.6	75
2	≤10	55	38.9	5.8	2.7	20.8	36
3	≥53.3	11	_[c]				

[a] No farmers fall in the category between 10 and 53.3 ha.
[b] Persons of age 16 and over, including nonagricultural laborers.
[c] No data.
Source: Sano and Romero (2010).

Table 10.5. *Jatropha plot size distribution by mode of management*

Group	Management mode	Total area under management	Number of corresponding households[a] (persons)	Average jatropha area per household (ha)
1a	Commissioned by the Forestry Department	53.4	37	1.4
2a	Commissioned by company	1,940.0	9	216
3a	Commissioned by village or other	101.7	3	34[b]
4a	Commissioned work on granted land	27.0	14	1.9

[a] Households without jatropha production (17 households) and incomplete data (3 households) were excluded.
[b] The average of 0.3, 1.3, and 100 ha.

(approximately 60% of total households) were engaged in small plots (2.5% of the total jatropha area) commissioned by the Forestry Department (Table 10.5).

4. Implications for sustainability from the village-level data

Sustainable jatropha production requires a sustainable supply of production inputs such as land, water, and labor. Most surveyed villages have a large share of forest or meadowland, with the majority of jatropha being planted on the unutilized part of such lands (Table 10.2). This indicates that the principle of "no jatropha on farmland" is actually being observed. Jatropha production also adds a modest contribution to forest cover. However, since a large share of the previously unutilized forest and meadowland is now utilized by the recent introduction of jatropha into these villages, land for additional jatropha production will have to be found elsewhere in the province to meet future targets.

Wu et al. (2010) found that a moderate amount of suitable land exists in the southwestern region (Yunnan, Sichuan, and Guizhou provinces). Nevertheless, the amount of suitable land is limited (Wu et al., 2010). This implies that some jatropha would be planted under less advantageous conditions (resulting in lower yields) or could potentially compete with other forest crops. In either case, it would not be easy for farmers to agree to adopt jatropha as a way of generating extra income for sustainable livelihoods unless they were guaranteed to get a relatively high price for jatropha fruits. Water resources are primarily dedicated to staple crops (rice), and irrigation is available mainly in areas with dry, arable land, but only on a limited scale (Table 10.6). Therefore jatropha needs to be planted in areas with adequate rainfall.

While land and water do not seem to immediately constrain production, labor may do so in the future. Currently the majority of households in these villages are

Table 10.6. *Jatropha production inputs in surveyed villages*

| | Labor | | Water resources | | Infrastructure | |
| | | | Ratio of | | | |
Village	Share of nonagricultural and forestry income (%)	Ratio of migrant work in total labor (%)	paddy areas in arable land (%)	Ratio of irrigation in dry land (%)	Distance from county capital (km)	Transportation cost to town for foodstuff[a] (USD/ton)
1	9	20	61	7	140	18
2	9	13	43	56	90	23
3	4	15	12	0	28	12
4	10	–[b]	34	0	20	15
5	25	18	32	5	0.5	3
6	10	2	18	0	18	15
7	10	13	29	41	27	44
8	5	59	24	0	27	15
9	30	44	66	0	6	2
10	7	6	53	0	6	4
11	60	6	1	100	3	9
12	20	18	69	6	27	10
13	9	15	16	7	12	12
14	20	34	95	0	2.5	6
15	21	4	27	8	31	29
16	10	8	63	13	3	3
17	10	1	25	0	8	7
18	10	8	100	–[b]	24	7

[a] Exchange rate as of December 2008 (USD 0.146/RMB; http://www.x-rates.com/).
[b] No data.
Source: Sano and Romero (2010), modified by authors.

engaged in agricultural activities,[5] but many households also engage migrant workers and receive income from nonagricultural–forestry activities (Table 10.6). It is worth noting that the number of migrant workers is increasing in China. Although the rate of decrease in the agricultural population of Yunnan Province was still less than 1 percent (between 2004 and 2005), higher rates of agricultural population decrease were found in neighboring provinces (2.1% in Sichuan and 1.6% in Guizhou) and in the nation as a whole (2%) (NBS, 2007).

In addition, high transportation costs may soon inhibit jatropha production. Some villages are located a long distance from the county capital and suffer from associated high transportation costs (Table 10.6). For example, if the unit transportation cost observed in village 7 (the highest of the 18 villages) is applied to jatropha transportation, it would correspond to 7.3 percent of the sales price of jatropha fruit.[6] Therefore,

[5] The share of agriculture population in total population is over 95% in the 14 villages.
[6] Assuming a unit transportation cost of 4 RMB/kg (fresh/wet-base, 60 cents/kg).

to be economically feasible, the jatropha market price must be high enough to cover such a high transportation cost.

5. Sustainability implications of the household-level data

5.1. Labor input

The labor input recorded for jatropha production between 2006 and 2008 was analyzed using aggregated household data (Table 10.7). Total labor demand sharply increased in 2007 when the Forestry Department began a strong jatropha promotion campaign. The most labor-intensive activity for jatropha cultivation by far is hole digging for planting, followed by planting and weeding. A considerable portion of work (approximately 40%) depends on hired labor. In general, jatropha production was observed to be extensive with little fertilizer and water use. Since jatropha production had just started at the time of field observations, little labor was used for pruning and harvesting. However, when jatropha trees mature, a significant amount of labor is needed for harvesting.

5.2. Production cost and profitability

Production costs and revenues were compared between 2006 and 2008 using aggregated household data (Table 10.8). The majority of costs came from purchasing seedlings and seed, followed by agricultural chemicals (mostly compound fertilizers). Although mostly hired labor was employed for production (Table 10.7), this cost was not included because no payment for hired labor was recorded.[7] The labor cost of engaging the producing households in jatropha agriculture was also excluded. This means that the associated implicit costs (in the form of potentially significant forgone opportunities if, for example, the household members could receive better wages as migrant workers) were not considered in the analysis.

Though production costs increased between 2006 and 2007, revenues jumped in both 2007 and 2008. This resulted in a more than sevenfold increase in the revenue-to-cost ratio between 2006 and 2008. These aggregate data indicate that jatropha production appears profitable in general. However, this ignores variance among the surveyed individual households. In fact, this increase in the revenue-to-cost ratio was evident due to the high revenues from selling harvested fruits for their seeds because seeds are marketed at considerably higher prices compared to the prices obtained from selling them for making oil.

Currently there are only a few seedling providers in Xinping County. The prices of seedlings and seeds that were sold to the Forestry Department were 0.5 RMB/seedling

---

[7] Hiring farm labor is a relatively new phenomenon gradually emerging since the late 1990s in China (Wang, 2007). Thus recording labor payments was not planned in the survey.

Table 10.7. *Labor input for jatropha production (aggregate) in 2006–2008*

Year		Land clearing	Hole digging	Planting	Applying agrochemicals	Weeding	Pruning	Irrigation	Harvest	Hired labor	Relative increase in total labor 2006 = 100
		Labor input (man-day/year) (top) and the share of each activity in total labor (%)									
2006		303.5	4,248.0	1,538.0	686.0	1,037.0	0.0	3.5	150.0	5,153.0	100
		2.3	32.4	11.7	5.2	7.9	0.0	0.0	1.1	39.3	
2007		4,642.0	53,974.0	18,271.0	5,360.0	3,447.0	289.0	468.5	1,568.0	51,024.0	1,060
		3.3	38.8	13.1	3.9	2.5	0.2	0.3	1.1%	36.7	
2008		1,379.9	9,304.5	3,660.5	489.5	5,463.5	63.0	34.5	1,909.0	14,910.0	284
		3.7	25.0	9.8	1.3	14.7	0.2	0.1	5.1	40.1	

Source: Sano and Romero (2010).

Table 10.8. *Costs and revenues of jatropha production (aggregate) in 2006–2008*

Year	Seedling, USD (%)	Seed, USD (%)	Agrochemicals, USD (%)	Transport, USD (%)	Relative increase in total cost (2006 = 100)	Relative increase in revenue from fruits (2006 = 100)	Relative increase in revenue/ cost ratio (2006 = 100)
2006	25,820 (86.0)	10 (0.0)	4,182 (13.9)	22 (0.1)	100	100	100
2007	132,448 (84.7)	18 (0.0)	23,870 (15.3)	30 (0.0)	540	3,189	591
2008	17,137 (82.8)	0 (0.0)	3,546 (17.1)	19 (0.1)	75	5,723	7,664

Source: Sano and Romero (2010), modified by authors.

(7.3 cents/seedling) and 60 cents/kg (wet fruit base), respectively, in 2008 (Figure 10.3). Seedlings were grown at a density of 375,000–450,000 seedlings/ha. This means that seedling growers could potentially receive payments between to 27,375 and 32,850 USD/ha, which are extremely lucrative. Yet the provision of seedlings and seeds is only a short-term business opportunity since jatropha is a perennial crop. Most households responded in the questionnaire that the main purpose of growing

Figure 10.3. Jatropha seedling preparation in Xinping County, Yunnan Province. (See color plate.)

jatropha was to get fruits for seeds, not for oil.[8] This is consistent with the fact that the Yunnan Forestry Department needed more seedlings to achieve its target levels of jatropha production, a need despite which it made no official commitment to buy up all the jatropha seedlings. In summary, these households appeared to be taking advantage of a quick speculative opportunity.

The main sources of uncertainty in the profitability of jatropha production include labor requirements for harvesting, transport costs to ship harvested fruits, and the market price for the fruits at harvest time. Jatropha harvesting is expected to be labor intensive because farmers need to pick only ripened fruits from trees by hand (leaving the unripened ones on the tree) and carry them across hillsides. This would most likely require hired labor, just as planting does. Transportation costs can be significant in remote production areas, as mentioned in the previous section, and there is evidence to suggest that these costs are already increasing (Table 10.8). Most critically, there is yet no well-functioning jatropha market. This is mainly because the harvest volume is very small, if there is anything to harvest at all, and because there are no established links with buyers (processing plants). As of yet, there is also no biodiesel plant set up in the province to utilize the produced jatropha, and there is no jatropha–biodiesel production chain. Therefore the long-term economic viability of jatropha production for biodiesel remains very uncertain.

Under these circumstances, both the government officials and the farmers interviewed appeared to have decided to quietly wait and see how things would turn out. Government officials did not seem to be rushing the introduction of jatropha production, while farmers did not appear desperate to grow additional or alternative crops. In Xinping County, farmers are well informed of their options. They make decisions with a healthy skepticism and compare jatropha with tested alternatives such as sugarcane, walnut, or grassland for raising water buffalo. While jatropha production would generate only about 400 RMB/mu (approximately 880 USD/ha) when fully matured,[9] sugarcane production could yield 700 RMB/mu (approximately 1,500 USD/ha). Walnut, an already existing cash crop, which the Forestry Department is also promoting as an alternative forestry product on unutilized hillsides in conjunction with their afforestation efforts, is believed to be more profitable. Yet farmers need to wait five years until the trees bear fruit, a longer time than is required for jatropha production.

5.3. Perceptions and concerns regarding jatropha production

This section discusses perceptions and concerns regarding jatropha production expressed by interviewed households. The analysis compares perceptions based on different production scales and management modes.

---

[8] The questionnaire offered respondents the following choices as a description of the purpose(s) of growing jatropha: for seed, as fencing, to retain water and soil, for cultivation cuttings, or "other."

[9] The estimate of revenue from jatropha production implied a rather low yield of jatropha. This estimate falls below the lower end of the reported yield range, i.e., 110–140 kg/mu in barren land (ICRAF China, 2007), if the same price is applied in the calculation.

These perceptions and concerns are summarized in Table 10.9 and organized by scale of production. Small-scale jatropha-producing households (group 2 in Table 10.9) showed the highest interest in expanding production. Moreover, 70 percent of the households in group 2 expressed their willingness to expand production even without subsidies. Large-scale jatropha-producing households (group 3) showed relatively lower interest, possibly because they were already engaged in large-scale jatropha production and had yet to reap profits. Non-jatropha-producing households (group 1) also showed interest in possible jatropha production.

The majority of interviewed households believed in the environmental benefits of jatropha (Table 10.9). About 75 percent of households believed that jatropha production would help to improve water and soil retention, and about 50 percent believed that it would be beneficial to plant jatropha to increase land cover on hillsides with few or no trees. However, at the same time, more than half of the interviewed households did not seem to be aware of the toxicity of jatropha, while more than 80 percent incorrectly responded that jatropha was not toxic to animals.[10] This implies that the information provided on jatropha to households somewhat emphasized its environmental benefits and downplayed other impacts.

The interviewed households were generally more concerned with the lack of a well-established jatropha market in Yunnan Province (Table 10.9). Similarly, relatively small concerns were expressed about yields, reflecting the fact that the full harvest had not yet been completed.

While cultivation technologies were important to large-scale jatropha-producing households (group 3), small-scale jatropha-producing households (group 2) were more concerned with land and labor availability. Since jatropha was planted in unutilized forest areas and not on farmland, land availability would not be an issue to an individual household. However, the survey results also revealed that almost half of the total land managed by these households came from newly developed farmland converted from unutilized forestland and hills. This indicates that larger land areas for agricultural crops were needed by these farmers for their subsistence or economic activities. Because many of these households (groups 1 and 2) were interested in expanding their jatropha production, a food–fuel conflict could occur if jatropha production were to become more attractive than other basic food crops and be introduced outside of the government's oversight. A food–fuel conflict could be worsened if it were coupled with a labor shortage, about which small-scale jatropha-producing households also expressed concern.

In summary, the survey results indicate that households are generally interested in jatropha production and have a positive image of it, but they are also aware of challenges such as the absence of an established market and associated uncertainty about the market price, recognizing that jatropha production is only in its initial stages. These challenges may become more serious as the scale of jatropha production increases.

---

[10] This may be because they did not commonly use jatropha as live fences, as is the case in other countries.

Table 10.9. *Perceptions and concerns about jatropha production by production scale*

Jatropha area (ha)	No. of samples/ households (persons)	Ratio of households who (%)			Ratio of households with concerns of lack of (%)							
		Consider expanding jatropha	Consider expanding jatropha without subsidies	Believe in environmental benefits from jatropha	Capital	Land	Labor	Seedling	Cultivation technique	Yield	Market	Other, better options
0	17	18	53	88	18	12	18	0	6	9	53	29
≤10	55	24	71	80	15	29	27	9	7	9	40	11
≥53.3	11	9	55	100	9	18	9	0	36	9	36	27

Perceptions and concerns about jatropha production by different management modes are summarized in Table 10.10. Among the four different management modes, the households that expressed the greatest interest in expanding jatropha production were those that benefited from subsidies from the Forestry Department (group 1a). The amount of subsidies provided to households was 5–60 RMB/mu or an average of 36 RMB/mu (11–131 USD/ha or 79 USD/ha, on average) in 2008.[11]

In contrast, more than half of the households that were engaged in contract work by PetroChina (group 2a), which also had the largest jatropha area per household of the four groups, did not show interest in expanding production. This was despite the fact that households in this group received around 150 RMB/mu[12] (330 USD/ha) in annual reforestation commission fees on average and typically spent only around 2 RMB/mu (4.4 USD/ha) on average for maintenance. This amount was significantly larger than what was given to the households in group 1a. These households may think that they have already received large enough commissioned work and appear to be more concerned about the cultivation technique in delivering the products (Table 10.10). Jatropha may have been one of the best options currently available for them, although not the only one. It is interesting to note that households in this group also strongly believed in the environmental benefits of jatropha.

In summary, the commission system with payments to individual households appeared to be successful in securing a large jatropha production area. Yet which management mode is more economically sustainable remains uncertain at this point, especially because little information was available regarding how these different management modes handled other critical activities such as pest management, harvesting, transportation, and technical support.

5.4. Other factors relating to the sustainability of jatropha production

Another factor that could affect the sustainability of jatropha production is the choice of jatropha variety for seeds and seedlings and maintenance of soil fertility. Jatropha is not an entirely foreign plant to the region, which implies that the surrounding natural environment is more or less suitable, and local households have some level of familiarity regarding how to manage it. However, there was no science-based process reported for the selection of jatropha variety during the field survey. Since jatropha is a perennial plant, a selected variety with superior properties should be introduced right at the beginning to minimize any potential negative environmental and socioeconomic impacts.

Regarding the processing of jatropha fruits, there are several steps that must be followed before oil can be extracted from the seeds. Conventional harvesting requires

---

[11] Criteria for subsidies are unknown.
[12] The estimated agricultural subsidy was reported at 160 RMB/mu (ICRAF China, 2007).

Table 10.10. *Perceptions and concerns about jatropha production by different management modes*

Group	Management mode	Ratio of households who (%)			Ratio of households with concerns of lack of (%)							
		Consider expanding jatropha	Consider expanding jatropha without subsidies	Believe in environmental benefits from jatropha	Capital	Land	Labor	Seedling	Cultivation technique	Yield	Market	Other, better options
1a	Commissioned by the Forestry Department	27	73	89	11	32	27	14	8	5	38	14
2a	Commissioned by company	11	44	100	11	22	11	0	44	11	33	22
3a	Commissioned by village or other	0	67	33	33	0	33	0	0	0	0	0
4a	Commissioned work on granted land	21	64	57	21	14	21	0	7	21	50	7

picking the mature fruits by hand since the fruits on each plant ripen at different times. This requires skilled labor allocation at appropriate harvest times and efficient transportation of jatropha seeds from the field to the processing mills. A jatropha sheller, a simple machine to open the fruit to separate the seeds, can be used to increase efficiency, but it was not yet available in the surveyed villages. If jatropha yield and oil extraction are poor, making cultivation unprofitable to farmers, labor resources could be redirected to other profit-generating activities. In addition, the long-term health impacts of jatropha oil on humans, which are still unknown today, should be taken into consideration.

For jatropha oil production to become a sustainable local business model (or regional industry), it should improve the participating households' cash flow and avoid stressing the labor allocation necessary for basic subsistence farming activities. To improve the profitability of the operation, owners of processing mills should consider the potential utilization of all possible by-products or secondary products that can be produced from the waste created by the processing of jatropha oil (see a similar discussion in the context of sugarcane in southern Africa in Chapter 12). Examples include organic fertilizers, which can be produced from seed cake, or biogas produced from jatropha waste. Establishing good vertical coordination among farmers, middlemen, and processing plants would help to stabilize the market price of jatropha (seeds, seedlings, and fruits), which can be one of the most significant incentives to farmers. The government has a role in this by either directly getting involved in establishing this vertical coordination or encouraging local enterprises to participate in it. Finally, it is also possible to utilize straight jatropha oil to fuel simple machines locally rather than further processing and blending it as biodiesel (see Chapter 1). However, none of these options are fully realized yet.

6. Discussion

The promotion of jatropha production in Yunnan Province is aimed at increasing China's renewable energy supply and promoting local communities' economies in conjunction with afforestation efforts, while avoiding the food–fuel conflict. Yet production had not fully taken off during the field survey in late 2008 (only a few years after its introduction to the region), and thus the analysis presented in this chapter is preliminary. However, despite this limitation, the snapshot taken in 2008 provided many useful insights to evaluate the impacts of existing jatropha production and their potential implications for sustainability.

In general, the observations made during the field survey indicate that natural resource endowments, such as land availability, soil conditions, water availability, and climate conditions, appeared adequate for jatropha production. Electricity is already available, and roads already run through remote villages where the crop is grown. This suggests that there is no evidence of poverty-stricken agricultural areas.

Jatropha was not the only available alternative crop from which farmers could choose, and no clear case was observed in which jatropha was competing with other crops for farmland. In this regard, direct food–fuel conflicts seemed to be avoided, as specified by the government's policy.

However, the analysis of the village-level survey data showed that the local economies were dependent on agricultural activities and migrant work, implying that there could be a labor shortage as villagers continue to migrate to cities. Although this survey only recorded partial labor for harvesting (because the jatropha trees are not yet mature and therefore not yet productive), the household data showed an increase in labor demand for harvesting. With the current labor-intensive manual harvesting practices, labor availability could become a critical factor for the long-term sustainability of jatropha production in these villages. In addition, the village-level survey data showed that transportation costs in remote villages are high. Although no increase in transportation costs was observed in the household survey data (because only a small amount of jatropha was harvested and/or sold locally as seeds or for seedlings), an increase in the transportation cost for harvested fruits is inevitable and will become a significant determinant of the profitability of jatropha production as trees mature.

At the moment, jatropha production appears to be accepted by households. First, this might be because households were given incentives, both cash and in kind (free seedlings), by PetroChina and the Forestry Department. The two most common modes of jatropha business identified in the survey were those of the Forestry Department and PetroChina. Each mode used different methods of commissioning households to engage in jatropha production. The former had the largest number of households enrolled (approximately 60% of total households), although the planted area accounted for only 2.5 percent of total area planted with jatropha recorded in the survey. Some of these households received small monetary subsidies. The largest portion of jatropha area (over 90% of the total) was commissioned by the petroleum companies with large subsidies. In some cases, households were granted land for jatropha production. However, this research found no special efforts to train households or build capacity for jatropha production. In this sense, knowledge transfer from the Forestry Department or PetroChina was limited.

Since China introduced the Household Responsibility System in rural areas, households have come to have more flexible time allocation and access to off-farm work opportunities. However, the land continues to be owned by villages (Wang, 2007). As Jacoby et al. (2002) found, the stringent land redistribution found in China could discourage farmers from longer-term investments for improving land, and the same may apply for jatropha production. Although the production modes currently found in Yunnan may prevent an unwanted rush for jatropha production by individuals, they could also fail to provide much incentive to produce quality products[13] or a long-term

---

[13] The jatropha price does not clearly reflect the quality of products due to the undeveloped market.

commitment given that decision making is left to village officials and not to the actual producers.

Second, a positive perception of jatropha production was observed among surveyed households. The direct involvement of the Forestry Department and/or the state-owned petroleum company (PetroChina) in jatropha production may have created this optimistic perception, especially regarding environmental benefits, while potential negative aspects, such as the toxicity of the jatropha plant, were not well recognized. Because Yunnan Province has a high ratio of forestry land[14] and the highest share of collectively owned forest land in China,[15] households are probably very interested in gaining access to this forestland.

Third, jatropha may be one of the easiest, though not the only, options available for additional income compared with other forest-based cash crops such as walnuts. So far, the profitability of jatropha production observed in the survey data only reflects the profitability of seeds and seedling sales, which is a one-time opportunity. Neither the Forestry Department nor the petroleum companies offered to buy future jatropha fruits from producing households. Moreover, jatropha oil processing plants are still under development and do not yet exist in the province. The Yunnan government proposed to build commercial-scale biodiesel refineries and has been trying to entice investors from the private sector without much success (ICRAF, 2007). Furthermore, the Chinese government has not yet provided any explicit subsidies to biodiesel producers similar to those provided to ethanol companies.[16] As a result, the jatropha industry is not economically sustainable without financial support from the government. Under these conditions, the future of the jatropha biodiesel industry is highly uncertain. Nonetheless, jatropha provided farmers with an opportunity that did not exist before. To cope with uncertainties, most households seemed to adopt a wait-and-see attitude toward future jatropha production, with healthy skepticism. Jatropha fruit can be left unharvested if the market price is too low. Yet the households' initial decision to engage in jatropha cultivation may have been based on expectations of short-term returns (subsides and seeds and seedlings sales) without much consideration of the longer-term costs, including their own labor (harvest of jatropha, other crops, or migrant jobs) and other opportunity costs.

On the basis of the survey data, the current environmental and socioeconomic impacts of jatropha production at the village level in Yunnan Province can be summarized as follows. Because jatropha production has been introduced only on unutilized

---

[14] Over 60% of the total agricultural and forestry land compared to the national average of 32%. In this respect, Yunnan is one of top five provinces in China (National Bureau of Statistics, China, 2007).

[15] Seventy-six percent by collective ownership and 24% by state ownership (ICRAF China, 2007). After the introduction of the Household Responsibility System in 1978, land was distributed to households according to family size, but landownership rights remained in the village (Wang, 2007).

[16] The retail price of diesel in northern China was reported to be in the range of 4.55–4.92 RMB/L (0.66–0.72 USD/L) in July 2007. The estimates of jatropha oil production costs in Yunnan were in the range of 4–11.5 RMB/L (0.58–1.68 USD/L) (ICRFA China, 2007).

hillsides without much land clearing, and few agricultural inputs (fertilizers, pesticides, irrigation water), no visible direct threat to the environment was observed. Subsidies to households may have caused improvements in welfare, but the impacts are limited and may last only a short period of time. No negative health impacts have been reported due to the very short history of exposure to jatropha production.

However, potential socioeconomic impacts could occur indirectly when the households begin full-scale harvest of jatropha production in the future. The possibility of a labor shortage is already indicated, and the situation could become more severe if more labor is diverted to jatropha in the future. This might result in a labor shortage for food production activities. Although direct land use conflicts are currently avoided by providing additional land from unutilized forested land, this type of two-step food–fuel conflict could be triggered by a labor shortage. Needless to say, if the assumption of no fuel from farmland is violated, jatropha production will compete with other cash crops, such as maize, sugarcane, and coffee, that are grown on more productive land, as indicated by the ICRAF China (2007). The potential negative impacts on the welfare of households could be significant, especially for those with lower incomes or with no access to their own food production.

7. Conclusions

Overall, the economic viability of jatropha production for biodiesel in Yunnan seems to raise more concerns when compared to potential environmental impacts. While environmental impacts of jatropha production appear minimal at present, there are important concerns about its future economic viability due to the lack of buyers for the feedstock (jatropha fruit) and the farmers' risk-averse behavior. The main limiting factors regarding the overall sustainability of jatropha production are the immaturity of the current jatropha oil market and the existence of several market uncertainties. Jatropha is a typical new perennial crop with a currently limited use. Once farmers commit to jatropha production, they will not have any other options but to sell their harvested fruits to the existing jatropha processing company (or companies). Moreover, there is no alternative market for jatropha and its derivatives, and even if there were, many of the farmers do not have their own means of transportation to take advantage of possible arbitrage opportunities. Often jatropha feedstock suppliers (farmers) are price takers, so their economic gain largely depends on the economic viability of the jatropha oil processing companies. Thus the location of the jatropha processing refineries is critical for growers as transportation costs of feedstock can be significant.

To overcome these challenges, the Forestry Department or jatropha fruit buyers (processing companies and/or final distributors such as PetroChina) should investigate possible additional markets and uses for jatropha production and by-products other than transportation fuel. This would bring more benefits for local people from long-term jatropha production. Government and university support for related R&D would be necessary to compensate for the inability of the private sector to conduct sufficient R&D of its own. At the same time, although contracts with the Forestry Department or PetroChina are well managed, along with land allocation in the village under local authorities, farmers would need an organized harvesting and transporting system in time for the large-scale harvest. Government support in these areas may reduce some of the economic uncertainties that farmers face.

At the national level, China introduced its biodiesel standard (B5) in October 2010, so the downstream development of the biodiesel industry is accelerating. Therefore it is likely that the introduction of a blending mandate will be considered in limited areas where sufficient supply is secured. At the same time, China's land tenure system and labor market are evolving. A full-scale sustainability assessment should be conducted when jatropha trees mature in the future, taking into consideration the impacts of these policy developments on households' decision making.

Acknowledgments

The research for this chapter was conducted as part of the research project "Biofuel Use Strategies for Sustainable Development," supported by the Global Environment Research Fund (E-0802) of the Ministry of Environment, Japan. Authors would like to thank Jikun Huang, director of the Center for Chinese Agricultural Policy (CCAP), Chinese Academy of Science; Zhurong Huang from the CCAP; Qingbo Liu from the Southwest Forestry College; and other staff at the CCAP for their cooperation in organizing and conducting the field survey in Yunnan.

Part Four
Africa

11

Biofuels and Africa: Impacts and linkages at the household level

SIWA MSANGI

International Food Policy Research Institute, Washington, D.C., USA

Abstract

This chapter addresses the issue of biofuels in Africa within the context of the potential effects major Organisation for Economic Co-Operation and Development biofuel producers might have on African households as well as what impacts fledgling biofuel programs in those countries might have on their own markets. A stylized model of household economic behavior is proposed to consider alternative ways in which consumption and production behavior interact to better understand the impacts on welfare that are transmitted through markets to the household level. The implications suggested by this analytical model's results are identified and related to findings from the empirical literature. In this analysis, the importance of off-farm labor or wage-earning opportunities for rural and urban households, in terms of the way in which they adjust to price shocks, is identified. Also pointed out is the importance of sequential labor decisions and the ways in which they might constrain households in their ability to adjust when proposed biofuel ventures fail and promised opportunities for wage earnings or on-farm production revenues do not materialize. These examples illustrate the kinds of risks to which foreign investment–driven, agribusiness-focused investments such as biofuels might expose vulnerable smallholder households in sub-Saharan Africa. Some key entry points through which policy can act to mitigate the risks that households face include the strengthening of institutions that regulate foreign investments and guarantee the rights of local citizens and their access to key resources. By looking at the key linkages between household decisions and market dynamics, a broader picture is gained of where energy and other national- and local-level government policies can make the most impact on household welfare and broader rural economic development.

Keywords: agricultural production, risk, rural development, welfare

1. Introduction

In the past decade, the growth of biofuel production has increased worldwide in the effort to reduce dependence on imported fossil fuels, boost market opportunities for domestic producers of key crop-based biofuel feedstocks, or reduce greenhouse gas (GHG) emissions from fossil fuel consumption in the transportation sector (see Chapter 1). While many countries promote their national biofuel programs, at a political level, it is difficult to achieve all these goals or make them mutually compatible. The United States has overtaken Brazil in its production of biofuels, and they both comprise more than 90 percent of the world's production. The United States' current national policy, embodied within its Renewable Fuel Standard, has articulated some goals regarding GHG reduction but is solely focused on biofuels rather than a broad array of renewables. Furthermore, it exempts corn ethanol from these GHG reduction goals, effectively allowing for its unchecked expansion (Sperling and Yeh, 2010). Some have pointed out that the combination of high tariffs and blending mandates might even encourage gasoline consumption (de Gorter and Just, 2008), while not properly incentivizing the reduction of GHG emissions (Khanna et al., 2010; Searchinger et al., 2008). The sugar-based ethanol program in Brazil, by contrast, was launched with the primary objective of reducing energy imports (see Chapter 6) and has grown steadily over time, thanks to the widespread availability of flex-fuel vehicle technologies and steady improvements in ethanol feedstock production and conversion technologies (Moreira, 2006).

In terms of biodiesel production, the European Union (EU) has led global production, which is concentrated mostly in Germany, France, and Italy (see Chapter 1). The EU's policies are focused mostly on GHG emission reductions, as reflected by the ambitious blending targets of the EU Renewable Fuel Directive. While U.S. national policy has begun to consider GHG emission levels, only the U.S. state of California has actually incentivized the reduction of the carbon intensity of transport fuels directly into its policy framework (Sperling and Yeh, 2010). Given the global efforts toward climate change mitigation, renewable fuels have been heavily scrutinized for their potential in reducing GHG emissions from the transportation sector. This focus on climate mitigation, however, is not a strong motivation for many of the countries within the developing world. Such countries are more concerned with the impact that high oil prices are likely to have on their vulnerable economies as well as on reducing their import dependence on fossil fuels, considering the large burden such a dependence represents to their national balance of payments.

Because of this pressure, there has been a growing interest in biofuel production expressed by developing nations (FAO, 2008a). The faster-growing economies among these nations (e.g., China and India) are expected to have a growing share of biofuel production in the coming decades (Fulton et al., 2004). In China, for example, the government's support for biofuels was expected to reach a level of at least

USD 1.2 billion by 2020, starting from the 2006 level of USD 115 million in subsidies to the sector (GSI, 2008). This has been curtailed, however, by concerns over the links between biofuel production and food prices (Qiu et al, 2010).

Within sub-Saharan Africa, a number of countries have been identified as having high potential for biofuel feedstock production (Cai et al., 2011; FAO, 2008a), even though their own capacity to process and absorb the final fuel product might be limited and warrant an export-oriented focus. Many African countries are motivated by the desire to reduce their dependence on imported foreign oil, which constitutes a sizeable share of their import bills (Mitchell, 2011). These countries range from relatively well-resource-endowed countries, such as Mozambique and Angola, to some of the more resource-constrained Sahelian countries within the West African Monetary Union (e.g., Mali), which are trying to balance the prospect of starting up national biofuel programs with the pressing needs of other important development objectives (UEMOA, 2008). The degree of infrastructure development in these countries, however, varies widely. This may facilitate the large-scale production of biofuels in certain countries, while leaving other countries noncompetitive (e.g., Malawi) (Peskett et al., 2007). While some of these countries are willing to offer generous concessions to outside commercial interests to encourage them to establish biofuel programs within their countries, the fiscal implications of forgoing tax revenues on fuel need to be carefully considered in light of the net benefits that will actually accrue to the country once other necessary expenses for supporting infrastructure for the biofuel industry are made (Arndt et al., 2010c).

This chapter highlights some key aspects of biofuel production and use that are pertinent to African economies and to welfare impacts at the rural, household level. There are, essentially, two dimensions along which this issue is addressed: (1) the effect of Organisation for Economic Co-Operation and Development (OECD) production of biofuels on Africa that is transmitted through international markets and (2) the effects that arise from the production of biofuels that could happen within Africa itself. Both these effects are mediated by markets (either international or local) and have different implications for various households, ranging from productive farm households to landless (or severely land limited) households that depend on wages from off-farm activities.

This inquiry resonates with a large body of literature that has investigated the rapidly growing biofuel production and demand within both the developed and developing world and the potential for adverse impacts on global food economies. Both the micro- and macro-level linkages that connect biofuel to agricultural markets are discussed, and the implications that exist for households in terms of price changes for outputs and key productive inputs, such as labor, are identified. Some stylized and theoretical economic models are used to characterize the varied nature of household-level impacts that can arise from biofuel-driven market shocks and to illustrate how the different cases apply to either urban or rural households (with or without access to land). While not all policy issues involved are resolved or addressed, there is a conscious effort to put the key issues within the context of what matters most to policy makers in Africa, as they contemplate alternatives and pathways for energy and food security.

2. Household-level linkages: Market-driven welfare impacts

2.1. Households as individual production units

This section provides a conceptual overview of the important linkages that tie together productive economic activities, resource availability, market conditions, and their implications for human welfare at the household level. The pathways through which biofuels affect human welfare in developing countries mostly follow on these linkages and can range from binding resource constraints within the household itself to the market-mediated ties that connect the expansion of biofuels production in OECD countries to food and fuel markets in less developed countries. But even in the case of biofuel production within the developing world (such as that being undertaken within several countries in sub-Saharan Africa), there are still market-based linkages at work, including those markets for labor, land, productive inputs, and food. To understand the nature of these linkages and how they relate to human well-being, we take a broader view of the household as a productive unit that uses its own labor to produce goods for consumption and sale. At the same time, such productive units also rely on markets for purchasing inputs that they either use in their own production or that they consume after some kind of transformation process (that can in itself be thought of as a within-household production process).

Figure 11.1 conceptualizes the household as a productive unit that draws from both natural and human resources to produce goods that it uses for its own consumption or for generating income, through the mediation of markets. The human resources from which it draws, such as labor, can be sourced internally or can be obtained through market-mediated wages (where such markets exist).

On the consumption side, households also make a number of consumption decisions, which are based on their own production as well as on the goods that are available through exchange or market interactions. If external institutions exist to assist the household in making savings and investment decisions (or in receiving remittances from household members that have migrated), then it will further enhance its ability to accumulate wealth and improve its own welfare. Otherwise, the household will have to accumulate assets in nonfinancial forms to build up its buffer against future income and other shocks. The ability of the household to smooth its own consumption over time and provide a stable environment for nurture, care, and the provision of healthy food and other inputs to personal well-being allows it to thrive and increase its own productivity in various endeavors. This, in a very compact form, summarizes the linkages and feedbacks that are most relevant to household-level productive and consumptive behavior and the outcome of health and overall well-being.

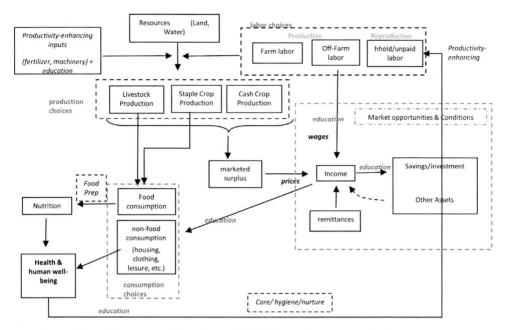

Figure 11.1. Schematic of key household linkages between agriculture, resources, and welfare.

However, it is difficult to fully appreciate the impacts of biofuel production and use for household-level welfare in sub-Saharan Africa while holding such a complicated view of all the possible linkages going on simultaneously (see Figure 11.1). Therefore, in the following section, the discussion of biofuel impacts is divided in a way that can allow the consideration of different types of households. As a result, a more complete understanding of how household-level heterogeneity complicates the picture of the positive or negative effects that biofuel growth can have at the household level is obtained. On this basis, we will be in a better position to provide the needed level of nuance for our observations and policy recommendations and will avoid lumping all cases together in a way that tempts a simplistic view of the prospects for (and the impacts of) biofuels in Africa.

2.2. Taking consideration of household differences

While it is challenging to present in the following analysis the full array of diversity that exists across households in sub-Saharan Africa, those stylized classes that have the most relevance to how food and energy markets affect urban and rural households are illustrated in Table 11.1.

Table 11.1 illustrates three cases in which households have a differing level of economic activity as well as a different degree of exposure to the dynamics of relevant input and output markets. It is shown that the degree of exposure that households

Table 11.1. *Alternative cases most relevant to biofuel-driven impacts on households*

	Household-level activities	Implications
Case 1	Households that sell their own labor to the market and buy all of the food they consume	Applies to most urban households Applies to landless rural households
Case 2	Households produce food, which is consumed and/or marketed	Applies mostly to rural households with land
	The same food commodity is used as a feedstock for biofuel production and is either sourced from the household or drawn from other marketed surplus	Price effects that are driven by biofuel production affect the production and consumption decisions of the household (whether they are net sellers or buyers of the good)
Case 3	Households that produce crops that are for food and for (potential) feedstock use separately and only consume the food crop, while both can be sold to the market	Applies mostly to rural households with land The key resource constraints are that of land (which must be divided between both types of crops) and labor (for which the household might be constrained, if labor markets are thin or missing)

have to biofuel-driven market forces (as felt through prices or more directly through production and consumption choices) is quite different across these cases. If Figure 11.1 is modified to reflect the stylized characteristics of households falling under case 1, it becomes evident that the main linkages to markets are through either providing labor for a wage or the purchase of food commodities (Figure 11.2).

Any changes in domestic availability of the food item (that are caused by market influences relating to changing levels of imports or exports coming from international markets, i.e., from OECD countries' biofuels policies) have a direct impact on the consumption side. This can cause changes in the allocation of consumption between goods or between the allocation of labor among different alternative activities.

Cases 2 and 3, conversely, are more aligned with Figure 11.1, in which household welfare is linked to the availability of food either from own-production (or from purchases) or from income that is earned either directly on farm or from off-farm work. Separate figures for these cases are not provided but are discussed within the context of the mathematical models that are presented subsequently.

2.3. Conceptualizing households under case 1

This household conceptualization only addresses the consumption side, in which decisions of consumption and labor are taken into account and are abstracted from explicit production activities. Following the neoclassical approach to modeling consumer

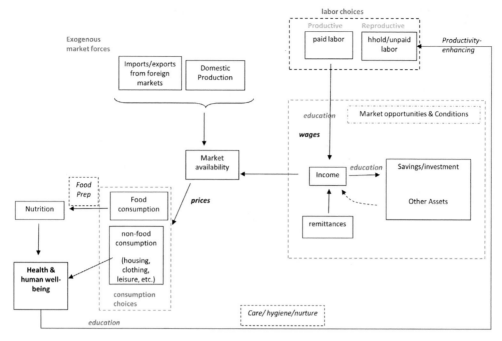

Figure 11.2. Characteristic household linkages for urban and rural-landless households.

behavior, the utility maximization problem of the (unitary) household consumption and labor decision problem can be written as follows:

$$\max_{c_f, l} \quad u(c_f, l) = \gamma_0 \log(c_f) + \gamma_1 \log(T - l), \quad \gamma_0, \gamma_1 > 0,$$

$$s.t. \tag{11.1}$$

$$p \cdot c_f \leq w \cdot l + \overline{I},$$

where the concave utility (or felicity) function (shown as a simple monotonic log function) contains consumption of food items (c_f) and labor (l) as arguments, and where leisure is the time left over when time spent in labor is subtracted from the total time available to the household (i.e., $T - l$). In this case, labor takes away from leisure (the residual of total time T that is left) and therefore enters negatively into overall household utility.

In this simple problem, the budget constraint $p \cdot c_f \leq w \cdot l + \overline{I}$ represents the only resource constraint that the household faces and is defined by prices (p), wages (w), and the income that is fixed and invariant according to household decisions ($\overline{I}$), which can be in the form of a remittance, fixed returns on assets, or transfers (e.g., from the government). If it was assumed that the maximizing behavior of the household would cause the constraint to bind (and all resources to be exhausted, in the absence of savings or asset carryover decisions), this problem could be reformulated as

$$\max_l u(l) = \gamma_0 \log(\tilde{w} \cdot l + \tilde{I}) + \gamma_1 \log(T - l), \tag{11.2}$$

where the only decision is on labor, due to consumption being replaced by the equation $c_f = (w \cdot l + \bar{I})/p = \tilde{w} \cdot l + \tilde{I}$. Taking the derivative of this expression and manipulating the resulting first-order necessary condition for utility maximization[1] leads to a very simple function that expresses the labor supply of the household, which responds positively to increases in wages and negatively to increases in fixed income (as expected):

$$l = \frac{\gamma_0 T - \left(\dfrac{\tilde{I}}{\tilde{w}}\right)\gamma_1}{\gamma_0 + \gamma_1}. \tag{11.3}$$

2.4. Conceptualizing households under case 2

For the cases in which the household is also a producer of the consumption good, the conceptual model presented earlier can be expanded to include the production-level decisions that also need to be made to maximize household welfare. Drawing from a long tradition of literature, starting with the seminal work of Singh et al. (1986), rather than treating them as separable, the production and consumption decisions of household-level agents are connected (more truly reflecting rural, smallholder farm households). As has been noted by other scholars who have built on this type of household producer–consumer model (e.g., de Janvry et al., 1991), the separability of consumption and production decisions cannot be treated separately on these kinds of peasant, household farms because such a large share of their consumption (both direct and purchased through farm income) is related to their allocation of labor and other inputs to on-farm production. In this case, the relevant utility maximization problem for the (unitary) household would have to be written, in its simplest form, as

$$\max_{c_f,l} \quad u(c_f, l_f, l_{nf}) = \gamma_0 \log(c_f) + \gamma_1 \log(T - l_f - l_{nf}), \quad \gamma_0, \gamma_1 > 0,$$

$$s.t. \tag{11.4}$$

$$p \cdot c_f \le p \cdot Q(l_f, A) + w \cdot l_{nf} + \bar{I},$$

where now a distinction is made between the labor used for on-farm food production (l_f) and the labor used for earning off- or nonfarm wages (l_{nf}). In this recursive household model, the on-farm food production now becomes part of the full-income constraint, where prices are relevant to both consumption and production decisions.

---

[1] Which can be simplified to $\tilde{w} \cdot \gamma_0(T - l) = \gamma_1 \left(\tilde{w} \cdot l + \tilde{I}\right)$.

In the case of a small household, the constraint on available land becomes relevant ($A \leq A^{\max} = \overline{A}$) and can be assumed to bind in the optimal solution, which simplifies the production decision to the use of labor, according to $Q(l_f, A) = \alpha\sqrt{l_f} \cdot \overline{A}$. If it is also assumed that the income constraint would bind at the optimum consumption level, then it can be substituted for consumption in the utility function, while also normalizing for prices, so that $p \cdot c_f = p \cdot Q(l_f, A) + w \cdot l_{nf} + \overline{I} \rightarrow c_f = \alpha\sqrt{l_f} \cdot \overline{A} + \tilde{w} \cdot l_{nf} + \tilde{I}$. This allows the household farm decision problem to be reformulated as

$$\max_{l_f, l_{nf}} \quad \gamma_0 \log(\alpha\sqrt{l_f} \cdot \overline{A} + \tilde{w} \cdot l_{nf} + \tilde{I}) + \gamma_1 \log(T - l_f - l_{nf}). \qquad (11.5)$$

This reduces the problem into one of labor allocation decision over farm and off-farm activities. The first-order necessary conditions for the maximization of the condensed utility function can be simplified to the following equations:

$$
\begin{aligned}
(l_f) \quad & \alpha\overline{A}\gamma_0(T - l_f - l_{nf}) = 2\gamma_1\sqrt{l_f} \cdot [\alpha\overline{A}\sqrt{l_f} + \tilde{w} \cdot l_{nf} + \tilde{I}] \\
(l_{nf}) \quad & \gamma_0\tilde{w}(T - l_f - l_{nf}) = \gamma_1 \cdot [\alpha\overline{A}\sqrt{l_f} + \tilde{w} \cdot l_{nf} + \tilde{I}].
\end{aligned}
\qquad (11.6)
$$

On the basis of further manipulation of these conditions (discussed in more detail in the appendix), the impact of exogenous factors such as wages on labor response can be evaluated. In one case, the decrease in the allocation of on-farm labor if off-farm wages were to rise (which is to say, $\partial l_f / \partial \tilde{w} < 0$) is an evident, intuitive result. Conversely (and just as intuitively), it can be shown that nonfarm labor (l_{nf}) responds positively to a change in off-farm wage according to the result $\partial l_{nf} / \partial \tilde{w} > 0$. This could relate to the demand for work in either food- or non-food-related activities outside the household farm. Other decisions that relate to production and land are treated in Section 2.5.

2.5. Conceptualizing households under case 3

In the case in which the farm household divides production between the food and dedicated biofuel feedstock crop, the maximization problem can be rewritten in a way that both activities are reflected in the full income constraint, that is,

$$\max_{c_f, l} \quad u(c_f, l_f, l_{bf}) = \gamma_0 \log(c_f) + \gamma_1 \log(T - l_f - l_{bf}), \quad \gamma_0, \gamma_1 > 0,$$

$$s.t. \qquad\qquad\qquad\qquad\qquad\qquad\qquad\qquad\qquad\qquad\qquad\qquad (11.7)$$

$$p_f \cdot c_f \leq p_f \cdot Q^f(l_f, \overline{A} - A_{bf}) + p_{bf} \cdot Q^{bf}(l_{nf}, A_{bf}) + \overline{I},$$

where (in place of off-farm labor[2]) a distinction is now made between the labor used for on-farm food production (l_f) and the labor used for on-farm production of biofuel feedstock (l_{bf}). In this formulation of the recursive household model, there are two on-farm production functions (one for each of the crops), and the allocation of land is made between them so that total available land ($\overline{A}$) defines the right-hand side of the land resource constraint ($A_f + A_{bf} \leq \overline{A}$), which is treated as binding (as is the case for time allocation).

If the production technologies of each of the crops are generalized in a way to behave, in terms of the use of labor and land, according to $Q^f(l_f, A_f) = \alpha_f(l_f)^{\beta_f} \cdot A_f$, $Q^{bf}(l_{bf}, A_{bf}) = \alpha_{bf}(l_{bf})^{\beta_{bf}} \cdot A_{bf}$, then, in a similar way as before, it can be assumed that the income constraint becomes binding at the optimum. This means that we can substitute for consumption in the utility function and normalize according to the food price:

$$c_f = \alpha_f(l_f)^{\beta_f} \cdot (\overline{A} - A_{bf}) + \frac{p_{bf}}{p_f} \cdot \alpha_{bf}(l_{bf})^{\beta_{bf}} \cdot A_{bf} + \tilde{I}. \tag{11.8}$$

This allows the reformulation of the household farm decision problem as

$$\max_{l_f, l_{bf}, A_f} \quad \gamma_0 \log\left[\alpha_f(l_f)^{\beta_f} \cdot (\overline{A} - A_{bf}) + \frac{p_{bf}}{p_f} \cdot \alpha_{bf} t(l_{bf})^{\beta_{bf}} \cdot A_{bf} + \tilde{I}\right]$$
$$+ \gamma_1 \log(T - l_f - l_{bf}), \tag{11.9}$$

which summarizes the problem of allocating land and labor across the two on-farm activities under time and resource constraints. From these, three first-order necessary conditions for the maximization of the household farm's problem can be obtained, which are (with respect to the labor decisions for the food and nonfood crop as well as for land devoted to the biofuel feedstock)

$$(l_f) \quad \alpha_f\beta_f\gamma_0(l_f)^{\beta_f-1}(T - l_f - l_{bf}) \cdot (\overline{A} - A_{bf})$$
$$= \gamma_1\left[\alpha_f(l_f)^{\beta_f} \cdot (\overline{A} - A_{bf}) + \frac{p_{bf}}{p_f} \cdot \alpha_{bf}(l_{bf})^{\beta_{bf}} \cdot A_{bf} + \tilde{I}\right],$$

$$(l_{bf}) \quad \alpha_{bf}\beta_{bf}\gamma_0(l_{bf})^{\beta_{bf}-1}\left(\frac{p_{bf}}{p_f}\right) \cdot (T - l_f - l_{bf}) \cdot A_{bf}$$
$$= \gamma_1\left[\alpha_f(l_f)^{\beta_f} \cdot (\overline{A} - A_{bf}) + \frac{p_{bf}}{p_f} \cdot \alpha_{bf}(l_{bf})^{\beta_{bf}} \cdot A_{bf} + \tilde{I}\right],$$

$$(A_{bf}) \quad \alpha_f(l_f)^{\beta_f} = \alpha_{bf}\left(\frac{p_{bf}}{p_f}\right) \cdot (l_{bf})^{\beta_{bf}}. \tag{11.10}$$

---

[2] The off-farm labor decision could have been kept the same, as in Case 2, for greater generality. However, it is dropped here simply to help focus the discussion on the key trade-off between food and biofuel-related allocation decisions

In this case, trade-offs in the allocation of land and labor to the food and biofuel feedstock production activities depend critically on the difference in price and productivity between the two activities, namely, the ratio p_{bf}/p_f, and on the relative magnitudes of β_f and β_{bf}. The on-farm activity that has the higher per-unit productivity per unit labor will tend to bias the allocation of labor toward it, whereas the more favorable price will cause the allocation of land to tilt toward it.

2.6. Summary across cases

The characterization of household-level resource allocation decisions across these three distinct cases is summarized in Table 11.2, which describes the nature of the key decisions and constraints as well as the way in which the prices for household consumption goods (limited to food, in these simple examples) affect the intrinsic trade-offs embedded in the resource allocation decisions. In the case in which all food must be purchased on the market – using income gained from paid labor – the household faces a trade-off between labor and leisure, where leisure must be forgone if an increase in food prices will erode the purchasing power of the paid wages. This characterizes the problems of urban and landless urban households that cannot directly produce food for their own consumption. In the cases in which own-consumption of household production is possible, the trade-offs center more around the allocation of labor across economic activities. This leaves households with marketed surplus from on-farm production at an advantage over net consumers (in case 2) or those with larger land holdings with a greater degree of freedom in balancing food production (and exposure to food price risk) with the production of a cash crop (which can be generalized beyond the case of the biofuel feedstock crop that we consider here).

It is recognized, at this point, that a unitary view of the household has been adopted, in which important differences, such as gender, are not explicitly taken into account. In the seminal work of Fontana and Wood (2000), the importance of unpaid household labor is acknowledged, and it is that the time spent by women in reproductive activities (such as cooking, providing fuel and other resources, and providing overall nurture and care to other household members) is essential in enabling the wage- or revenue-earning activities of the household to take place. For example, if the time needed to collect firewood can be avoided by providing alternative fuels such as a form of biofuel amenable to household use (e.g., ethanol gels) (Kammen et al., 2003), then this would be a way to relax the total time constraint and allow the household to engage in more income-earning production activities or in more leisure. This dimension of biofuels is not explicitly treated in this chapter, although it is acknowledged that it has emerged in the discussion on biofuels' potential in Africa (Ewing and Msangi, 2009; Mitchell, 2011; Rossi and Lambrou, 2008).

Section 3 considers other dimensions through which the welfare of households might be affected by biofuels-related market forces and what these dimensions might

Table 11.2. *Summary of key household characteristics for the three cases*

	Key decision variables	Maximization problem	Key trade-offs	Effect of food price changes
Case 1	Nonfarm labor allocation (l)	$$\max_{l} \quad \gamma_0 \log(\tilde{w} \cdot l + \tilde{I}) \\ -\gamma_1 \log(T - l)$$	Trade-off in time use between leisure and paid work	Affects purchasing power of labor wages
Case 2	On- and off-farm labor allocation (l_f, l_{nf})	$$\max_{l_f, l_{nf}} \quad \gamma_0 \log\left(\alpha\sqrt{l_f} \cdot \overline{A} + \tilde{w} \cdot l_{nf} + \tilde{I}\right) \\ -\gamma_1 \log(T - l_f - l_{nf})$$	Trade-off between leisure and the returns from two competing economic activities, one of which is defined by return per unit wage and the other by returns per unit land	Affects net return (if household markets food) or compensating power of off-farm wage (if household is net food buyer)
Case 3	Allocation of land and labor between food and nonfood crops (l_f, l_{bf}, A_{bf})	$$\max_{l_f, l_{bf}, A_f} \quad \gamma_0 \log \left[\begin{array}{l} \alpha_f (l_f)^{\beta_f} \cdot (\overline{A} - A_{bf}) \\ + \dfrac{p_{bf}}{p_f} \cdot \alpha_{bf} (l_{bf})^{\beta_{bf}} \cdot A_{bf} + \tilde{I} \end{array} \right] \\ -\gamma_1 \log(T - l_f - l_{bf})$$	Trade-off in returns per unit land of two activities, which also compete for limited labor time; therefore productivity per unit labor is key	Determines the degree to which the household is exposed to food price risk and should compensate with feedstock production

242

imply for the design of policies that aim to increase the well-being of rural households in Africa.

3. The dynamic nature of household allocation decisions

3.1. Resource allocation as a sequential series of decisions

In previous sections, the trade-offs that face different types of households, within a static framework in which all decisions are made simultaneously, and in which the competition for various resources occurs within the same period as the allocation decision, were identified. Starting from the insight gained from looking at household resource allocations as a sequential series of decisions, as in the seminal paper of Fafchamps (1993), it gradually becomes obvious how the dimension of time enters into the maximization problem of the household-level decision maker. Fafchamps (1993) addressed the way in which households are prevented from fully realizing the returns to certain investments if they face constraints in certain key resources such as labor.[3] Within the context of the cases considered, one can appreciate how the time dimension might enter into the labor and land allocation decisions of the various household types, especially for cases 2 and 3.

For case 2, where on- and off-farm labor decisions encompass the key trade-offs the household faces, if one labor decision must be made before another can be taken, then the choice of the labor allocation in the second period rests purely on the trade-off between labor in that one activity and overall leisure. In the case of biofuels, if the off-farm labor opportunity is related to the production or processing of biofuel feedstocks on a plantation or at a processing facility, then certain on-farm production activities might have to be decided on first to ensure that adequate time is available to allocate to the off-farm activity. In the case of farming, this could involve preparation of land for sowing or purchasing inputs that will be needed for the growing season. If the off-farm labor opportunity were to fall through and not materialize as expected, then there may not be sufficient time for the household to readjust its on-farm production plans in time to make up for the lost wage revenue or to find alternative off-farm employment. This case would represent one in which the household is negatively affected by a missed opportunity rather than by an adverse market price shock that we would normally consider within a static framework.

For case 3, the sequential nature of the decision might lie more in the allocation of land area between on-farm production activities. If the opportunity to market on-farm production of the nonfood or cash crop occurs in the future (after the land allocation has already been made), then any change in the prospects of realizing the sale of the nonfood feedstock crop in the future could not be directly compensated by a reallocation of land if the necessary preparation needed to be done prior to the point

---

[3] Especially for those labor-intensive activities like weeding which are important in determining the productivity of any yield-improving farm input, like improved seed or fertilizer.

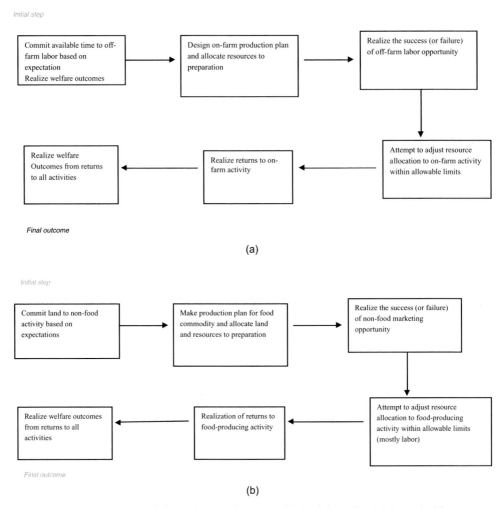

Figure 11.3. Sequential dependency of household decisions for (a) household conceptualized as "Case 2" and (b) household falling conceptualized as "Case 3".

at which the realization of a failed market opportunity was made. Despite the fact that labor could be reallocated completely to the alternative activity, the allocation of land could not be adjusted so readily.

A general representation of this kind of sequential decision making, and the dynamic dependency of the returns realized from one decision on the outcomes of preceding ones, is shown in Figure 11.3. It is obvious that the welfare outcomes depend on the degree of adjustment that is possible after an anticipated opportunity has fallen through.

3.2. *Household risk and failed biofuel ventures*

This illustration becomes relevant when one considers the risks that households might be facing when making sequential decisions in anticipation of biofuel ventures that may or may not be successfully implemented in developing countries, especially in sub-Saharan Africa. Taking case 2, we can think of higher off-farm wage opportunities as being offered by large feedstock plantations to cultivate and harvest sugarcane or other high-value energy crops. If it turns out that the on-farm activity is the cultivation of staples, then the household would be electing to forgo household labor (and food production) for the prospect of gaining nonfarm wages that offer a higher return to effort. While this has the potential of being welfare increasing, it depends on the opportunities actually being materialized.

There are a number of cases (Hufstadter, 2010) in which the promise of paid wages from biofuel plantations did not actually materialize because of the venture falling through, putting the household in a difficult position because it had already reduced its labor allocation to certain key farming activities (such as planting) in anticipation of the other opportunity coming through. Similarly, the setting aside of land for the planting of jatropha, as is being done in both Tanzania (FAO, 2010c) and Mozambique (Arndt et al., 2010a), can represent the situation for case 3, in which the household has to lock a certain amount of land into a perennial crop that may not realize a sufficient yield or have a suitable market price or outlet when it is actually harvested. This makes the forgone food production on that land a missed opportunity and increases the exposure of the household to potentially volatile market prices and availability for food.

These examples illustrate a transference of risk onto smallholders that might go beyond what the normal community risk-sharing strategies are able to cope with, given that most traditional, reciprocal arrangements for insuring risk between rural households relies on the principle that risks be idiosyncratic and not affect more than a handful of households within a community at a time (Coate and Ravallion, 1993; Fafchamps, 1992; Ligon et al., 2001). The failure of an announced commercial venture within a targeted region will likely affect a large number of households, both in terms of missed opportunities for off-farm labor and in terms of their ability to produce feedstock as an outgrower within the larger production scheme. The only kind of plausible insurance arrangement that could offset the effects of such an occurrence could be for the central government to provide compensation through a guaranteed rural employment scheme that would offer paid work for a limited time to the households within the affected target area. This kind of employment scheme is used in India to boost rural household income and is not conditional on the occurrence of such events but might prove to be expensive for most African governments to undertake on a continual basis. Exceptions are richer African countries, such as South Africa, which has a national program to combat the large-scale problem of unemployment with which it has struggled during the post-apartheid era (Antonopoulos, 2008). Therefore these governments might require companies that are coming in to make investments to pay a bond up front that would help to finance such a scheme in the event of commercial failure. The company would be repaid if the venture were successful and went forward for at least a couple years, and this would serve as a down payment on the success of the venture and transfer some of the risk of failure back to the enterprise.

Paying a large bond up front might, however, act as an additional barrier to entry for some firms, potentially discouraging some of the investments that could otherwise occur in such countries. This barrier would be lower if there were an existing rural household compensation scheme that was co-financed through the contributions of other commercial activities beyond just biofuel-related ventures. In this case, this payment would be a contribution to a larger pool and could be lowered in value, as it would be in the case of paying an insurance premium. Alternatively, the involvement of multilateral organizations, such as the International Finance Corporation, could be useful as they could act as the guarantors, co-financiers, or reinsurers of this kind of commercial risk, possibly helping small countries vet the merits of such commercial ventures ahead of time so that the bad ventures could be identified early and weeded out, before on-the-ground commitments are made and costs are incurred. Such a scheme could take a number of possible configurations, with each having its own merits and weaknesses. Even though an exhaustive discussion on this is not possible within the scope of this chapter, this is identified as a fruitful area for further research and exploration.

4. Policy implications and relation to the empirical literature

A number of papers in the academic literature have tried to address the potential impacts of biofuel expansion on households in sub-Saharan Africa, from the perspective of expansion of production in OECD countries and from the perspective of the possible expansion of biofuel programs within sub-Saharan African countries themselves.

Regarding the impacts on Africa coming from the expansion of biofuel production in OECD countries, a number of studies have taken a global perspective in an attempt to infer the impact that growth in first-generation biofuel production has on commodity prices and what the implications are for welfare. The analyses done by the International Food and Policy Research Institute (IFPRI) (Msangi et al., 2007; Rosegrant et al., 2008) illustrated the linkage between increasing commodity prices, decreases in calorie availability (driven by negative demand response), and the inferred increase in levels of child malnutrition. Although their analysis did not consider individual household characteristics, the implications that they draw cover the household types that fall into the case 1 description (Section 2.3). Because much of this ex ante assessment of the impact of biofuels was done within the context of attempts to understand what was driving the 2007–2008 spike in food prices, and what role biofuels had to play in it, there was not enough hindsight to be able to test whether the causal linkages between biofuel growth, price increases, and decreases in welfare would actually play out in the way that was projected. The analysis done by Ivanic and Martin (2008) on the likely impacts of food price increases (regardless of driver) on household welfare showed the importance of taking into account not only the price effect on consumption (as the IFPRI analysis did) but also the implications for wages and for sales revenue for the households that are net buyers of agricultural commodities. In light of those additional inferences, the gains seen by net milk producers in Latin America or net rice producers in regions of Asia (e.g., Vietnam) could lead to increases in welfare in the face of higher regional food prices – with a sharp contrast drawn between urban and rural households (with urban households mostly losing out) (Ivanic and Martin, 2008). Their framework extends to those households that are described by case 2 and case 1. Even in their analysis, though, sub-Saharan Africa still fares badly under higher prices and sees increases in poverty for both urban and rural populations. A recent World Bank study (Timilsina et al., 2011) also shows decreases in welfare for sub-Saharan Africa under higher growth in OECD biofuel production and higher food prices compared to other regions.[4] Presumably, the representative household that is modeled for sub-Saharan Africa should also respond to consumption effects as well as labor opportunities, so it would certainly cover the case 1 descriptive category. The linkage of production and consumptive behavior is not as direct as the household farm structures of cases 2 and 3, although there is repatriation of producer revenues that does accrue to households to the extent that is captured by the regional social accounting frameworks. So, in an indirect way, the linkages made explicit during the description of case 2 households are implicitly picked up in the global, macro-level computable general equilibrium (CGE) modeling framework that they use as well.

As far as the impacts of biofuel development within sub-Saharan Africa on household welfare are concerned, there is very little actual biofuel production within Africa on which to base ex post analysis of welfare impacts. Nonetheless, a number of studies have done ex ante analyses on what the poverty and welfare effects might be if certain countries were to follow through on their national plans for scaling up ethanol and biodiesel production, using targeted feedstock crops such as sugarcane and jatropha (see Chapters 12 and 13).

The analysis of Arndt et al. (2010a) points out the trade-off that is likely to occur in a land-abundant but labor-scarce country like Mozambique if the country was to pursue an export-oriented domestic biofuel program, as has been proposed and planned since the early 2000s. The fact that a new agribusiness sector is likely to draw labor away from the production of agricultural staples and cash crops is contrasted with the additional income that would accrue to the producers of the feedstock as well as to the laborers who are hired to produce it. They highlight the effect that a biofuel-induced rise in the exchange rate is likely to have on the competitiveness of more traditional, nonbiofuel export commodities as well as the changes in the poverty head count of

[4] It is not clear however, whether this study made the same distinction between urban and rural households and net-buyers and consumers of key commodities that was made by the World Bank study of Ivanic and Martin (2008).

urban and rural households. They show that urban consumers lose when prices of food items increase, due to the decrease in production as labor is diverted toward biofuel production. It also demonstrates the positive effects that accrue to rural households as their wage earnings and on-farm incomes increase, especially under the adoption of the outgrower-oriented feedstock production schemes. As we demonstrated in our theoretical models, the implications of wage and income changes differ across alternative types of households, with those households seeing wage benefits falling into cases 1 and 2, whereas those that engage directly in the production of biofuel feedstock on-farm would fall into case 3. The analysis of Arndt et al. (2010b) for the case of Tanzania also demonstrates a trade-off between rural and urban households that results from food price changes. They find that biofuel production could actually increase food production because it requires less intensive labor inputs compared to some of the more traditional, long-standing export crops, such as cashews, by freeing up labor that can go toward food production. This analysis was used within a larger case study of Tanzania conducted by the Food and Agriculture Organization (FAO, 2010c).[5] In the scenarios run for the Food and Agriculture Organization analysis, it was found that even modest increases in crop yields could more than offset the negative food security impacts coming from biofuel-driven price increases on international markets and could even improve the potential competitiveness of the nascent biofuels sector in Tanzania itself. Using the framework of cases 2 and 3, this would mean that the yield improvements that boost on-farm production, captured by $Q(l_f, A)$ (equation (11.4)) and $Q(l_f, \overline{A} - A_{bf})$ (equation (11.7)), would overcome the expenditure increases realized on the left-hand side of the full-income constraint, represented by $p \cdot c_f$.

Other potential trade-offs that should be considered when scaling up to a national biofuel program at the policy level relate to the protection that African countries want to give to their economies against volatile fossil fuel prices and the implications that this might have for the allocation of resources to other export-oriented and foreign exchange–earning sectors. Indeed, the cost that fossil fuel imports represents for the national balance of payments of many non-oil-producing African countries is considerable and serves as a powerful motivation to seek alternative, renewable forms of energy that can be produced locally (Mitchell, 2011). While scaling up domestic biofuel production has the potential to reduce fossil fuel imports, the implied savings have to be balanced with the loss of foreign exchange that might result from decreasing output in traditional exports. This decrease in the output of traditional exports would be due to their lowered competitiveness on the world market as massive foreign investments in the local economy begin to have macroeconomic effects on exchange rates. This potential effect was noted in both the cases of Tanzania and Mozambique by Arndt et al. (2010c), who pointed out that offering generous concessions (in the

---

[5] This study also attempted to show the linkages to international price effects, and the land use requirements that country policy makers should consider in the design of the national policies.

form of reduced fuel taxes) to foreign companies who are contemplating investments might also represent a forgone source of government revenue that would need to be accounted for in the national budget balance.

Given that very few African countries are actually producing biofuels on a large scale, it is not possible to compare the hypotheses raised in earlier literature or generated from our theoretical analysis to the reality happening on the ground. Nonetheless, the long research experience and large body of evidence that have been gathered from observing the effects of other kinds of market-mediated shocks on African agricultural economies point to the same channels of transmission that have already been highlighted in this chapter. The first of these relates to the market prices for agricultural production outputs, which face international trade dynamics and various degrees of price linkages with other countries that produce and consume the same good. The other is through key input markets for agriculture, such as labor markets, which function at different levels of efficiency in various parts of sub-Saharan Africa but which remain a key linkage to household-level changes in income and welfare.

Recent literature on land acquisitions in sub-Saharan Africa has also raised the concern of whether agribusiness ventures financed by foreign interests (such as those providing investments for biofuels production) could cause more vulnerable households to lose access to important (although untitled) areas used for grazing or other types of agricultural commons, especially in countries with weak institutions for protecting farmers (Cotula et al., 2009; Friends of the Earth, 2010). These arguments, although not relating exclusively to the case of biofuels, speak to a larger need for countries in sub-Saharan Africa to design policies that more carefully accommodate agribusiness ventures to foreign interests. These policies would prevent already disadvantaged populations from being further constrained by the reallocation of key resources such as land. Such policies would accomplish this by ensuring that adequate compensation, documentation, and regulation are exercised in the event of the execution of leases as well as in the transferring or granting of use and access rights for land and other natural resources. Where institutions are failing to oversee and regulate these activities, urgent policy reform is needed to provide contract enforcement and oversight of foreign-funded agribusiness enterprises. Policy efforts should also be made to provide social protection to avoid the transference of risk to households that might bear the loss of promised employment or earned on-farm income from failed or underperforming ventures, as was discussed in the previous section.

5. Conclusions

This chapter highlighted some key issues that are relevant to the impacts of biofuels on African agricultural economies and households. In particular, the implications of biofuel growth were identified and discussed from (1) the perspective of likely impacts on households and consumers, (2) the micro-perspective of the household farm, and

(3) from the perspective of the linkages between those micro-level effects and the macro-level drivers of change that have been addressed in recent literature. By using a simplified and abstract representation of the household farm, key points of intuition about what the impacts (negative or positive) might be on consumers and producers of biofuel-related goods (or of the goods that compete with biofuel feedstocks for key resources such as land and labor) were articulated.

Additionally, the household-level effects in terms of specific cases, which span the spectrum of urban and rural household types that are observed in sub-Saharan Africa, were discussed. These household types, or cases, differ in the degree to which both consumption-side and production-side responses are linked to each other and undergo different types of welfare changes under increases in prices of consumption goods, production outputs, and productive inputs. Tying this discussion to ex ante policy simulations done in the literature, the key role that yields have to play begins to emerge and serves as an entry point into how the impacts of biofuels in less-developed countries can be ameliorated or attenuated (in the case of negative impacts) as well as elaborating their importance in improving the performance of agriculture (for the purpose of producing biofuel feedstock crops or even food crops).

At the household level, the importance of the labor response to market forces was also highlighted. Moreover, the division between on-farm and off-farm labor allocations highlights the different responses and welfare impacts that rural households are likely to face when new agribusiness opportunities, such as biofuel production, appear on the economic horizon. Those (case 1) households that are landless (or have very little access to productive land) and depend mostly on wage labor will react differently to new forces in the economic environment – manifested through wage changes and increases in off-farm employment – compared to households that derive most of their income from (and devote most of their labor efforts to) their own farm production (case 2). The false hope that some biofuel ventures have given to rural households who are counting on proposed projects to succeed, and on promised jobs or purchases of on-farm production to appear, can have important consequences for other labor allocation decisions and might result in forgone opportunities that cannot easily be recovered once they are past. In this case, it is the riskiness of waiting for opportunities that may not materialize in their own countries that is likely to affect these poor households, more than the price shocks that they are likely to receive from international markets, as a result of biofuel production happening elsewhere in the OECD.

In light of the analysis conducted in this chapter, it might be argued that the food-for-fuel trade-off that some policy analysts use to characterize the prospects of large-scale expansion of biofuel production need not always occur. This is true if the appropriate investments and efficiency improvements are made in advance and/or if the moderation of biofuel targets (in light of their implications for agricultural markets) is taken into account at the policy decision–making level. The empirical assessments of Tanzania (Arndt et al., 2010b; FAO, 2010c) illustrate the benefits of technological spillovers as well as the potential that increasing the productivity of labor in agribusiness sectors such as biofuels might have in actually releasing labor for other activities such as food production. This can actually lead to enhanced food security and welfare outcomes.

Undoubtedly, market-level price effects will result in the event of a large-scale increase in the production of a feedstock commodity that also has sizeable food and feed use value. Those who are most vulnerable to price increases could be adversely affected. But this should also be considered within the context of specific countries where agricultural (and nonagricultural) wages might improve in the face of such changes and where there might be possible beneficial spillovers into food production, as was illustrated in the studies of Arndt et al. (2010a, 2010b, 2010c). Therefore the need for continued micro- and macro-level economic and policy analysis in this area is clearly evident and should remain a priority for researchers.

Appendix

The derivation of the comparative static results for the household can be derived from equation (11.6). If the total derivatives of both sides of this equation system[6] are taken, then after some rearrangement, equation (11.A1) is obtained:

$$
\left[\alpha \overline{A} + \frac{\tilde{w} \cdot l_{nf} + \tilde{I}}{2\sqrt{l_f}} + \frac{\alpha \gamma_0 \overline{A}}{2\gamma_1} \right] dl_f + \left[\tilde{w}\sqrt{l_f} + \frac{\alpha \gamma_0 \overline{A}}{2\gamma_1} \right] dl_{nf}
$$
$$
= \left[\frac{\alpha \gamma_0}{2\gamma_1}(T - l_f - l_{nf}) - \alpha l_f \right] d\overline{A} - \sqrt{l_f} d\tilde{I}
$$
$$
\left[\frac{\alpha \overline{A}}{2\sqrt{l_f}} + \frac{\tilde{w}\gamma_0}{\gamma_1} \right] dl_f + \tilde{w}\left(1 + \frac{\gamma_0}{\gamma_1}\right) dl_{nf}
$$
$$
= \left[\frac{\gamma_0}{\gamma_1}(T - l_f - 2l_{nf}) \right] d\tilde{w} - \alpha \sqrt{l_f} d\overline{A} - d\tilde{I}. \qquad (11.A1)
$$

This can be rewritten as the following linear system, where the endogenous (left-hand side) and exogenous (right-hand side) variables are separated:

$$
\begin{bmatrix} \left(\alpha \overline{A} + \dfrac{\tilde{w} \cdot l_{nf} + \tilde{I}}{2\sqrt{l_f}} + \dfrac{\alpha \gamma_0 \overline{A}}{2\gamma_1} \right) & \left(\tilde{w}\sqrt{l_f} + \dfrac{\alpha \gamma_0 \overline{A}}{2\gamma_1} \right) \\[2ex] \left(\dfrac{\alpha \overline{A}}{2\sqrt{l_f}} + \dfrac{\tilde{w}\gamma_0}{\gamma_1} \right) & \tilde{w}\left(1 + \dfrac{\gamma_0}{\gamma_1}\right) \end{bmatrix} \begin{bmatrix} dl_f \\ dl_{nf} \end{bmatrix}
$$

---

[6] Neglecting those parameters which we will assume as fixed, namely T, α, γ_0, and γ_1.

$$= \begin{bmatrix} 0 & \left(\dfrac{\alpha\gamma_0}{2\gamma_1}(T - l_f - l_{nf}) - \alpha l_f\right) - \sqrt{l_f} \\ \dfrac{\gamma_0}{\gamma_1}(T - l_f - 2l_{nf}) & -\alpha\sqrt{l_f} & -1 \end{bmatrix} \begin{bmatrix} d\tilde{w} \\ d\overline{A} \\ d\tilde{I} \end{bmatrix}. \quad (11.A2)$$

To do comparative statics, based on this linear system, the sign of the principal determinant has to be determined first:

$$\mathbf{D} = \begin{vmatrix} \left(\alpha\overline{A} + \dfrac{\tilde{w} \cdot l_{nf} + \tilde{I}}{2\sqrt{l_f}} + \dfrac{\alpha\gamma_0\overline{A}}{2\gamma_1}\right) & \left(\tilde{w}\sqrt{l_f} + \dfrac{\alpha\gamma_0\overline{A}}{2\gamma_1}\right) \\ \left(\dfrac{\alpha\overline{A}}{2\sqrt{l_f}} + \dfrac{\tilde{w}\gamma_0}{\gamma_1}\right) & \tilde{w}\left(1 + \dfrac{\gamma_0}{\gamma_1}\right) \end{vmatrix}$$

$$= \tilde{w}\left(1 + \dfrac{\gamma_0}{\gamma_1}\right) \cdot \left(\alpha\overline{A} + \dfrac{\tilde{w} \cdot l_{nf} + \tilde{I}}{2\sqrt{l_f}} + \dfrac{\alpha\gamma_0\overline{A}}{2\gamma_1}\right)$$

$$- \left(\dfrac{\alpha\overline{A}}{2\sqrt{l_f}} + \dfrac{\tilde{w}\gamma_0}{\gamma_1}\right) \cdot \left(\tilde{w}\sqrt{l_f} + \dfrac{\alpha\gamma_0\overline{A}}{2\gamma_1}\right). \quad (11.A3)$$

It can be argued to be positive in sign, given the relative size of the first product of terms relative to the second product that is subtracted from it. On this basis, the key derivatives of interest (i.e., the impact of exogenous factors like wages on labor response) can be evaluated as shown:

$$\dfrac{\partial l_f}{\partial \tilde{w}} = \dfrac{\begin{vmatrix} 0 & \left(\tilde{w}\sqrt{l_f} + \dfrac{\alpha\gamma_0\overline{A}}{2\gamma_1}\right) \\ \dfrac{\gamma_0}{\gamma_1}(T - l_f - 2l_{nf}) & \tilde{w}\left(1 + \dfrac{\gamma_0}{\gamma_1}\right) \end{vmatrix}}{\mathbf{D}} \quad (11.A4)$$

$$= \dfrac{-\left(\tilde{w}\sqrt{l_f} + \dfrac{\alpha\gamma_0\overline{A}}{2\gamma_1}\right) \cdot \dfrac{\gamma_0}{\gamma_1}(T - l_f - 2l_{nf})}{\mathbf{D}} < 0.$$

This shows that in response to higher off-farm wages, the allocation of on-farm labor decreases, and this is an intuitive result. Conversely, and just as intuitively, it can be shown that nonfarm labor (l_{nf}) responds to a change in off-farm wage according to the following expression (equation (11.A5)), which suggests a positive response to higher wages:

$$\frac{\partial l_{nf}}{\partial \tilde{w}} = \frac{\begin{vmatrix} \left(\alpha \overline{A} + \dfrac{\tilde{w} \cdot l_{nf} + \tilde{I}}{2\sqrt{l_f}} + \dfrac{\alpha \gamma_0 \overline{A}}{2\gamma_1} \right) & 0 \\ \left(\dfrac{\alpha \overline{A}}{2\sqrt{l_f}} + \dfrac{\tilde{w}\gamma_0}{\gamma_1} \right) & \dfrac{\gamma_0}{\gamma_1}(T - l_f - 2l_{nf}) \end{vmatrix}}{D} \tag{11.A5}$$

$$= \frac{\left(\alpha \overline{A} + \dfrac{\tilde{w} \cdot l_{nf} + \tilde{I}}{2\sqrt{l_f}} + \dfrac{\alpha \gamma_0 \overline{A}}{2\gamma_1} \right) \cdot \dfrac{\gamma_0}{\gamma_1}(T - l_f - 2l_{nf})}{D} > 0.$$

12

Energy security, agroindustrial development, and international trade: The case of sugarcane in Southern Africa

BOTHWELL BATIDZIRAI
Utrecht University, Utrecht, Netherlands

FRANCIS X. JOHNSON
Stockholm Environment Institute, Stockholm, Sweden

Abstract

For most of the Southern African Development Community (SADC) countries, energy security is a key developmental issue, given the limited capacity and supply of modern energy services. Fortunately, the region is bequeathed with abundant natural resources that can potentially be developed to support a thriving biomass energy industry. The development of modern biomass energy is likely to contribute to solving energy security concerns, improving rural livelihoods, and mitigating a number of environmental and socioeconomic impacts of current energy systems. This chapter explores the numerous opportunities and challenges associated with an expansion of biofuel production from the sugar industry as well as potential international trade implications. Current analysis shows that land is not a limiting constraint to bioenergy production from sugar resources. This chapter discusses possible implementation mechanisms to maximize the benefits of sugar resources through multiproduct strategies. One of the key issues to emerge from the analysis is the implementation of regional biofuel strategies to take better advantage of the complementarities in local, regional, and global biofuel markets.

Keywords: energy security, regional trade, southern Africa, sugarcane ethanol

1. Introduction

Energy security has become a significant concern in many regions of the world in recent years, especially for aid-dependent, oil-importing developing countries in Africa. Oil supply disruptions, high and/or volatile oil prices, power blackouts, and fuelwood shortages are all issues of major concern in the region (GNESD, 2007; UNECE, 2007). For many countries in the southern African region, the ability to meet a growing demand for energy is a top national priority. Traditional biomass

Table 12.1. *Biomass energy indicators in selected Southern African Development Community (SADC) countries (2008)*

Country	Biomass as % of TPES[a]	Fuelwood consumption (m³ per capita)	Wood charcoal consumption (kg per capita)
Angola	63	0.24	16.20
Botswana	22	0.36	34.65
DR Congo	93	1.28	29.08
Mozambique	82	0.89	5.30
South Africa	10	0.27	0.91
Tanzania	88	0.57	33.63
Zambia	81	0.67	96.28
Zimbabwe	65	0.63	0.70

[a] TPES, total primary energy supply.
Sources: FAO (2006a; IEA (2010b); Johnson and Rosillo-Calle (2007).

energy accounts for more than 80 percent of national primary energy supply in several countries (Table 12.1), and the lack of access to affordable and reliable modern energy services stifles economic growth and income generation, especially in rural areas. The per capita consumption of fuelwood and charcoal is, in some countries, many times higher when compared to South Africa (Table 12.1).

Estimates indicate that by 2020, traditional biomass energy use will increase with population growth in sub-Saharan Africa, and thus the heavy reliance on traditional biomass in total final energy supply may not change significantly unless new policies and institutions aimed at improving energy access and energy security are put in place (IEA, 2010c). Owing to the specific needs of household cooking and other end uses, the substitution of traditional biomass with modern bioenergy (including first-generation biofuels) generally does not occur directly. Thus the energy security perspective adopted in this chapter is associated not with domestic energy access per se but with the biomass resource base more generally and how it can contribute to growth and development.

Modern biomass energy represents a strategic resource that the region can utilize to support development and improve energy security. Sub-Saharan Africa has substantial potential to develop modern biomass energy, given its tropical climate, low population density, and natural resource endowments (Hoogwijk et al., 2003; Batidzirai et al., 2006; Johnson and Matsika, 2006; Smeets et al., 2007; Wicke et al., 2011). The development of modern biomass energy can address energy security concerns (Mathews, 2007; Goldemberg, 2007), improve rural livelihoods (Dar, 2007; PAC, 2009; Rutz and Janssen, 2008; UN-DESA, 2007), and mitigate some of the environmental and socioeconomic impacts of current fossil-based energy systems (Woods et al., 2008; Goldemberg et al., 2004b; AGAMA Energy, 2003; Mathews, 2007; Edwards et al., 2007b; Macedo et al., 2004) (see Chapter 1). There are real opportunities for rural farmers in Africa to produce modern biomass energy feedstocks to supply national, regional, and even international bioenergy commodity markets (de Fraiture et al., 2008) (see also Chapter 11).

Despite the potential socioeconomic benefits of modern bioenergy and biofuels, there are also risks that large-scale bioenergy programs can result in food–fuel conflicts (Brown, 2006; Runge and Senauer, 2007; UN, 2007b) (see Chapters 2 and 11) and ecosystem degradation (de Fraiture et al., 2008; Dufey et al., 2004; GEF-STAP, 2006) (see Chapters 3, 4, 5, and 9). Given these fears, there is a strong lobby for a strict sustainability regime in biofuel production and supply (Dufey, 2007; FAO, 2008b; Knauf et al., 2007; RSB, 2010; van Dam et al., 2008). Such schemes could be implemented with national or, preferably, regional institutional support.

Greater economic integration within the region could allow biomass resource development to be better allocated to those areas where it is most productive, rather than being overly constrained by national priorities and policies (Johnson and Matsika, 2006). The increased trade in bioenergy and associated industries that accompanies economic integration can help to address the gap between bioenergy potential and utilization by directing resources and investment toward more promising areas and improving the distribution and transport infrastructure. International collaboration could also benefit local and regional initiatives, given the growing global experiences in biofuel production, supply, and utilization (Mathews, 2007).

Sugarcane is the most promising feedstock for biofuels in Southern Africa in the near term (Cornland et al., 2001; Takavarasha et al., 2006; Watson, 2011). The sugar industry is thus one of the key sectors with the potential to contribute significantly to national and regional energy security, especially through the supply of bioethanol for transportation as well as bagasse for electricity generation. Traditionally, sugar has been the main output of the sugar industry, but its by-products and coproducts (both energy and nonenergy) have been gaining in importance. Molasses and cane juice are valued for their fermentable sugars, which can be converted into ethanol. Cane residues, namely, bagasse and cane trash, are valued for their fibre content and organic residues as well as their use as fuel in cogeneration plants (see Chapter 6). Bioethanol and bagasse cogeneration of heat and electricity is the most commercially significant coproduct option.

This chapter explores these issues and assesses the numerous opportunities and challenges associated with increased biofuel production, focusing on the case of sugarcane because it is a commercially viable near-term option in southern Africa. Initially, an overview of the resource base is provided, with special reference to sugarcane and the key renewable energy options, ethanol and electricity cogeneration, that can be obtained (Section 2). An analysis is provided of the energy security implications of expanded sugarcane energy (Sections 3–4) and their relation to trade and development policies (Section 5). It should also be mentioned that there are significant opportunities for climate mitigation based on sugarcane bioethanol because it is among the most efficient of all energy crops. However, climate change is not a focus of this chapter, as it is discussed in greater detail in Chapters 1, 3, and 6.

Table 12.2. *Land statistics for SADC countries (2008)*

Country	Total land area (1,000 ha)	Agricultural area (% of total land area)	Cultivated area (% of total land area)	Population density (Persons/km^2)
Angola	124,670	46	2.9	14.5
Botswana	58,173	46	0.7	3.4
DR Congo	226,705	10	0.42	28.3
Lesotho	3,035	77	11.0	66.5
Madagascar	58,704	47	6.1	32.9
Malawi	11,848	47	27.5	151.8
Mauritius	204	56	52.2	625.0
Mozambique	80,159	62	5.8	27.7
Namibia	82,429	47	1.0	2.6
South Africa	122,104	82	12.9	40.1
Swaziland	1,736	81	11.2	67.9
Tanzania	88,580	39	1.4	48.0
Zambia	75,261	47	7.1	17.0
Zimbabwe	39,075	53	8.7	32.2

Source: World Bank (2010).

2. Biomass resources and sugarcane in Southern Africa

SADC[1] is a diverse region, ranging from sandy deserts in the southwest to tropical dense forests in the north. Lying in the tropics, the climate generally supports high biomass growth, although erratic rainfall patterns affect many parts of the region. Extending over an area of 9.3 million km^2, and similar in size to the United States or China, the region has considerable forest and woodland resources (covering 40% of the area), which provide a rich source of biomass fuels (Nyoka, 2003). Abundant agricultural land area and low population density in many countries suggest the potential for expanding agroindustrial development (Table 12.2). The average area under cultivation is 6 percent, which is low by international standards. Additionally, the agricultural productivity is low due to the subsistence type of agriculture in many SADC countries (see Chapter 11).

2.1. Current sugarcane production

The SADC region produces nearly 50 million metric tons of sugarcane per year on some 850,000 ha, accounting for somewhat less than 3 percent of the world total

[1] The Southern African Development Community (SADC) is an intergovernmental organization whose goal is to promote socioeconomic cooperation and integration among its 14 members, which include Angola, Botswana, Democratic Republic of the Congo, Lesotho, Madagascar, Malawi, Mauritius, Mozambique, Namibia, South Africa, Swaziland, Tanzania, Zambia, and Zimbabwe.

Table 12.3. *Sugarcane production in Southern Africa (2008)*

	Area (1,000 ha)	Production (1,000 metric tons)	Share of world total	Yield (t/ha)
Angola	10	360	0.02	38
DR Congo	40	1,550	0.09	39
Madagascar	82	2,600	0.15	32
Malawi	23	2,500	0.14	109
Mauritius	62	4,533	0.26	73
Mozambique	180	2,451	0.14	14
South Africa	314	20,500	1.18	65
Swaziland	52	5,000	0.29	96
Tanzania	23	2,370	0.14	103
Zambia	24	2,500	0.14	104
Zimbabwe	39	3,100	0.18	79
Southern Africa	**849**	**47,464**	**2.73**	**56**
Africa, Total	**1,478**	**85,392**	**4.92**	**58**
India	**5,055**	**348,188**	**20.05**	**69**
Brazil	**8,140**	**645,300**	**37.17**	**79**
World	**24,257**	**1,736,271**		**72**

Source: FAO (2011).

(see Table 12.3). The soil is amenable and the climate is quite suitable to sugarcane cultivation in many areas due to a well-defined growing season with an extended period of dry, sunny conditions. According to reported yields (FAO, 2011), and as discussed in Woods et al. (2008) and Innes (2010), the region is home to some of the most efficient sugar industries in the world. Malawi, Tanzania, and Zambia have consistently had yields above 100 metric tons/ha, although other regions have much lower yields under current conditions (Table 12.3). Whereas development in the past has been driven by differing national priorities, a regional approach to sugarcane expansion in the future would naturally emphasize higher-yielding areas and better link these areas to regional and international markets.

2.2. Sugar markets in SADC

The possibility to expand renewable energy production from sugarcane (mainly bioethanol and electricity cogeneration) depends significantly on the nature and structure of the market for sugar. The SADC region is a net exporter of sugar, due in part to the existence of preferential markets. Several countries in the region have preferential access to European Union and U.S. sugar markets, in which prices are higher than the world market price. As shown in Box 12.1, some countries rely heavily on such preferential markets.

Box 12.1. SADC sugar markets

The SADC region is a net exporter of 1.6 million metric tons of sugar per annum. Around 63 percent of SADC's sugar exports are sold to the European Union (EU) and the United States under preferential access arrangements during 2006–2008, compared to only 40 percent during 2003–2005. SADC producers export sugar to the European Union and the United States under three preferential access arrangements:

- Economic Partnership Agreements, which replace the previous Sugar Protocol
- The EU Everything but Arms Initiative (EBA), which granted LDC sugar producers duty-free access to the EU sugar market (the qualifying SADC countries are Angola, DR Congo, Malawi, Mozambique, Tanzania, and Zambia)
- The United States, where market access is granted through a tariff-rate quota

The following table presents an overview of the sugar supply–demand balance in SADC and South African Customs Union (SACU) countries for the three-year period 2006–2008. The table shows that 90 percent of the 1 million metric tons that SADC sells under preferential access arrangements is sold to the EU. Some countries rely heavily on preferential EU and U.S. markets (e.g., Mauritius), whereas others, such as South Africa, have no preferential access to EU markets.

SADC supply–demand balances: Average 2006–2008 (in 1,000 metric tons)

	Production	Consumption	Preferential trade			Surplus (deficit)
			EU EPA/EBA	U.S. quota	Total	
Botswana	0.0	51.0	0.0	0.0	0.0	−51.0
Namibia	0.0	57.0	0.0	0.0	0.0	−57.0
South Africa	2,395.2	1,700.4	0.0	21.6	21.6	673.2
Swaziland	639.4	112.0	173.5	11.9	185.4	342.0
SACU	**3,034.6**	**1,920.4**	**173.5**	**33.5**	**207.0**	**907.2**
Angola	0.0	256.0	0.0	0.0	0.0	−256.0
DR Congo	62.8	85.0	0.0	0.0	0.0	−22.2
Madagascar	18.6	140.0	0.0	0.0	0.0	−121.4
Malawi	273.3	170.0	63.6	6.7	70.3	33.0
Mauritius	482.3	41.2	478.8	2.8	481.6	−40.5
Mozambique	245.5	161.6	77.0	8.5	85.5	−1.6
Tanzania	270.0	312.5	15.4	0.0	15.4	−57.9
Zambia	231.1	116.2	43.1	0.0	43.1	71.8
Zimbabwe	361.9	223.0	62.4	11.2	73.6	65.3
Non-SACU SADC	**1,945.5**	**1,505.5**	**740.3**	**29.2**	**769.5**	**−329.5**
SADC	**4,980.1**	**3,425.9**	**913.8**	**62.7**	**976.5**	**577.7**

Sources: Innes (2010); ISO (2010)

There are several ways in which the sugar market affects the incentives to expand into renewable energy options and constrains the scale of such expansion. First and foremost, ethanol producers look for a higher return on their investment. With access to preferential markets, producers will only make ethanol from molasses as they will have little incentive to forgo sugar production by using raw cane juice for ethanol. Second, because the factory size and level of efficiency are based on sugar production, investment in cogeneration facilities depends on accompanying investments in the factory performance and scale (Seebaluck et al., 2008). Finally, the overall package of coproducts and services associated with a sugarcane agroindustrial complex are interrelated. This means that sugar producers will be interested in a higher overall return but at the same time cannot extend their marketing and distribution to cover too many products. Consequently, the development of new markets for ethanol, cogeneration, and other coproducts (beyond the various sugar products themselves) takes time and requires some investment in technical and marketing infrastructure.

2.3. Bioenergy and sugarcane potential

A number of studies suggest that the region has a great potential to produce bioenergy from energy crops using available, unutilized, and suitable land after satisfying competing land use needs (conventional agricultural cropland and human settlements) and without degrading ecosystems (Batidzirai et al., 2006; Johnson and Matsika, 2006; Smeets et al., 2007; Watson, 2011) (see also Chapter 2). Watson et al. (2008) use Geographic Information Systems (GIS) analysis to overlay regional mapping by IGBP-IHDP (1995) to provide a regional distribution of areas that are potentially suitable for sugarcane production. Their methodology is based on remote sensing techniques and delineates land surfaces in southern Africa that have a spectral signature corresponding to sugarcane and sugarcane-like crops (Figure 12.1).

In a more recent study, Watson (2011) identified at least 6 million ha of land as available and suitable for rain-fed sugarcane production in selected southern African countries[2] using GIS to interrogate 1 km^2 grid cell resolution, taking into account protected areas, land cover, climate, elevation, and soil data sets (see Table 12.4). To promote sustainable land use expansion, sensitive areas were excluded, including nationally and internationally protected areas, closed canopy forests, wetlands, and all areas under crop production. It should be noted that such an analysis does not provide any indication as to whether local economic and institutional conditions can facilitate access to this land for development. At the same time, the analysis excludes areas currently under crop production, and because some of those areas have low productivity, sugarcane might offer a much better overall return compared to the existing crops,

---

[2] Some countries (DR Congo, Mauritius, South Africa, and Swaziland) were excluded due to limited expansion options associated with land resources, biophysical constraints, and investment climate.

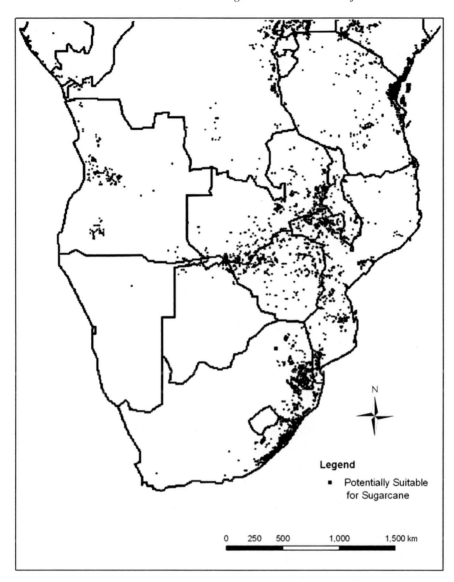

Figure 12.1. Areas in southern Africa potentially suitable for rain-fed sugarcane production. *Source*: Watson et al. (2008).

thus presenting an opportunity for growth and development. Therefore the analysis is focused first and foremost on physical potential by identifying areas that are both suitable and available.

For the selected countries, the proportion of available and suitable land varies from about 1 to 3 percent of total national land areas. Mozambique has the highest potential, with at least 2.3 million ha of land for expanding sugarcane production, followed by Zambia (1.2 million ha) and Angola (1.1 million ha), Zimbabwe (620,000 ha), Tanzania (467,000 ha), and Malawi, which is the smallest of the six countries

Table 12.4. *Land availability and suitability for rain-fed sugarcane production in selected SADC countries (in 1,000 ha)*

	Angola	Malawi	Mozambique	Tanzania	Zambia	Zimbabwe
Total land area	124,670	9,408	78,409	87,869	74,339	38,667
Potentially suitable land area	1,626	742	4,906	1,694	3,546	2,935
Total land area available and suitable for rain-fed sugarcane production	1,127	206	2,338	467	1,178	620

Source: Watson (2011).

(206,000 ha). The availability of 6 million ha of suitable land for rain-fed sugarcane production in these six countries suggests that land is unlikely to be the limiting factor in sugarcane expansion. By way of comparison, Brazil uses about 8 million ha to produce all its sugar and ethanol. Given that similar, or in some cases higher, yields than those in Brazil can be achieved in southern Africa, the raw potential is comparable to that of Brazil.

2.4. Productivity and competitiveness

Many producers in the SADC region are highly cost competitive by world standards, reflecting not only excellent growing conditions for sugarcane but also efficient sugarcane milling operations. Milling performance in the region is high in terms of sugar extraction compared to other parts of the world. In Mauritius, the average mill extraction is 97.2 percent, while for South Africa and Zimbabwe, it is 97.6 percent and 96.9 percent, respectively. This is comparable to some of the leading sugar-processing countries, such as Brazil and India, where the average mill extraction is 96.5 percent and 96.0 percent, respectively (Seebaluck et al., 2008). The crushing capacity is also high in a number of the southern African sugar mills, reaching 313 metric tons of cane per hour (TCH) in Swaziland and 455 TCH in Zimbabwe. However, facilities in some countries are in very poor condition due to civil unrest and lack of investment, for example, Angola, DR Congo, Madagascar, and Mozambique.

It is interesting to note that the potentially most productive areas for future sugarcane expansion do not lie in South Africa, where most of the sugarcane is now produced. Nor is there scope for expansion in Mauritius, which has an advanced industry but comparatively poor soils and limited land. Therefore any significant expansion, if it is to be efficient in technical terms, would occur elsewhere and would lead to greater economic development in the poorer countries of the region. Consequently, sugarcane expansion would have redistribution impacts, which are related fundamentally to the resource potential of the region rather than to the industrial market access that has defined previous development paths (Johnson et al., 2008).

Yet it nevertheless appears that this shift would only occur if there was greater regional cooperation and less emphasis on national priorities.

3. Economic drivers and constraints

Availability of affordable and reliable energy services is a key factor for economic growth and development (Cluver et al., 1998; IEA, 2004b; WCED, 1987). Attainment of the Millennium Development Goals similarly hinges on improved access to modern energy services (UNDP, 2005). With traditional biomass energy accounting for up to 90 percent of national energy supply in some countries, the region needs to develop its domestic energy resource base in a more efficient manner. Renewable energy from sugarcane can thus be viewed in terms of how it addresses some of the constraints to economic growth and development in the region.

3.1. Regional energy infrastructure and economic growth

The World Bank (2006) estimates that high oil prices, together with domestic capacity constraints and limited export demand, have reduced growth among oil-importing developing countries by up to 7 percent between 2002 and 2006. Lower-income countries dependent on oil imports are the worst affected by oil price shocks. When prices rise, the energy security of the poor is threatened as their ability to purchase sufficient fuels for basic cooking, heating, and lighting needs is greatly diminished. Also, the International Energy Agency estimates that a sustained USD 10 per barrel increase in oil prices would result in a loss of more than 3 percent of gross domestic product in the ensuing years in Africa (IEA, 2004a).

Apart from dampening economic development, energy shortages also discourage investment. According to the World Bank, firms in developing countries lose about 5 percent of their annual sales due to power outages. Its Investment Climate Surveys have consistently found that unreliable or unavailable electricity services are a "major obstacle to doing business" for 44 percent of firms in Africa (Saghir, 2006: 2).

Oil import dependency is high among SADC countries, except for Angola, which is a major oil exporter, the DR Congo, which is also a net exporter, and South Africa, where coal-to-liquid fuel technology provides a significant share of transport fuels (Table 12.5). For the landlocked countries in southern Africa, the landed price of oil is even higher due to more complicated logistics and higher transport costs (Johnson and Matsika, 2006; Scurlock et al., 1991).

The different taxation regimes in different countries also give rise to wide variation in prices, as shown in Table 12.5. For example, petrol or diesel prices are much higher in landlocked Zambia and Malawi. As an oil producer, Angola keeps its fuel prices among the lowest in the world and thus does not address the social and environmental costs of petroleum consumption. Several oil-importing countries,

Table 12.5. *Oil prices and import dependency in selected SADC countries*

	Gasoline price (U.S. cents/L)	Diesel price (U.S. cents/L)	Oil import dependency
Angola	53	39	n/a
Botswana	88	102	1.25
DR Congo	123	121	n/a
Lesotho	79	93	1.07
Madagascar	155	143	0.91
Malawi	178	167	1.10
Mozambique	171	137	0.92
Namibia	78	88	0.89
South Africa	87	95	0.26
Swaziland	86	93	1.01
Tanzania	111	130	0.98
Zambia	170	161	0.92
Zimbabwe	130	105	1.08
Average	107	101	

Note: Bunker fuels used in shipping are not included; oil dependency can vary by 10% or more in a given year based on transfers, storage, etc.
Source: GTZ (2009).

including South Africa, also tax petroleum fairly lightly and thereby reduce the incentives for renewable energy investment or for improving energy efficiency. The vast geographical extent of the region and the undeveloped infrastructure also mean that supply lines are vulnerable to disruptions in the case of civil unrest, natural disasters, or trade measures.

Electricity infrastructure is inadequate and/or unreliable in several countries, result-ing in lower access and widespread blackouts, which dampen productivity[3] and invest-ment. Payments for electricity and imported petroleum fuels continue to drain scarce convertible currency, thereby reducing available resources for investment in other important sectors and in overall infrastructure development. Energy imports in SADC countries account for up to 28 percent of total export revenues and for more than 20 percent of import expenditures in some countries (IMF, 2005, 2006a, 2006b, 2006c, 2006d). The African Development Bank estimates that oil imports make up between 10 and 25 percent of total imports of at least 28 African countries. Therefore increas-ing and/or volatile oil prices will constrain future economic growth and development in the region (AfDB, 2006).

---

[3] Foley (1990) valuates the costs of supply outages according to studies done in different developing countries. These range from USD 1.00–7.00 per kWh for industrial users to USD 1.50–1.65 per kWh for commercial users to USD 0.05–1.50 per kWh for domestic users. Thus the supply outages result in some cases in costs that can be 1 or even 2 orders or magnitude greater than electricity prices.

Table 12.6. *Current and planned fuel ethanol production capacity in selected SADC countries*

Country	Stage of development	Ethanol annual distillery capacity (million L)
Malawi	Existing	30
South Africa	Planned (corn-based)	155
Zambia	Planned	36.5
Zimbabwe	Existing	40

Source: Batidzirai and Wamukonya (2009).

3.2. Bioethanol production in Southern Africa

Bioethanol production from sugarcane is not a new phenomenon in the region. There is a significant production experience in several SADC countries, although not at a large scale, as in Brazil and elsewhere. Historically, much of the ethanol produced in the region has been used for industrial and potable applications and exported to the world market. These nonfuel ethanol markets can command a higher price than fuel ethanol and thus offer attractive options when the production scale is smaller. However, the tendency to pursue such markets means that the associated benefits of substituting ethanol for fossil fuel are forgone, including energy security, health impacts (e.g., lead removal), and climatic and environmental benefits.

Zimbabwe pioneered ethanol production from sugarcane molasses and blending with petrol in 1980, followed by Malawi in 1982 and Kenya in 1983. These national bioethanol fuel programs were aimed at reducing dependency on oil imports, saving foreign exchange, and developing a new domestic industry (Habitat, 1993; Scurlock et al., 1991). Ethanol blending is still practiced in Malawi, it has been reintroduced in Zimbabwe, and it is likely to be introduced in South Africa in the near future (Batidzirai and Wamukonya, 2009; DME, 2007). New ethanol programs have been announced in several African countries, including Mozambique, Zambia, Swaziland, and Mauritius.

Table 12.6 shows the existing and planned bioethanol production capacity in selected SADC countries. Diversification of sugar producers into bioethanol is in part a response to competitive pressures in southern Africa due to the gradual phasing out of subsidies and/or preferential markets for sugar in the European Union (Yamba et al., 2008). Some SADC sugar-producing countries will still qualify for preferential markets and/or price supports in the European Union for some years to come, and therefore most plans for ethanol production also involve some type of expansion in sugar supply (Jolly, 2011). South Africa does not qualify for European Union preferential markets, although its major sugar companies, Illovo and Tongaat-Hulett, control sugar operations in other SADC countries that do qualify.

Table 12.7. *Electricity export potential from bagasse energy in SADC countries*

Country	Cane crushed annually (1,000 metric tons)	Bagasse availability (1,000 metric tons, at 50% moisture content)	Power at 20 bars, 325°C (GWh)	Power at 45 bars, 440°C (GWh)	Power at 82 bars, 525°C (GWh)
Angola	360	108	9.0	27.0	46.8
DR Congo	1,669	536	41.7	125.2	217.0
Malawi	1,796	630	44.9	134.7	233.5
Mauritius	5,800	1,560	145.0	435.0	754.0
Mozambique	397	120	9.9	29.8	51.6
South Africa	22,103	6,126	552.6	1,657.7	2,873.4
Swaziland	4,103	1,350	102.6	307.7	533.4
Tanzania	1,289	600	32.2	96.7	167.6
Zambia	1,600	540	40.0	120.0	208.0
Zimbabwe	4,535	1,360	113.4	340.1	589.6
Total/average	**43,652**	**14,395**	**1,091**	**3,274**	**5,675**

Source: Seebaluck et al. (2008).

3.3. Cogeneration in the sugar industry

Bagasse is another important sugarcane coproduct and is the main energy source for the production of electricity and process heat in sugar factories (see Chapter 6). With recent reforms in the electricity sector (UN Energy/Africa, 2007) and the possibility of independent power producers generating and selling electricity to national public electricity grids (Marandu and Kayo, 2004), sugar factories now have an opportunity and incentive to optimize their cogeneration facilities and earn additional income through electricity trade (Seebaluck et al., 2008). Mauritius has successfully developed a sugar-based electricity industry in which about 22 percent of the national electricity supply is based on bagasse (MEPU, 2010).

From the 40–50 million metric tons of cane crushed each year in the SADC region, the potential electricity generation from SADC sugar mills for export to the public national grid is estimated to be 600 GWh using high-pressure boilers (82 bars) and high-temperature (525°C) technology (Table 12.7). Most of this potential is found in South Africa, Mauritius, Swaziland, and Zimbabwe. At lower operating parameters, the potential electricity dramatically decreases by 42 percent (at 45 bar and 440°C). The analysis suggests excellent potential for expanding bagasse cogeneration plants in the region, but realizing this potential would require investment in factory upgrades for the cogeneration plants to be able to operate efficiently. Also required are the appropriate infrastructure and policy measures to facilitate the export of surplus electricity from the factories.

Except for Mauritius, no SADC country has invested significantly at the national level in bagasse electricity cogeneration. Most sugar mills concentrate on satisfying their own electricity consumption needs and have somewhat low efficiency, mainly due

Table 12.8. *Voltage regulation performance with and without an embedded cogeneration unit*

Substation	Light load without cogeneration	Peak load without cogeneration	Light load with cogeneration	Peak load with cogeneration
Chiredzi Town	32.7	29.5	34.6	33.6
Chiredzi East	32.5	29.0	33.7	31.6
Hippo Valley Pumps	32.7	29.3	34.7	33.5

Note: Units are in kilovolts.
Source: Batidzirai (2002).

to low steam pressure and temperature of the boiler house. A number of pilot projects have been initiated with some success. For example, Zimbabwe's two sugar factories produce electricity from bagasse to meet electricity requirements for irrigation, sugar plant machinery, and equipment and for the local community (Mbohwa and Fukuda, 2003). Replication of the Mauritian experience would require a cohesive plan to upgrade factories but also to create supporting legislation.

3.4. Reliability of supply and avoided costs

Independent power production from facilities such as bagasse cogeneration plants can contribute significantly to improving reliability and avoiding costly investments in electric utility generating capacity, thereby addressing another aspect of energy security. Batidzirai (2002) illustrates the economic and technical benefits of embedded generation of electricity in a sugar mill at Hippo Valley Estates (HVE) in Zimbabwe. Chiredzi is one of the furthest load centers from the main power generation plants in Zimbabwe, and thus providing reliable and quality power supply is a challenge to the utility, especially voltage control. Electricity demand in the Chiredzi area is 25 MVA, and the export of up to 11 MW from HVE's cogeneration plant to the local grid vastly improves the reliability of supply in the area. Since HVE feeds the local area in an island mode configuration, supplies in Chiredzi are firm during the milling season. Thus, even if the national grid were to collapse, the network supplied from HVE would not be affected, and a total blackout of Chiredzi could be avoided. Quality of supply also improves with an embedded cogeneration plant as good voltage regulation can be achieved, as shown in Table 12.8. Poor voltage regulation can easily lead to system instability, and technical fixes, such as line drop compensation, may not be effective. With improved voltage regulation, it is possible to increase line loading, extend the network, and serve new consumers.

The HVE plant also provides economic benefits to the local utility by means of avoided investment in new generation capacity. Such economic benefits also trickle down to the local economy through productivity improvements and tariffs. In addition, network reinforcement and uprating of feeders to carry higher loads can be avoided or delayed, with significant cost savings (Foley, 1990). Also, the locally generated power can displace costly imports and associated interconnector infrastructure.

Costs of meeting peak demand can be very high if a utility has to dispatch inefficient thermal plants. The load flows conducted for the HVE plant showed significant reduction in line losses with cogeneration in the Chiredzi network (peak power losses are reduced from 16.1% without cogeneration to 8.3% with cogeneration). Thus distributed generators embedded in the national grid can play a significant role in stabilizing power in the local network and reducing overall system losses. Furthermore, the construction period for such plants is shorter as they require smaller capital investment, necessitate lower running costs, and enjoy quicker returns on investment compared to an equivalent fossil fuel–based power plant.

4. Implementation strategies

A number of different implementation options might be pursued by individual producers in expanding renewable energy from sugarcane in the region. These options can be viewed in terms of various coproduct strategies that are based on differing levels of sucrose and fiber utilization (Yamba et al., 2008). The more sucrose that is directed toward sugar, the less there is available for ethanol production, while fiber utilization depends mainly on efficient factory configurations and bagasse and residue management strategies (as well as economic markets for cogenerated electricity). There are other energy-related coproduct options, such as biogas production using the stillage (waste stream) from ethanol production, and there are also many nonenergy-coproducts that might be included. These implementation strategies are discussed in this section.

4.1. Ethanol strategies

Since the sugar industry has traditionally focused on extracting as much crystalizable sugar (sucrose) from the cane as is economically possible, any sustainable strategy to co-produce ethanol and raw sugar is driven by market factors to apportion some of the sucrose resources to ethanol production. To avoid negative economic and social impacts associated with a food–fuel competition, local demand for crystalline sugar must be met before any sugars can be diverted to ethanol production.

The traditional strategy of only producing sugar is not economically sustainable unless preferential export markets are guaranteed, which is no longer the case (Innes, 2010). The alternative strategy of producing only ethanol in an autonomous distillery has been pursued in Brazil and elsewhere and has the advantage of significant savings in capital investment costs because only juice preparation and extraction facilities are needed rather than a complete sugar factory. It is nevertheless difficult economically for producers to forgo sugar production. A study carried out in Zambia (Cornland et al., 2001) concluded that ethanol-only strategies are not viable unless sugar prices drop significantly. Furthermore, the feasibility of an autonomous distillery depends significantly on both the size of the market and the size of the sugarcane estate. The plant size and feedstock production would need to be optimized accordingly. An autonomous distillery would need to be operated at a reasonable scale and would require a stable market at this higher level. Such stability is difficult in the near term, given that no regional ethanol market currently exists. The introduction of high ethanol blending mandates, E85, or flex-fuel vehicles coupled with policy measures to stimulate a regional market could improve the feasibility of ethanol-only options (via autonomous distilleries) in the future (see Chapter 6 for a discussion on the Brazilian context). The low rate of personal car ownership (and resulting low consumption of petrol) essentially means that the small scale of the national markets (except for South Africa) makes these markets less competitive in comparison to a regional market (Johnson and Matsika, 2006).

A third approach considers producing sugar and ethanol in fixed quantities by maximizing extraction for sugar production following the standard 3-massecuite raw sugar production process and leaving only C molasses (final molasses) available for ethanol production.[4] This approach is based on annexed distilleries to make use of the surplus molasses stream while prioritizing sugar production. At Triangle Estates in Zimbabwe, C molasses were at some stage imported from neighboring HVE and from as far as Zambia to supplement the B molasses from the sugar mill (Scurlock et al., 1991). This approach, based on fixed quantities, remains viable if sugar prices are competitive, sugar markets are active, and ethanol and oil prices are low. Where sugar markets become saturated and prices decline, and where ethanol markets emerge, a fixed sugar–ethanol strategy would fail to capitalize on changing economic opportunities.

The fourth strategy involves producing sugar and ethanol in flexible proportions. In this scenario, sugar is extracted up to the first or second stages, resulting in the production of A or B molasses, respectively. These molasses streams will have fermentable sugars that can still be economically extracted. The presence of additional fermentable sugars increases the efficiency of ethanol conversion. Consequently, if ethanol is expected to have a market value close to or greater than that of sugar, then it makes economic sense to prioritize ethanol production over some sugar production by using A or B molasses as the ethanol feedstock. If market prices are fluctuating over time, a producer can benefit from having the flexibility to switch among these alternative balances of molasses use. The capital and operating costs of the additional

---

[4] The three strikes (A, B, and C) used at sugar factories extract sugar (sucrose) from the syrupy mixture (massecuite) until further extraction is not economically viable. Each subsequent mixture is known as A, B, or C molasses. Economic viability is normally exhausted by the third strike (producing C molasses).

processing stations for B and C molasses are not significant compared to the overall production costs (Seebaluck et al., 2008). Consequently, the decision whether to emphasize sugar or ethanol production can be made at the margin. Such an approach is already adopted by producers in the mature sugar–ethanol market of Brazil.[5]

In the case of expected increases in ethanol market demand, the approach would involve a progression in the use of feedstocks for ethanol production from C molasses to B molasses to A molasses to cane juice, as required over time to meet increasing ethanol demand. Thus, on an annual basis, the first step would be to compare ethanol yield from C molasses to target ethanol demand. If ethanol demand exceeds ethanol yields from C molasses alone, then B molasses is chosen as feedstock (for ethanol production in as many mills as required, starting with the largest mills, but only if national sugar demand is still met). If raw sugar demand is not met, then additional feedstock needs to be developed. But as long as sugar demand is met, ethanol continues to be derived from higher-sugar feedstocks (up to A molasses and from cane juice). Where ethanol and sugar demand are not met, there is a need to assess additional sugarcane production required to meet demand and ascertain whether this would be feasible. This may include considering the need for additional factories. If expanding the cane-growing area is not feasible, alternative feedstocks (e.g., sweet sorghum, maize) or imports may be considered. The other crops could, in most cases, provide equivalent feedstock to the same ethanol distillery, although there are always some additional feedstock preparation costs when using multiple sources of raw material. Sweet sorghum is not suitable for crystalline sugar production but is suitable for ethanol production and therefore can be viewed as a potential supplementary feedstock where ethanol production is valued more than sugar at the margin (Woods, 2001).

4.2. Bagasse-based electricity generation strategies

For given electricity production targets, possible options for meeting these targets can be determined using a decision flowscheme. Following the same logic used for ethanol strategies, the largest mills are considered the most suitable for early modification to produce surplus electricity. The first step involves assessing electricity production from the sugar industry under existing conditions and comparing it with demand targets. If this production is not adequate, the second step is to assess electricity production, assuming that all factories optimize their existing boilers and back-pressure steam turbines, and improved efficiency in the use of steam. If demand is still not met, the next step is to assess potential electricity generation in a scenario where the largest mill is retrofitted with state-of-the-art condensing extraction steam turbines (CEST)

[5] In Brazil, ethanol producers generally rely on a broth of B molasses and cane juice that has optimal biochemical properties for the distillation process, while also allowing some flexibility in diverting additional sucrose to ethanol production when sugar prices are low, and vice versa (Macedo, 2005).

and the highest boiler pressures deemed technically and economically viable for the mill. If demand is still not met, power production from the next largest CEST-powered mill is estimated until demand is met or the mill set is exhausted. If after all viable mills are converted to using CEST and demand is still not met, possible new cane production and factories are considered.

Two main strategies for selling surplus electricity from a sugar factory are possible. The first option can be to sell to local off-grid customers, such as local industries or rural electricity cooperatives, thereby providing electrical services without the costs that accompany grid connections. For areas without a guaranteed market and low local demand, an investment in a cogeneration plant with the intention of selling electricity to the neighboring communities may not be viable.

The second option is to sell surplus electricity to established utilities or distributors as an independent power producer (IPP). This requires appropriate national policies that allow IPPs to generate and supply the public network. However, in most SADC countries, the national utilities are not offering competitive feed-in tariffs to encourage this option. As an important input in the productive sector, many governments still maintain some form of control of electricity pricing to lower the cost of industrial production. This has generally stifled development of cogeneration in the sugar industry in the region, despite sectoral reforms in the electricity industry. However, the rapidly changing electricity tariffs in the region will gradually make cogeneration economics more attractive in the short to medium term.

4.3. Other coproduct markets

Besides production of electricity and ethanol, there are many other marketable by-products or coproducts that can be produced in annexed plants to sugarcane factories. The establishment of a centralized and integrated sugarcane processing complexes creates further economic opportunities by increasing the viability for exploiting additional cane-based coproducts. Whether this is done through extensions to existing factories or through entirely new facilities depends crucially on scale and market structure for the intended coproducts. The initial investment costs will be lower via the expansion of existing factories, whereas a fully optimized production that is related to the specific product mix would require changes already at the design phase (Seebaluck et al., 2008).

By definition, a sugarcane complex or sugar agroindustrial complex is a cluster of industries that rely in some way on sugarcane and its many coproducts as the raw biomass feedstock material. The sugar factory serves as the nucleus to produce sugar, while other production units can process as many value-added products as the market allows, while also modulating the input requirements (especially energy and water) in various forms (live steam, exhaust steam, vapor at various pressures and temperature, and electricity) from the combustion of the fibrous fraction of cane (bagasse) in an integrated manner.[6]

Downstream products that are already produced in South Africa include furfural (used mainly in lube oil refineries for the purification of oils), furfural alcohol (used mainly to produce a resin in the foundry industry as a binder for foundry sands), diacetyl and 2,3-pentanedione (both used as high-quality natural flavorings), Agri-guard (an agricultural nematocide), and lactulose (a natural laxative). These products are marketed as high-value sugarcane-derived biochemicals in niche markets (Illovo, 2009). Such higher value-added products that are produced under high-quality control and modest volumes can serve as economic complements to the production of high-volume energy coproducts. The high efficiency of sugarcane and the competitiveness of sugar industries in southern Africa offer opportunities for a green chemicals industry to emerge on a larger scale.

5. International trade and policy options

The current setup of the sugar industry in southern Africa is based largely on historical patterns for production, consumption, and trade of sugar, with a number of countries depending on the preferential sugar markets, as mentioned earlier. With the reform of the European Union sugar regime and the reduction in the price support system, sugarcane-based renewable energy could become more valuable in improving the competitiveness of the sugarcane agroindustry in southern Africa. In this section, we assess the different scales of production, consumption, and market access in relation to the options for regional and international trade.

5.1. Local and national markets

Local and national markets for sugar determine the scope for expansion into other coproducts and other agroindustrial options as well as constraining (or creating) opportunities for regional and international trade rather than national markets. The national markets for sugar will generally need to be satisfied in most cases to avoid the higher cost of imports. In some rural areas, sugar is used as a way to deliver greater nutrition to poorer households by adding vitamins (e.g., vitamins A and D) to the sugar because these households sometimes have access to sugar but not all other staple foods (Cornland et al., 2001). It is also important to consider the options for alternative products and alternative local markets, including various agricultural applications of animal feed, fertilizers, and other small-scale and relatively low-tech

---

[6] The Jiangmen Sugar and Chemical Complex in China is among the best examples of an advanced sugar agroindustrial complex, producing more than 28 distinct coproducts (Rao, 1997).

end uses. Even where bioethanol markets may not be feasible for the transport sector due to insufficient scale, the use of other coproducts and applications might be used to strengthen competitiveness.

A possible alternative market for bioethanol is for household cooking. The use of ethanol stoves has been championed in Ethiopia by the award-winning Gaia Association, a nongovernmental organization connected to the U.S.-based Project Gaia (Takama et al., 2011). Other countries have also already experimented with options for expanding the household ethanol market (Takama et al., 2011). The decrease in emissions compared to traditional biomass or charcoal (see Chapter 3) results in health and environmental benefits, while the high efficiency can provide savings in annual fuel costs, especially in urban or per-iurban areas where wood and charcoal are purchased. The use of bioethanol for cooking is an option that can simultaneously improve energy access, health, and environment (emissions). The main barrier is the higher up-front cost of the ethanol stoves, although there is also evidence to suggest that consumers are willing to pay more for the reduced emissions because of the associated health and safety benefits (Tsephel et al., 2009).

5.2. Regional trade

The SADC countries are gradually moving toward greater economic integration at the same time that the preferential sugar markets are being transformed. A number of member countries will continue to be able to export sugar at preferential prices, but there will be greater price constraints and quotas compared to the previous sugar regimes (Innes, 2010). Nevertheless, economic integration has progressed slowly and is complicated by the fact that there are various overlapping economic trade zones in Africa. In addition to SADC, there is the East African Community, in which Tanzania participates, and the Common Market for Eastern and Southern Africa, in which several SADC members participate. There is also the African Union, which addresses especially political cooperation and external relations for the entire continent and thus has some influence on the strategic direction of market developments.

Expanded regional trade in sugar, ethanol, molasses, and other sugarcane coproducts could have a number of economic advantages over the persistent tendency for national strategies. First, the countries that currently dominate the industry (Mauritius and South Africa) do not have good agricultural conditions for expansion compared to many other possible expansion areas, especially Mozambique and Zambia. Directing future expansion into the most agriculturally suitable areas will therefore contribute to the longer-term sustainability of the industry. Second, the lack of a cohesive regional market results in some resources (e.g., molasses) being exported outside the region that could otherwise be employed in higher value-added production of ethanol, biochemicals, and other products for use within the region. Finally, the regional trade can be linked to the changes in international markets so that ethanol exports might effectively replace or supplement sugar exports to the European Union and elsewhere as preferential prices for raw sugar are lowered in international markets.

5.3. International biofuel markets

International biofuel trade has been developing fast amid growing bioenergy demand during the past few years and is expected to increase further in the coming decade (Junginger et al., 2008). In OECD countries, especially the European Union, renewable energy targets and greenhouse gas (GHG) abatement targets are creating a global market for biofuels such as biodiesel and bioethanol. With crude oil prices reaching historically high levels, the global prospects for biofuel use are growing. These emerging markets for biofuels are likely to create a worldwide demand for these commodities and create unique opportunities for regions that can produce biofuels competitively. The high cost of biofuel production and the shortage of land in Europe and other world regions means that an import strategy would probably be more cost effective in meeting Europe's bioenergy requirements.

Southern Africa has therefore emerged as a potential region for expanded production and supply of biofuels for the international market. Exporting value-added products rather than raw materials to supply the world's energy markets could provide a more stable and reliable market and an incentive for greater rural investment in the region. Trade is therefore an instrument for modernizing bioenergy and contributing to rural and agricultural development, with the practical monitoring of sustainability as a key factor for long-term security (Woods et al., 2008). It is nevertheless important to acknowledge that competing in a nascent global ethanol market where Brazil is already the dominant player (see Chapter 6) poses a considerable challenge. At the same time, the agricultural conditions in some parts of southern Africa are actually better than those of Brazil (in terms of yield), and furthermore, a truly international market for any product will always require several producers in various world regions to be viable. In this respect, SADC countries can become the additional producing countries that can help consolidate the formation of an international ethanol market (see Chapter 6).

5.4. Domestic versus export markets

Historically, biofuel programs in Africa targeted primarily local and national markets to substitute expensive imported oil from the world market, similar to the manner in which the Brazilian bioethanol program began. At the national level, it may be easier to make a policy choice on whether to use biofuels locally or export depending on the economics, especially if biofuel prices on the international market are lucrative. In the case of sugar trade, some producers in the ACP region even sell beyond their own consumption needs and then import sugar at lower world market prices (Seebaluck et al., 2008; Yamba et al., 2008). However, at global level, policy decisions to promote local use compared to international trade are more complex. While there are calls to allow market forces to control trade, generally, it is more rational, from an energy efficiency point of view, to use the biomass primarily locally and only export the surplus. However, the actual energy balances and environmental impacts depend strongly on the reference energy systems in both the exporting and importing countries (Junginger et al., 2006).

5.5. International trade barriers

International trade in biofuels is currently limited by the fact that countries with large markets (the United States, Japan, the European Union) maintain tariffs on these fuels to protect their domestic industries and/or to ensure that their substantial domestic subsidies are not used to support the industries of other nations. For instance, the United States applies an extra USD 0.54 to each gallon (about USD 0.14/L) of imported ethanol on top of a 2.5 percent tariff, bringing the cost of imported bioethanol in line with that produced domestically (IEA, 2004a). Moreover, the tariff escalation systems that prevail in many industrialized countries encourage developing countries to export feedstock, such as unprocessed molasses and crude oils, while the final biofuel conversion (and associated value addition) takes place in the importing country. In addition, the lack of technical specifications for biomass and biomass import regulations can be a major barrier to trading. For example, in the European Union, denaturized ethanol above 80 percent attracts an import levy of 102 EUR/m^3, representing substantial additional costs (Junginger et al., 2006). The denatured ethanol can be seen as a higher value-added product compared to nondenatured ethanol since the former is designed to facilitate broader commodity trade.

The current lack of a clear classification of biofuels within the multilateral trading system constrains effective trade. At present, there is no broad agreement on whether biofuels are industrial or agricultural goods. On one hand, biofuels are traded as other fuels or as alcohol (in the case of ethanol) and are subject to general international trade rules under the World Trade Organization. Biofuels are potentially classifiable as environmental goods, as discussed in trade liberalization talks under the Doha Round, and such classification would facilitate their expanded use in environmental policy objectives as well as in the current emphasis on energy security goals (Dufey, 2007).

5.6. Standards and certification

It is critical that bioenergy be produced in an ecologically and socially sustainable manner; otherwise, its contribution as a renewable energy resource is compromised. It is important to distinguish between standards and certification. Although both can be related to sustainability concerns, the former establishes specific technical guidelines and measurement procedures (see Chapter 1), whereas the latter is more concerned with the institutional mechanisms through which an external third party can inspect production and trade to ensure that it meets particular criteria (see Chapter 7). With respect to standards, the International Standards Organisation (ISO) has recently initiated a major effort to establish sustainability standards, which will take several years to complete (ISO, 2010).

Systems of certification can include criteria that stipulate sustainable production methods. One approach could involve agreements between the major bioenergy producers and consumer countries concerning mandatory sustainability criteria that aim to minimize the negative environmental and socioeconomic impacts of biofuel production and use, which would be enforced by national legislation. The participating nations would then exclude unsustainable bioenergy from their markets and find ways to make this WTO compatible (Knauf et al., 2007). For example, the European Union Renewable Energy Directive has established sustainability criteria for biofuels, including default values for GHG emissions, while the Better Sugarcane Initiative has established a detailed set of criteria aimed at achieving best practice in sugarcane production in terms of environmental and socioeconomic standards.

6. Conclusions

This chapter explored the intersection between energy security, agroindustrial development, and international trade in SADC countries, taking the case of sugarcane and its various coproducts, particularly bioethanol and bagasse electricity cogeneration. Increasing and/or volatile oil prices and a lack of reliable electricity supply pose threats to energy security in the region. The sugar industry is undergoing global restructuring as price supports are reduced, and those areas that can improve their competitiveness stand to benefit. The increased production and use of modern bioenergy from sugarcane, in combination with expanded international trade, offers one approach to simultaneously address agroindustrial competitiveness while also improving energy security. Products that have low value in exports (e.g., molasses) could be used instead for value-added production (of ethanol and biochemicals) within the region. A regional expansion strategy would require much greater investment in the physical and institutional infrastructure for technology development and regional trade.

Modern bioenergy options such as sugarcane ethanol can, in some cases, directly replace traditional biomass (in the case of household cooking), while sugarcane bagasse can substitute for coal and otherwise diversify the sources of electricity supply, thereby improving the long-term reliability and sustainability of energy supply. The contributions to energy security and energy access are more likely to be indirect, however, resulting from the economic linkages associated with an agroindustry such as sugarcane, which requires a considerable amount of labor and local investment to be effective. At the same time, in the large region of southern Africa with fairly low population density, the availability of dispersed but efficient sources of domestic energy has strategic value for the region's long-term economic development.

The potential for rain-fed sugarcane in the region is comparable to the current production levels in Brazil. If areas of subsistence agriculture were also considered, the potential would be higher still, under the assumption that a commercial crop with higher value added, such as sugarcane, is preferred to subsistence farming (see Chapters 1 and 11). A more competitive global market for sugar means that expansion of sugarcane can no longer be predicated mainly on expanded markets for sugar alone but rather on the whole feasible set of coproducts, including renewable energy in the form of ethanol and bagasse electricity cogeneration, but also the many non-energy products, such as biochemicals that can be obtained using sugarcane biomass as a feedstock. A regionally integrated expansion of the sugarcane agroindustry offers opportunities for improving energy security and competitiveness, while also posing considerable challenges in terms of the significant investment required in physical infrastructure as well as in political and economic institutions.

13

Environmental and socioeconomic considerations for jatropha growing in Southern Africa

GRAHAM P. VON MALTITZ
Council for Scientific and Industrial Research, Pretoria, South Africa

ANNE SUGRUE
University of Johannesburg, Johannesburg, South Africa

MARK B. GUSH
Council for Scientific and Industrial Research, Stellenbosch, South Africa

COLIN EVERSON
University of KwaZulu-Natal, Pietermaritzberg, South Africa

GARETH D. BORMAN
University of the Witwatersrand, Johannesburg, South Africa

RYAN BLANCHARD
Stellenbosch University, Stellenbosch, South Africa

Abstract

Since about 2005, there has been a growing enthusiasm around biofuel development within southern Africa, with almost all countries in the region initiating biofuel projects. The key driver for biofuel expansion in the region is its potential to boost rural development and national energy security and to improve national trade balances, given that imported fuel contributes up to 25 percent of national foreign expenditure in some countries. *Jatropha curcas*, in particular, has captured the imagination of developers in the region, despite the fact that as a commercial crop, it is totally untested in the region. Even though jatropha growing is a new industry in southern Africa, there are already a number of projects being established in Mozambique, Swaziland, Zambia, Tanzania, and Namibia. This chapter discusses the results of different field experiments and observations from existing jatropha projects in an attempt to better understand the social and environmental impacts of jatropha introduction within southern Africa.

Keywords: jatropha, outgrower schemes, southern Africa, yield

1. Introduction

Jatropha curcas L. (jatropha) is a shrublike, small tree that produces a seed with approximately 35 percent oil content (Achten et al., 2008; Openshaw, 2000). Though jatropha originates from Central America, it is found growing throughout the tropics and subtropics, where it was introduced as a hedge or for its oil, originally for applications such as lamp oil or soap making (Henning, 2006; Jongschaap et al., 2007). The seeds are toxic and the plant is unpalatable, which allows it to be introduced into areas with high grazing pressure, but at the same time, it carries a health risk. The potential to use jatropha oil as a biodiesel feedstock, together with claims of high yields, drought tolerance, and the ability to grow in dry areas with poor soils, has created huge expectations around jatropha (e.g., Fairless, 2007).

Since the mid-2000s, there has been a growing enthusiasm around biofuel development within southern Africa (see Chapter 12). A number of factors have led to this, including high petroleum prices, the potential for economic and rural development, insecurity of the fuel supply, and the potential savings of foreign exchange expenditures. However, it appears that the perceived potential of a large market created by the biofuels incentives in the European Union's Renewable Energy Directive of 2008–2009 was a principal driver of biofuel expansion. This is substantiated by the fact that most of the early investment in biofuels was from foreign investors seeking to sell into the European market (Schut et al., 2010; von Maltitz et al., 2009). Two main crops have dominated proposed biofuel investment in most southern African countries: sugarcane (see Chapter 12) and jatropha. Jatropha has been widely recommended as a potential income generator for both small-scale farmers and large-scale commercial farming (Schut et al., 2010; von Maltitz et al., 2009).

At present, most sugarcane projects are still in the planning phase, whereas over 54 jatropha-based projects have been identified as being in the process of establishment in southern Africa (GEXSI, 2008d; von Maltitz et al., 2009), though this is almost certainly an underestimate and far less than the land requests received. In Mozambique alone, Schut et al. (2010) have identified 17 projects. Though some countries, such as Mozambique, initially welcomed biofuel investment (von Maltitz and Setzkorn, 2012), relatively limited land has so far been allocated to projects, with estimates being about 114,000 ha planted in southern Africa and 2.3 million ha anticipated to be planted by 2015 (GEXSI, 2008d).

The rapid rise in interest from investors in biofuels caught most southern African countries by surprise in that there was no existing policy framework to regulate biofuel expansion, yet investors were actively seeking access to land for biofuel plantations. South Africa was the first country to put in place a formal biofuel policy (2007), followed by Mozambique (2009) and Angola (2010). Tanzania and Zambia have completed policies, but they have not yet been made publicly available. With the exception of South Africa (which banned jatropha), Angola (which is focusing on palm oil), the Democratic Republic of Congo (for which no data are available), and Lesotho (which is climatically unsuitable), all other SADC countries are actively promoting jatropha as their principal biodiesel crop (Lerner et al., 2010).

South Africa's biofuel trajectory has differed substantially from its neighboring African states. There was initially a high interest in jatropha growing from some provincial departments and private investors, with KwaZulu Natal's department of agriculture actively promoting jatropha and one large project initiated in the North-west Province. However, in 2007, the national department of agriculture banned the planting of jatropha as it considered it a category 2 weed based on an assessment of its invasive potential. Jatropha was ripped out in some areas where it had been planted, and its promotion ceased. D1 Oils, a commercial company that had done initial assessments, including impact assessments, for specific planting locations, was forced to move its plantations to Swaziland. The Northwest Province project collapsed due to social issues, with planted trees dying due to the area having an unfavorable climate for jatropha. A detailed scientific trial funded through the Water Research Commission and planted by the Council for Scientific and Industrial Research (CSIR) on the University of KwaZulu Natal's experimental farm in Pietermaritzburg was, however, allowed to remain, and results from this trial are discussed in detail subsequently.

Claims about jatropha include that it has low water use, grows on marginal and degraded lands, has high oil yields, has low labor costs, and tolerates pests and diseases. However, there is little scientific evidence to support many of these claims. The success of jatropha projects to date has fallen far short of the initial expectations as envisaged five years ago. A review of project planning proposals, company statements, and media releases gives expected yields in the 5–8 t/ha range for seeds, or 1,000–2,000 L/ha of oil from dryland plantings, with even higher yields expected under irrigation (e.g., Henning, 2006; Jongschaap et al., 2007; Schut et al., 2010). The suggestion is also made that yields will start being economically viable from two to three years after planting (Henning, 2006). Though most southern African plantations are still relatively young and it can be argued that they have not as yet reached productive maturity, it would appear that yields are falling far short of initial expectations (Borman et al., in press; Loos, 2009; Wahl et al., 2009). This chapter brings together a diversity of results from experiments and observations around jatropha introduction in an attempt to better understand the social and environmental impacts of jatropha introduction within southern Africa.

2. Greenhouse gas emissions of jatropha production

One of the key global drivers for biofuels is that they reduce greenhouse gas (GHG) emissions and thus help climate change mitigation attempts. Biofuels are by no means a carbon-neutral replacement for fossil fuels, though most life cycle assessments suggest that there is some saving (see Chapter 3). The level of these savings is situation specific, depending on the crop used, the way it is managed, and the processing and post-processing emissions. Land use change due to biofuel feedstock production can have potentially disastrous effects on carbon emissions (Fargione et al., 2008; Searchinger et al., 2008), but this is typically excluded from life cycle assessments, and both these aspects are explored in more detail later. Indirect land use change resulting from growing jatropha has not been assessed for the region but is currently assumed to be relatively small since most jatropha is planted on land not currently used for agriculture, abandoned agricultural land, or land with very low subsistence yields. There is a concern, however, that jatropha is often planted on good-quality land that could have been used for food crops.

2.1. Land use change impacts on carbon

Predictions are that global biofuel production will quadruple within the next 15–20 years (IEA, 2004a). Southern Africa estimates a 20-fold increase in the area under jatropha cultivation from 2008 to 2015 (GEXSI, 2008d). During land clearing, or land use change, biotic stores of carbon are released through combustion and/or decomposition, and these emissions may not be recouped from bioenergy production for many years, thus creating a carbon debt (Fargione et al., 2008). It has been reported that the potential for a carbon debt as long as several hundred years results from replacing high-carbon stock land (e.g., peatlands or rainforests) with biofuel feedstocks (Fargione et al., 2008; Gibbs et al., 2008).

It has been argued that direct land clearing can be avoided by cultivating jatropha on fallow or degraded lands, that is, lands low in accumulated carbon (Achten et al., 2007). Part of jatropha's sustainability acclaim is that it performs in marginal environments (Heller, 1996; Henning, 2006; Jones and Miller, 1992; Openshaw, 2000), avoiding the risk of competition with agricultural land and indirect land use change. However, experience with jatropha suggests that its cultivation will likely not be limited to poor lands as such a practice will often give uneconomic yields (Achten et al., 2010b; Borman et al., in press; Montobbio et al., 2010).

Approximately 50 percent of the African landmass lies within the savannah and miombo bioregions. Woody biomass production ranges from 15 t/ha in dry savannahs to greater than 90 t/ha in moist miombos. Miombo and wetter areas of savannah are most likely to be targeted for jatropha expansion. However, the actual land targeted for cultivation to date is mixed. In Mozambique, it is often arable land on abandoned tobacco and cotton plantations that has been allocated, whereas in Tanzania, there is evidence of the clearing of more pristine savannah vegetation (von Maltitz and Setzkorn, 2012).

The Century ecosystems model (Parton et al., 1987) was used to model the total carbon stock of different vegetation and soil carbon pools at two sites. The sites were in the region of Skukuza, South Africa, and Mongu, Zambia, which represent a semiarid savannah and miombo woodland, respectively. These sites were chosen as representative of potential jatropha growing areas along a rainfall gradient and, just as important, sufficient detailed data exist for parameterizing the Century model. In both locations, soils are sandy and production is nutrient limited (primarily nitrogen, then phosphorus). Mean annual rainfall at Skukuza and Mongu is 616 and 1,035 mm, respectively. Land clearing was simulated in the model, transforming the selected sites' land use to jatropha cultivation.

Biomass production from jatropha is poorly quantified. Henning suggests a total biomass of 80 t/ha after seven years for an irrigated plantation in Egypt, based on a nonempirical estimate of 200 kg biomass per tree, including belowground biomass (Benge, 2006). Jatropha has a very light wood density of between 0.29 (Achten et al., 2010c) and 0.37 g/cm^3 (Benge, 2006), which makes this value appear to be grossly overestimated, especially considering that the woody aboveground biomass of a mesic savannah has been observed at approximately 40 t/ha (Shackleton, 1997), with, in addition, just over half that value present in belowground tissue (Scholes and Walker, 1993). On the basis of available allometric equations (Achten et al., 2010c; Ghezehei et al., 2009) and comparing with known savannah tree biomass (Shackleton, 1997), jatropha rain-fed plantations are likely to have standing biomass of 6–9 t/ha (Borman et al., in press).

The Century model reaches an equilibrium aboveground biomass state for natural vegetation (including the herbaceous component) of 24–27 t/ha for the semiarid savannah site and 58–62 t/ha for the miombo woodland site. Figure 13.1 displays the respective magnitudes of the different carbon pools included in the study. The passive soil organic matter (SOM) pool includes physically and chemically stabilized matter, resistant to decomposition. The slow SOM pool includes resistant structural plant material and stabilized microbial products. The active pool includes soil microbes and products with short turnover rates (one to three months). Belowground biomass comprises both fine and coarse roots. Aboveground biomass has been split into the herbaceous component, including grasses, and the woody canopy.

Jatropha plantations result in a lower total carbon store than the two previous land uses. The approach taken in this research defines the carbon debt (or surplus) at any given point in time as the gap that exists between the amount of carbon stored in the jatropha plantation at that time and the amount that was stored in the previous natural ecosystem. The carbon debt payback time is the number of years it takes for the net carbon balance to be equivalent to the initial carbon stock, and this is strongly dependent on the jatropha yields obtained. The GHG emission reduction from combusting biodiesel rather than petro-diesel is what repays this debt. The dotted lines in Figure 13.1 display the best case scenarios of credits generated from the displacement of petro-diesel by various jatropha seed yields.

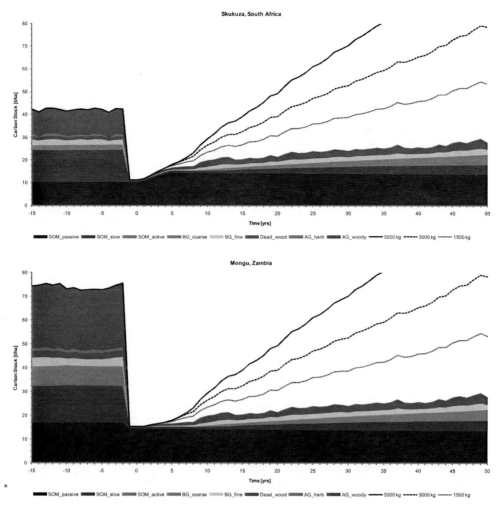

Figure 13.1. The net carbon balance of jatropha biodiesel production over time at the selected sites in semiarid savannah (Skukuza) and miombo woodland (Mongu). The black lines indicate the net carbon balance for different yield scenarios (kg/yr). (See color plate.)

2.2. CO_2 emissions compared to petro-diesel

The biofuel life cycle is not entirely emissions neutral, considering that biofuel production itself requires significant external energy inputs (see Chapter 1). Furthermore, the application of mineral fertilizers increases the emissions of nitrous oxide (N_2O), a powerful GHG due to its relatively long atmospheric lifetime, with a global warming potential about 310 times greater than CO_2 (IPCC, 2007). On the basis of Achten et al. (2010d), it was assumed that the life cycle CO_2 equivalent emissions from jatropha offer a 55 percent reduction from a petro-diesel reference system. This value falls within the range of 64 ±18 percent based on a review of several studies (Borman et al., in press) and also compares similarly with studies on other feedstocks such as oil palm, soya, and rapeseed when land use change is excluded from the life cycle analysis (see Chapter 3). The literature suggests that jatropha is one of the top performing feedstocks in life cycle GHG reductions (Dehue and Hettinga, 2008; Fritsche, 2008; Fritsche and Wiegmann, 2008; Reinhardt et al., 2007b; RIVM, 2008; van der Voet et al., 2008), in part due to predicted oil yields but mainly due to low nitrogen-fertilization requirements and it being a perennial crop. Perennial crops have been said to outperform annual crops in carbon terms (Ramachandran Nair et al., 2009). They require less input, such as fertilizers, and conserve soil carbon through annual belowground production and decay. Soils beneath perennial crops are also not tilled annually, and as tilling results in CO_2 emissions, this has an overall beneficial impact (Lal et al., 1995).

Cradle-to-grave, petro-diesel production and use emits 287 ± 62 g CO_2 equivalent GHGs per MJ of energy released during combustion in a car engine (Achten et al., 2010d). A 55 percent reduction in GHG from replacing petro-diesel with jatropha biodiesel avoids the release of half a metric ton of carbon into the atmosphere per hectare per year, based on an annual seed yield of 1,500 kg. Higher yields provide proportionally greater benefits, provided that more inputs, such as fertilizer, are not required (Borman et al., in press). Depending on the yields achieved, jatropha biodiesel can repay the carbon debt of land conversion from savannah and miombo within 17–36 years and 32–81 years, respectively (Figure 13.1). These are more conservative estimates than the range provided by Gibbs et al. (2008) for converting woody savanna into oil palm for the purpose of biodiesel production. Jatropha generates considerably greater carbon savings than nonperennial feedstocks for biodiesel such as groundnut and soybean (see Chapter 3).

It should be mentioned that about 30 percent of South African fuel is derived through a coal-to-liquid (CTL) technology that dates back to the 1920s, when two German scientists (Fischer and Tropsch) developed a process to convert coal into liquid petroleum to assist the German war effort. During the years of international isolation brought on by the apartheid era, South Africa invested heavily in developing and refining this process through its three Sasol plants. Today South Africa is the world leader in CTL technology, and annually, about 42 million metric tons of coal are converted to liquid fuel (150,000 barrels a day at a single plant). Despite its attractiveness, there are serious negative environmental concerns associated with CTL technology in terms of carbon emissions and water quality impacts. Liquefied coal emits twice as much carbon dioxide as burning oil and consequently will have larger impacts on CO_2 emissions than normal, oil-based fuels. CTL technology also emits large amounts of sulfur dioxide, resulting in respiratory problems for nearby communities. South Africa's CTL technology consequently produces syn-diesel with a higher CO_2 footprint than conventional crude oil, so the GHG benefits from jatropha, if compared to South African fuel, would be much greater.

3. Biodiversity

While biofuels might have a positive environmental impact through reducing global carbon emissions, there is a very real possibility that this will be at a cost to biodiversity (see Chapter 9 for a discussion on the impacts of palm oil on biodiversity). Biodiversity concerns relating to jatropha expansion in southern Africa are in two main areas, one relating to the potential invasiveness of jatropha and the other relating to land use change. Aspects such as water pollution or changes in streamflow that could potentially affect river biodiversity are currently seen as relatively minor biodiversity threats due to the relatively low fertilizer use and relatively low hydrological impacts (see Section 4). Insecticide use might be quite high and may in the longer term be an important biodiversity consideration.

3.1. Invasiveness

The potential threat that an introduced species may be invasive (i.e., the species is able to spread and establish itself outside the sites where it is cultivated) has led to the banning of the growing of jatropha in South Africa. In cases in which a species is invasive, its introduction into an area can have huge impacts on the indigenous biodiversity; South Africa experienced this with the Australian wattle, *Acacia mearnsii*, and other commercially useful tree crops (Rouget et al., 2002), and Kenya has experienced this with the purposeful introduction of *Prosopis* spp. (Witt, 2010). As these and many other examples have demonstrated, the cost of control can be disproportionately high in the African context, potentially negating the initial benefits from the specie's introduction (Le Maitre et al., 2002; Turpie and Heydenrych, 2000).

The specific plant traits, such as rapid growth rates, wide environmental tolerance, ease of establishment, low water demand, ability to resprout when harvested, and prolific seed production, that make a species an attractive biofuel feedstock are the same traits that are likely to lead to a species being invasive (Raghu et al., 2006; Richardson and Blanchard, 2011). Many of the nonfood plant species currently under consideration for biofuels are known to be invasive, at least under some circumstances in some parts of the world (e.g., *Arundo donax, Acacia saligna*, and *Jatropha curcas*) (Barney and DiTomaso, 2010; Buddenhagen et al., 2009; Crosti et al., 2010). Additionally, introducing nonnative species to a new region, especially if they are cultivated in large numbers, increases the likelihood of invasion. It does this by increasing the number of available propagules (Crosti et al., 2010; Křivánek et al., 2006; Richardson and Blanchard, 2011). The scientific community has responded to the challenge posed by increased trade in invasive species. This has seen the strengthening of screening tools, such as the Australian Weed Risk Assessment, to make them useful across a wide range of geographic regions (Crosti et al., 2010; Dawson et al., 2009; Gordon et al., 2008).

Specific biofuel guidelines and protocols have been developed and are being tested to minimize invasive risks (IUCN, 2009; Low and Booth, 2007).

In the case of jatropha, it is still uncertain as to whether it will be invasive in southern Africa. It is well known that there is often a long lag period between species introduction and the first indications of invasion (Křivánek et al., 2006). South Africa has chosen to take a precautionary approach in this regard, while most other African countries have chosen to allow jatropha to be grown.

3.2. Biodiversity impacts from land use change

Land transformation and the accompanying habitat loss are two of the biggest threats to terrestrial biodiversity (GBO3, 2010; MA, 2005; Sala et al., 2005). Currently there are limited studies indicating the likely impacts on biodiversity from changing land uses to accommodate biofuels in the landscape, and this is especially true of possible impacts from jatropha in southern Africa (Hellmann and Verburg, 2010; von Maltitz et al., 2010). Jatropha could potentially result in the following four categories of land use change: (1) the conversion of existing crops (food or industrial), (2) the conversion of abandoned agricultural land, (3) the conversion of degraded lands, or (4) the conversion of natural vegetation (von Maltitz and Brent, 2008). Each of these options is likely to have varying degrees of impact on biodiversity, with the conversion of natural habitats being the most severe. Meanwhile, the conversion of existing crops and degraded lands is expected to have the least direct impact (Tilman et al., 2009; von Maltitz et al., 2010). Potential impacts on biodiversity were modeled using the Biodiversity Intactness Index (BII) (Scholes and Biggs, 2005), as described in detail in von Maltitz et al. (2010). In Figure 13.2, the biodiversity impact varies considerably, depending on which current land use is allocated to biodiversity. The land allocation rules used in the modeled example allocate all cultivated land before allocating degraded land, and all degraded land before allocating lightly used land, in the scenarios where this is applicable. However, converting existing agricultural or rangeland may well result in indirect land use change as the agricultural activities may be displaced into new areas (Bird et al., 2010; von Maltitz et al., 2010).

From a theoretical perspective, there are a number of considerations regarding the impact on biodiversity resulting from land use change due to the production of biofuel crops such as jatropha. In addition to the obvious changes at the field scale, there are likely landscape and regional impacts. Species choice and planting configuration can influence landscape heterogeneity where large-scale plantings may act to create a more uniform landscape compared to small-scale schemes that may create a more diverse setting and increase fragmentation (Firbank, 2008). Changes to structural and functional composition within planted areas can negatively influence the delivery of goods and services. Firbank (2008) suggests that the impacts should be assessed separately at the field and crop, landscape, and regional scales. This approach

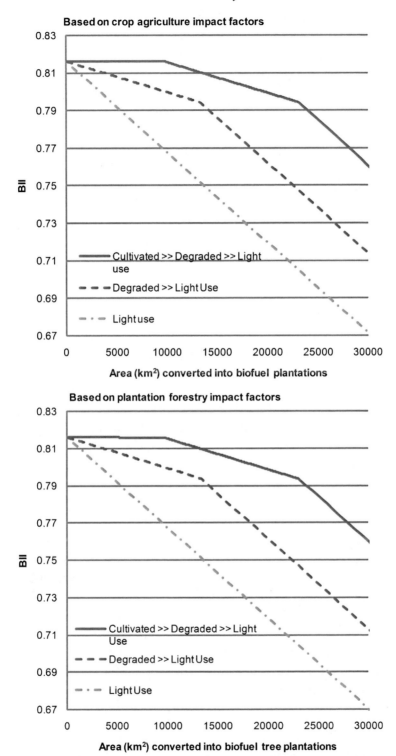

Figure 13.2. The influence of type of land allocated to biofuel on the BII impacts for annual and tree biofuel crops for Eastern Cape South Africa. *Source*: von Maltitz et al. (2010). (See color plate.)

allows for the separation of the pressures, impacts, and likely solutions within each scale to be assessed. Dauber et al. (2010) followed this approach in a meta-analysis of published results for biodiversity impacts from biomass production in temperate regions and concluded that the spatial layout and distribution of cropping systems required to meet intended targets was the biggest threat to biodiversity. They high-lighted the need to protect areas of high biodiversity importance from biofuel expansion and recommended the use of risk assessments to complement strategic landscape planning as a way to minimize overall costs to biodiversity.

A strong emphasis on avoiding deforestation in many biofuel certification schemes has made companies reluctant to plant jatropha in areas of forest or woodland, and developers in many instances have sought out degraded land or fallow or abandoned land (von Maltitz and Setzkorn, 2012). In some miombo areas, tree clearing is a necessity for the creation of jatropha plantations (Romijn, 2011). Many projects, such as TerraRossa Plantations in Madagascar, claim that their plantations will be cultivated on land that is already degraded (GEM, 2009). A potential unintended consequence of avoiding deforestation is that many plantations are on grasslands, a vegetation type that is both high in biodiversity and under extreme threats to biodiversity due to the extensive land transformation that is taking place (Gibbs Russell, 1986; Mucina and Rutherford, 2006; O'Connor and Bredenkamp, 1997).

3.3. Biodiversity impacts of biofuel versus fossil fuel

Separating out primary and secondary impacts of fossil fuels versus biofuels on bio-diversity is complex and poorly researched in general, particularly for jatropha. In the southern African region, there are relatively limited direct biodiversity impacts from fossil fuel because in most cases, the fuel is imported, and the extensive biodiversity loss that can occur from oil exploration (e.g., the BP oil spill in the Gulf of Mexico) is not experienced locally. The Fischer-Tropsch oil production in South Africa is a clear exception to this rule. The open cast coal mining involved has a clear biodiversity impact, as does the acid mine drainage into the river systems (see Section 4). The other exception is Angola, where there is extensive offshore oil exploration. Though not formally quantified, it is almost certain that biofuel has a higher primary impact on biodiversity than fossil fuels. The impact will be felt largely through an overall reduction of biodiversity when measured using an indicator such as the BII (Scholes and Biggs, 2005). Actual individual species extinction from biofuel expansion is less likely and can be prevented to a large extent through careful and strategic region-wide planning. Such planning could take the form of employing biodiversity-conscious management practices or determining the locations of plantations in such a way that conserves areas of high biodiversity importance are conserved (von Maltitz et al., 2010).

The secondary impact of fossil fuel on biodiversity is enormous. Thomas et al. (2004) predict that 20 percent of species may be driven to extinction by 2050 directly as a consequence of global warming (IPCC, 2007). The combined impact of land transformation and global warming is likely to be bigger than either on its own as it will inhibit adaptation through species migration (Hannah et al., 2000; von Maltitz et al., 2007). The relatively small positive benefits from biofuels in terms of reducing global warming are unlikely to compensate for the huge negative biodiversity impacts from land transformation. Though not quantified, there are likely many order of magnitude differences between the carbon benefits from biofuel (in terms of reducing global climate change and related extinctions) and the impacts on biodiversity caused by land use change (species extinctions being less likely to form a part of such land use change impacts). Crops that produce very high yields of biofuel per area planted will in general have fewer biodiversity impacts due to smaller land use change per unit of fuel produced. This does, however, assume that they are not grown in highly biologically diverse environments and that there are not secondary impacts on biodiversity through, for instance, high fertilizer use. In this regard, jatropha provides mixed benefits. It is good in that little fertilizer is used, but current indications are that yields are going to be low, and hence larger land areas will be required compared to crops such as oil palm or sugarcane to achieve equivalent fuel volumes.

4. Hydrology and yields

Large areas of southern Africa have limited water, and it is well established that plantations of introduced species (e.g., eucalyptus) can have a detrimental effect on catchment hydrology by reducing streamflow (Dye and Versfeld, 2007). Where large-scale changes in vegetation cover are proposed, differences in total evaporation (water use) between the current (or natural) and the proposed vegetation ultimately translate into changes in available streamflow from that catchment. Consequently, large-scale changes in land use could have significant hydrological implications. Furthermore, in South Africa, approximately 80 percent of arable land is classed as dryland agriculture, and there is concern that an expanding biofuel industry may negatively affect the already stressed water resources (Berndes, 2002). A good understanding of the water use of major land uses is thus fundamental to effective land use planning aimed at sustainable and integrated water resource management (Jarmain et al., 2008). Furthermore, the ability to relate some measure of the water use (e.g., transpiration or total evaporation) to some measure of production (e.g., seed yield) over a particular time period and spatial scale provides a measure of a crop's water use efficiency.

4.1. Hydrological research

To better understand the potential impacts of jatropha on catchment water resources, two separate experiments were established in different geographic regions within South Africa to investigate jatropha water use, growth patterns, and yields. The first experiment was carried out along the hot and humid subtropical eastern seaboard

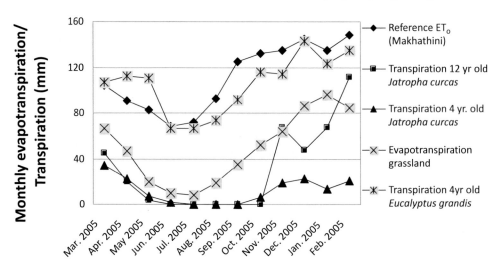

Figure 13.3. Comparative voluies of water use (transpiration or evapotranspiration) of a range of measured (*Jatropha curcas*, grassland, and eucalyptus plantation) and modeled (FAO56 grass reference ET_o) vegetation types.

in northern KwaZulu-Natal (KZN) Province, on young (4-year-old) jatropha trees at the Owen Sithole College of Agriculture near Empangeni and on mature (12-year-old) jatropha trees in the Makhathini flats (Gush, 2008). The second experiment was conducted in the KZN midlands (dry and temperate) on the research farm (Ukulinga) of the University of KwaZulu-Natal (Everson et al., in press).

In the northern KZN trials, the heat pulse velocity sap flow monitoring technique (Burgess et al., 2001) was used to monitor individual tree water use volumes of the 4- and 12-year-old trees. Total annual water use for the jatropha trees was measured and scaled up to plantation equivalent areas. The data were compared to the water use of indigenous vegetation (grassland), a eucalyptus plantation, and an FAO56-type (Allen et al., 2004b) reference potential evapotranspiration (ET_o). Evapotranspiration data from grasslands were sourced from Everson (2001), while transpiration data for four-year-old *Eucalyptus grandis* trees were taken from Dye (1996). Results confirmed that water use peaked during the warm, wet summer months but was negligible during winter due to the deciduous nature of the species. Compared to the other vegetation types, jatropha water use was less than for natural vegetation and significantly below that used by a typical Eucalyptus plantation (Figure 13.3).

To facilitate the application of the results across a wider (national) scale, monthly jatropha water use totals (mm) were divided by Penman–Monteith equivalent reference evapotranspiration (ET_o) totals (derived from weather station measurements) for the same period and location. In this way, a unique set of basal crop coefficient (K_{cb}) values (crop factors) for jatropha were calculated. The K_{cb} values were determined by only considering wet periods (soil moisture content >25%) so as not to violate the design applicability of the FAO56 method under free water conditions (i.e., under

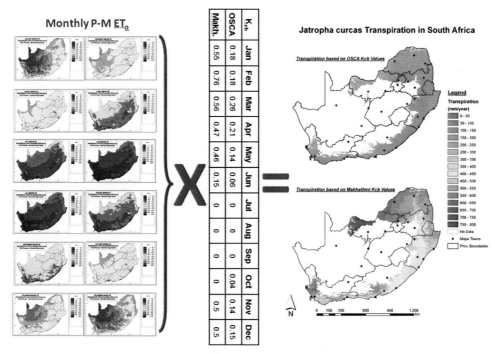

Figure 13.4. Illustration of the method used to estimate spatial water use patterns of *Jatropha curcas* in South Africa. *Source*: Gush and Hallowes (2007). (See color plate.)

conditions in which water uptake by the plant is not limited by soil water availability and usually where irrigation takes place). In this way, the hydrological impacts of the wide-scale planting of jatropha were assessed for South Africa by multiplying spatially explicit monthly totals of ET_o by monthly crop factors for jatropha (Hallowes, 2007). Spatial estimates of jatropha water use were compared against spatial estimates of water use by natural vegetation types in South Africa to determine if there would be a net increase or decrease in water use under jatropha plantations (Figure 13.4). The study concluded that in the majority of areas where jatropha can be planted, it is unlikely to use more water than indigenous vegetation types. It was thus unlikely to have a streamflow reduction impact as defined by the South Africa National Water Act of 1998.

In the second experiment in the KZN midlands, the water dynamics and productivity of jatropha were investigated in a silvopastoral experiment with the grass *Pennisetum clandestinum*.[1] Ukulinga receives an average of 680 mm over 106 rain days, with 23 percent of the mean annual precipitation (MAP) falling during the winter months (Camp, 1999). It is situated at an altitude of 721 m and experiences warm to hot

---

[1] The experiment was a randomized block design with three replicates of six treatments (*Pennisetum* only, control; jatropha only, standard spacing (3 × 3 m) with *Pennisetum*; single-row sets, double-row sets, and triple-row sets of jatropha with alleys of *Pennisetum*).

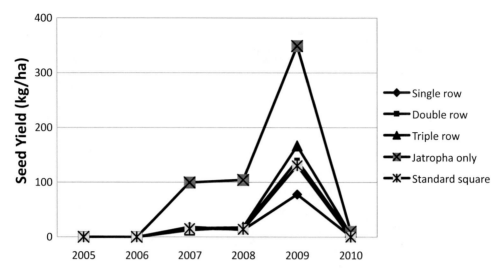

Figure 13.5. Seed yield (dehusked) from various *Jatropha curcas* experimental plots at Ukulinga research farm.

summers and mild winters with occasional frost (Camp, 1999). The mean annual temperature is 18.4°C (Camp, 1999). Although the cooler winter temperatures at Ukulinga may not be ideal for the growth of jatropha, the rainfall is considered sufficient.

Measurements of daily total evaporation rates during December to February (summer) on clear, hot days ranged between 3 and 4 mm/d. However, owing to the deciduous nature of the species, water use was negligible (<1 mm/d) during winter (May to August). The results confirmed those of the preceding study, namely, that two- to four-year-old jatropha trees were conservative water users. This study simultaneously provided some of the first recorded seed yield data for *Jatropha curcas* plantations in South Africa.

4.2. Regional experience with yield

The mean expected yield that some jatropha biofuel projects have anticipated is 2.64 t oil/ha or approximately 8 t seed/ha based on data from 12 jatropha projects in Mozambique (Schut et al., 2010). However, yields from the nonirrigated jatropha trial in KZN (described earlier) ranged between 1 and 2 orders of magnitude below that. The obtained yields were considerably lower than the yields commonly cited in jatropha literature and used in many plantation business plans (e.g., Heller, 1996; Henning, 2006). This finding is understandable given that high oil yields in jatropha plantations grown under dryland conditions are unlikely due to the low seed production linked to correlations between production and water availability. The best achieved seed yield at the Ukulinga trial was in 2009 (348.8 kg/ha) in the jatropha-only plots (Figure 13.5). The other treatments (where pasture competition was a factor) ranged between 77.8 and 166 kg/ha. Following heavy pruning in 2009, the 2010 yields dropped substantially (Figure 13.5).

Published yield data from the rest of southern Africa are scarce, as are scientifically rigorous trials. Companies tend to be sensitive around yield data, but no evidence has been found of high yields in the region, while there is extensive evidence of projects with low yields or even total project collapse (Haywood et al., 2008; Loos, 2009; Ribeiro and Matavel, 2010; Wahl et al., 2009). Current indications imply that yields are lower than expected, though it is likely that actual yields vary substantially between projects. This may be due to several reasons: plantations are still relatively young, most commercial plantations planted seeds collected from unimproved parent material, no rigorous species site matching has taken place, and management practices such as fertilizer use and pruning regimes are still being developed. D1 Oils, which had an extensive research program in Swaziland, have disinvested from Swaziland and only maintain a few plantations in southern Africa. D1, however, reports that it is making significant progress in breeding high-yielding jatropha varieties predicted to produce 2 t oil/ha (Volckaert, 2009). The accumulating local and global evidence is that commercially viable yields will require high rainfall; good soils; fertilization; extensive management, including weeding and pruning; and good genetic material. It is also becoming clear that jatropha plantations take a number of years to mature and that, contrary to early reports, economic yields are not available until one or two years after planting. At the Ukulinga trial, jatropha also showed a low tolerance to pests (specifically the golden flea beetle) and was prone to diseases. Anecdotal reports suggest that this pest problem is widespread. The prevention and treatment of these threats result in additional costs that need to be considered when evaluating viability.

Combining data on water use (irrigation plus rainfall) and energy produced from jatropha, some estimates of water use efficiency for jatropha have been calculated (see Chapter 4). Gerbens-Leenes et al. (2009a) calculated jatropha to have the highest water footprint of 12 biofuel crops considered at 396 m^3/GJ electricity. These calculations have been disputed by Maes et al. (2009) and Jongschaap et al. (2009). Jongschaap et al. (2009) suggests that for yields of 1.3 t/ha and a rainfall of 650 mm/yr, the water demand would be 8,281 L of water/L of oil (or 128 m^3 of water/GJ). However, since southern African yields are still speculative, it is difficult to reach firm conclusions about production efficiencies.

4.3. Jatropha water use compared to fossil fuel

The importance of CTL technology in providing fuel for South Africa has already been mentioned. From a fuel production perspective, it is therefore relevant to compare the water use impacts of such technologies against jatropha production. Standard production of fossil fuel from imported crude oil is assumed to have an insignificant water use impact in terms of quantities used, but this is not the case for CTL technology.

There are no published firm estimates of the water demand for CTL technology as it depends on many factors such as technology used and the quality of the coal. However, water demand is considered significant. For example, Sasol's total water use in 2010 was estimated to be approximately 151.3 million m^3, or about 4 percent of the storage capacity of the Vaal Dam, the main water storage facility for the industrial hub of Gauteng in South Africa. Of this, approximately 91 million m^3 of water were used in their Synfuels processes (Sasol, 2006).

While hydrological studies tend to focus on water quantity aspects, it is also pertinent to consider the water quality (pollution) aspects of jatropha production, particularly in comparison with the production of fossil fuels. Though there are minimal data regarding the impacts of jatropha production on water quality, relatively low impacts are expected due to the crop being perennial. As a result, jatropha provides good ground cover, which should result in reduced erosion compared to annual crops. In addition, low levels of fertilization and associated nutrient leaching are expected. Well-managed jatropha processing using good management practices is also not expected to have major water pollution impacts. By comparison, coal mining and processing (in this case, to produce fuel through CTL technology) have a large negative impact on water quality owing to acid mine drainage (AMD). AMD is highly acidic water, usually containing high concentrations of metals, sulfides, and salts. The major sources of AMD include drainage from underground mine shafts, runoff and discharge from open pits and mine waste dumps, tailings, and ore stockpiles, which make up nearly 88 percent of all waste produced in South Africa (Manders et al., 2009). From these observations, it may be concluded that while CTL technology uses relatively high volumes of water, jatropha has a higher impact on the quantity of water than fossil fuel production but a lesser impact on the quality of water and would similarly require considerably greater land areas for production.

5. Social issues

5.1. Types of projects

A literature review and direct visits to numerous projects in South Africa, Mozambique, Malawi, and Zambia were used to investigate the management models for jatropha production within the subregion (Haywood et al., 2008; von Maltitz and Setzkorn, in press). Within southern Africa, two distinctly different management models have emerged. In one model, feedstock comes from local small-scale farmers (1–10 ha) who are growing jatropha as a cash crop or for household use within a mixed farming enterprise. The other model is based on large-scale corporations that are owned mostly by foreign investors and funded through direct foreign investment. These projects tend to be thousands of hectares in extent. In some cases, outgrowers are linked to large plantations. It is also evident that there are two key motivations for growing

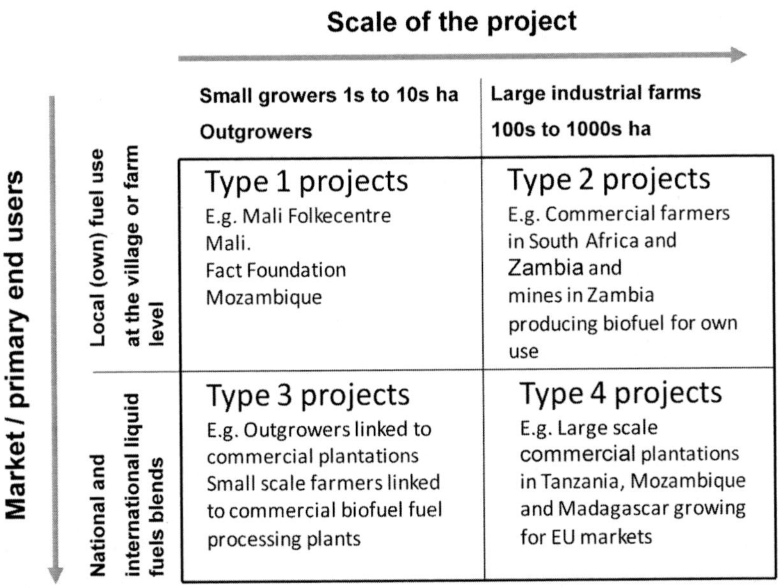

Scale of the project

	Small growers 1s to 10s ha Outgrowers	Large industrial farms 100s to 1000s ha
Local (own) fuel use at the village or farm level	**Type 1 projects** E.g. Mali Folkecentre Mali. Fact Foundation Mozambique	**Type 2 projects** E.g. Commercial farmers in South Africa and Zambia and mines in Zambia producing biofuel for own use
National and international liquid fuels blends	**Type 3 projects** E.g. Outgrowers linked to commercial plantations Small scale farmers linked to commercial biofuel fuel processing plants	**Type 4 projects** E.g. Large scale commercial plantations in Tanzania, Mozambique and Madagascar growing for EU markets

Market / primary end users (vertical axis label)

Figure 13.6. A typology of biofuel projects based on size of production unit and intended use of the fuel. Based on Haywood et al. (2008).

jatropha. The jatropha is either grown for the provision of local energy, or it is intended for sale into national or international transportation fuel markets (Figure 13.6).

Small-scale biofuel production is promoted both by development nongovernmental organizations and as commercial ventures. In the case of Marli Investment's plantations in Kabwe, Zambia, long-term (30-year) contracts are established with the farmers, who are contracted to grow 5 ha of biofuel on their 10 ha landholdings. The contract requires Marli to provide initial inputs and bridging finance until the jatropha starts seeding. In return, the farmers take care of the crop and harvest the seeds, which he or she is contractually obliged to sell to Marli. In practice, most farmers have only planted a small area of the farm to jatropha, with 1.6 ha being the mean jatropha field size in Kabwe (Haywood et al., 2008). In the case of Oval, Zambia, the company both supports the establishment of smallholder growers and establishes its own large plantations. Both Marli and Oval plan to sell jatropha oil into the national and international transport fuel markets. Prokon Tanzania reportedly has 12,400 ha under cultivation by 16,900 contracted farmers (Loos, 2009). By contrast, the FACT Foundation helps small-scale growers to cultivate jatropha in hedgerows to produce pure plant oil, which is intended for local energy provision (in diesel generators or as a fuel for lights and stoves) or to be used as a material in soapmaking. Though there are reports that fuels are being processed, to date volumes are small and yields are low. The farmers have a guaranteed market within the FACT projects as they sell their seeds to the project. For example, in Mozambique, the project pays 5 meticais/kg

(USD 0.7cents/kg) (personal communication). Some farmers in Zambia, Tanzania, and Mozambique (excluding those linked to the FACT Foundation) have reported that it is difficult to find markets for seeds, even when linked by contract to the local industries. It would appear that prices generally achieved for jatropha seeds tend to be below farmer expectations (German et al., 2010; Haywood et al., 2008; Schut et al., 2010).

Numerous large-scale plantations have been initiated in the region, such as Ecomoz and ESV in Mozambique; SUN biofuels in Mozambique and Tanzania; and D1 Oils in Zambia, Mozambique, and Tanzania (Schut et al., 2010; von Maltitz and Setzkorn, 2012). TerraRossa Plantations, Madagascar (previously called GEM Biofuels), represents the largest single large-scale plantation, with 57,700 ha planted out of a 490,000 ha land acquisition (GEM, 2009). Common to all of these projects is the fact that large tracts of land are acquired for the exclusive use of the biofuel venture. This means that previous formal or informal land use is largely or totally displaced. Though some projects are funded by private companies, most are funded through listed companies. The companies employ extensive labor for the planting and tend to pay slightly above the going agricultural minimum wages (though, in real terms, labor wages are low). Since all of the plantations are still in the establishment phase, it is too early to get reliable estimates on the labor opportunities that will be sustainably created over the long term. Company estimates in Mozambique range from 0.14 to 0.17 jobs per ha (Schut et al., 2010). The first jatropha oil is starting to come from plantations, with TerraRossa Plantations in 2010 claiming to be the first plantation to export jatropha oil. The quantities are small, and it would appear that it has taken at least some of the plantations longer than was initially envisaged to reach this point of initial oil production. The total area that governments have allocated to large-scale plantations is only a fraction of the area that was requested by companies. In addition, most companies have only planted a small proportion of the allocated land to date.

5.2. The economics of jatropha growing

When considering the economics of jatropha growing, it is useful to consider both the national and project-level perspectives. Obviously, the national perspective assumes that individual projects are financially viable, and for a new industry, this is not guaranteed. If the national economic perspective shows clear advantages for the country, then there might be a justification for the government to assist in enhancing viability at the project level.

An economy-wide computable general equilibrium model of Mozambique suggested that biofuels could have a significant impact on Mozambique's overall economy, contributing 0.37 percent to overall gross domestic product (GDP) from 550,000 ha of plantation, while generating 271,000 rural jobs (Arndt et al., 2009; EcoEnergy, 2008). Jatropha, when compared to sugarcane ethanol, had a far higher pro-poor

---

Box 13.1. ESV-Panda, Mozambique – A case study

ESV is a jatropha plantation located next to the village of Inhassune, in the district of Panda. The company was allocated 6,000 ha, but only 1,000 ha were planted when the plantation was visited for the first time in September 2008. At this visit, the plantation was thriving, the nursery was bustling, and the area had an atmosphere of activity and growth, with 1,200 of the 3,000 local villagers employed in the project. The management of ESV had supported the development of a literacy school, a soccer team, and a cultural group and was paying above the minimum wage. The community had a cell phone shop, a tailor, and electricity until 9:00 P.M. The local people were seen using the company's tractors for tilling their own fields. A year later, in August 2009, a return visit found that the plantation manager had left almost a year before, with instructions to his staff to continue looking after the plants. No salaries were paid for the 11 intervening months, but the staff was still trying to do their best under the encouragement of the district government. However, the plants were diseased (yellow beetle), they had not been pruned, and the weeds were growing prolifically around the plants, preventing good seed production. It was interesting to note that despite the loss of management support, the literacy project, the soccer team, and the cultural group were still going. Shortly after the second visit, another management team took over the farm, and reports of good recovery were received. Local people were interviewed during the second visit. The local chief, who has a thriving business growing vegetables for local sale, was not impressed with jatropha. The local people were still able to hire the company tractor, but without any wages, there was no money for the fuel. A new company was about to start working the fields but had not yet arrived. ESV announced the sale of its company for USD 4 million in late 2010 to SAB Mozambique SA, a company controlled by two Italian energy companies (Ribeiro and Matavel, 2010).

---

impact. The model assumed a 3t/ha jatropha yield, or 630 L/ha, which is relatively conservative but probably more realistic than company estimates. The model, however, assumes 3 ha per laborer on small-scale plantations, compared to the industry's estimate of 5.8–7.4 ha per person on large-scale plantations (Schut et al., 2010), so it may have inflated the overall labor benefits.

The Bioenergy and Food Security (BEFS) study conducted by the Food and Agriculture Organization developed a national-level model to investigate biofuel impacts for Tanzania. A key finding from the modeling data was that jatropha had the lowest cost of production of biodiesel if produced through outgrower schemes, which the report suggested as a more viable option than estate farming. However, the report also indicated that the risks of growing jatropha were too high at this point because of the uncertainties in jatropha production, which could not be accurately modeled (Maltsoglou and Khwaja, 2010).

The fact that many large corporations have engaged in jatropha growing clearly indicates that the private sector considers jatropha to be a financially viable venture.

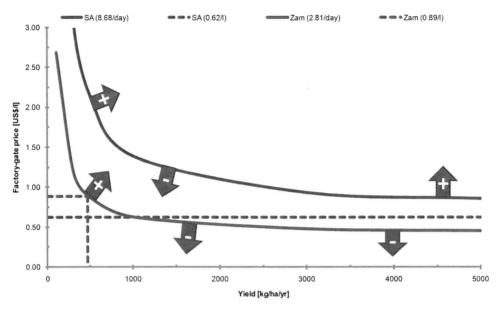

Figure 13.7. Break-even plot of complementary factory-gate prices for different yields that support the legal minimum wage payment in South Africa and Zambia. Areas above the curves indicate net profits, and areas below represent net losses. Values in parentheses for the respective countries represent minimum legal wage and basic fuel price in USD/d and USD/L, respectively. (See color plate.)

Underpinning this assumption are assumptions of yields and the costs of cultivation. As indicated earlier, original yield estimates may be overly optimistic; however, at this stage, no large-scale or small-scale schemes have reached full maturity, so a modeling approach is necessary in an attempt to cope with the uncertainties concerning profitability. The industry might also have underestimated labor requirements in these projects.

Data on the time taken to harvest and dehusk the seeds at the Ukalinga trial showed that 1 kg of seed took approximately three hours to pick and dehusk, suggesting that mechanical harvesting may be necessary to make seed production economically viable (Everson et al., in press). It is, however, clear that picking rates are influenced by yield and dropoff significantly when yields are low (Borman et al., in press). In an attempt to understand the interplay between labor needs during production and harvesting, seed yields per hectare, and fuel prices, a spreadsheet-based model was developed to ascertain the conditions under which profitability could be obtained (Borman et al., in press). Model results suggest that jatropha growing is most likely profitable where labor rates are low and fuel costs are high, as in Zambia. Under this set of conditions, jatropha may be profitable at yields of less than 1 t/ha. This result, however, excludes the establishment and overhead costs. In South Africa, under current fuel costs, the model predicts that jatropha will be unprofitable at any realistic yield (see Figure 13.7).

Grass competition was shown to significantly impact both growth and yields, indicating that weeding is required to guarantee good yields (Everson et al., in press). ESV in 2008 indicated that three weedings a year were required, particularly in the first two years (personal communication). In addition, there is known to be a strong link between yield and fertilizer, particularly in the early years, indicating an additional cost. Use of the seedcake as a fertilizer also enhances yields but reduces the profit that the processing company may obtain by selling the seedcake as a by-product (Borman et al., in press). Seedcake has good fertilizer properties (Openshaw, 2000), and D1 is also experimenting with methods to denature the toxins in seedcake so that it can be used as a chicken or cattle protein supplement (Volckaert, 2009). The FACT Foundation is using seedcake for biogas generation.

5.3. On-the-ground perceptions

In contrast to the concerns about environmental issues and climate change, biofuels have been promoted within southern Africa as an opportunity for the rural poor. However, perceptions of biofuels and, specifically, of jatropha are mixed. While the commercial sector and many small farmers see biofuels as an opportunity, civil society is less homogenous in its response (e.g., Ribeiro and Matavel, 2010). The Via Campesina is a global movement of peasant farmers working to promote food sovereignty and local self-determination and, in the tenth conference of the parties of the Convention on Biological Diversity Conference, they held a press release calling once more for an end to the expansion of what they term *agrofuels*. The União Nacional de Camponeses Mozambique (UNAC) is a key member of Via Campensina, but thus far, it has been reluctant to adopt a positive or negative stance on biofuels. Ribeiro and Matavel (2010: 6) state, "Friends of the Earth International (FoEI) believes that the dominant arguments used to promote jatropha – as a food security-safe biofuel crop, a source of additional farm income for rural farmers, and a potential driver of rural development – are misinformed at best and dangerous at worst." The report bases its concerns on the fact that jatropha is not as pest resistant as initially suggested, is not being grown on marginal land, is competing with food crop production as a result, and utilizes large amounts of water. The report recommends that further development of jatropha be halted in Mozambique, at least until some of the major development issues surrounding subsistence farming are addressed and rural communities obtain food sovereignty (Ribeiro and Matavel, 2010).

Our own unpublished research suggests that local concerns are largely about land rights and employment. In some community meetings held in Mozambique, the community was vociferously in favor of the proposed biofuel developments, and their main critique was that the projects were taking too long to develop. Most respondents, when questioned, indicated that they would prefer to get jobs with the biofuel company rather than grow biofuel themselves, indicating that this would mean that they can be assured of a monthly income and thus not have to deal with the the two to three months of food insecurity that accompanied their subsistence farming. However, on jatropha plantations, the employed labor and local people seemed less sure about growing jatropha, indicating that they thought it was not a good crop as they had not seen much seed production and they felt it was too easily attacked by pests. Local laborers and farmers hopeful to commercialize were far more interested when there was the possibility of entering into small-scale production of sugar, for instance. The research was carried out at a time when, generally, prospects were bleak in the jatropha biofuel sector as one of the jatropha projects visited was not operating (ESV) and another (ENERGEM) was about to close down.

5.4. Concerns over tenure and food security

Biofuel expansion might result in the loss of land rights by indigenous people in Africa (Cotula et al., 2008; Sulle and Nelson, 2009). Vermeulen and Cotula (2010: 900) suggest that "the internationally recognized right to food arguably requires that, at a minimum, land takings in contexts where people depend on land for their food security must be offset by alternative livelihood assets so as to ensure at least the same level of food security. Furthermore, land in Africa, as elsewhere, has important spiritual and social values, so that purely economic calculations are unlikely to capture local perceptions about proposed land deals. Global normative standards for consultation, consent, and recompense are framed by the principle of free, prior, and informed consent (FPIC)." Concerns about how biofuel expansion might result in the loss of land and how it might impact rural poor people in Africa are echoed in many reports by civil society and academic think tanks (Oxfam, 2008; Ribeiro and Matavel, 2010).

Jatropha supporters claimed initially that jatropha can be grown on marginal land in areas with medium rainfall. However, as noted elsewhere in this chapter, this is not the case. Richard Morgan, CEO of Sunbiofuels, at a conference in Maputo, put it succinctly when he said, "If you have good arable land and your rainfall is unpredictable, then grow jatropha; if the rainfall is good, rather grow a higher value crop." Jatropha, if grown on arable land, will compete with other crops, including food crops. However, it is possible to intercrop with jatropha, depending on how the spacing of the shrubs is done. The FAO's BEFS project (Maltsoglou and Khwaja, 2010), based on modeling of national data, suggests that food production would increase slightly under most biofuel investment scenarios. Additionally, it is more likely that the trade-off between biofuels and other crops would adversely affect other cash crops that are traded internationally. In the final conclusion, it was noted that "while all biofuel production scenarios improve household welfare, it is the small-scale outgrower schemes, especially for typical smallholder crops such as cassava and jatropha which are most effective at raising poorer households incomes" (Maltsoglou and Khwaja, 2010: 6). Schut et al. (2010), conversely, regard jatropha as a low-value crop, which presents a challenge in terms of purchasing inputs required for cultivating the crop.

Three factors underpin uncertainty around household benefits from jatropha: (1) uncertain yields, (2) the labor requirements needed to cultivate and harvest jatropha, and (3) the farm-gate market price for jatropha. In India, where there is an active market, jatropha seeds sell for USD 0.16/kg (Borman et al., in press). Using this rate, a yield of 1 t/ha will provide USD 160/ha/yr, while a yield of 5 t/ha will provide USD 800/ha/yr. Though seemingly low, this income is significant in a situation where other income-generating activities are scarce and limited markets exist for the sale of agricultural surplus. Considering that rural workers in Mozambique will earn about USD 50 per month, and that few formal jobs are available, this also is a seemingly reasonable return. However, currently, it would seem that farmers are not even achieving 1 t/ha yields, and in some instances, no markets exist (personal observation; Haywood et al., 2008; Loos, 2009). As pointed out earlier, the labor to weed, prune, pick, and dehusk is considerable, involving possibly as much as 3,000 hr/t (Everson et al., in press). So while theoretically, jatropha might be considered a good outgrower crop, it is possible that the benefits of being a farmworker, and the security given by a wage, may outweigh the benefits of a potentially better income but from a risky crop within a risky market (Ribeiro and Matavel, 2010).

Working on plantations does not need to negatively impact the ability of a farmer to grow subsistence crops to support his or her livelihood. For instance, in one project operated by Energem in Bilene, Mozambique, the plantation owner allowed his female workers to leave work at noon so that they still had daylight hours to work on their own land. At ESV (described earlier), the plantation manager provided the workers with company tractors at low rates to till their land, which meant that many hours of hard labor were avoided. If plantations facilitate their workers and local communities with their subsistence food growing, it is arguable that the benefits of a waged job can outweigh those of being a contract farmer in an outgrower scheme. Loos (2009) found very little evidence that small-scale jatropha farmers had less agricultural crop production, but Ribeiro and Matavel (2010) suggest that work on large-scale farms was suppressing smallholder production.

Globally, outgrower schemes are promoted as being the best model to uplift poor communities. Undoubtedly, the success with some high-value cash crops, such as sugar, coffee, and tea, in some parts of the world supports this view. When we are dealing with jatropha, however, these benefits may not extrapolate to the same extent. Jatropha is presently a low-value crop. When workers on jatropha plantations are receiving wages, however low, they appear to be satisfied. Criticisms of jatropha are justified, the crop is still an unknown, and accurate figures of expected yields seem to be slow in coming. Given the uncertainties, it is not wise to expect the rural poor to take the risk along with commercial operators by planting jatropha, considering that if the yield is low, then the payback will also be low.

5.5. *The move to certification*

Biofuel certification and policy development is starting in southern Africa. The Southern African Development Corporation (SADC) initiated a biofuel task force in 2008 with assistance from GTZ/ProBEC. It recently released its sustainability criteria for biofuels, and all member states are being encouraged to develop country-specific criteria (SADC, 2010). The SADC task force is also encouraging its member states to develop biofuel policies, and the SADC secretariat, in partnership with GTZ/ProBEC, recently released its biofuel policy support tool (Sugrue and Andrew, 2010). Most SADC member states have, to varying extents, policies or strategies that reflect their continued efforts to work on the issues. This initiative closely follows the work done in Mozambique using the Cramer Commission Criteria for sustainable biofuels. These criteria, which were developed by the Dutch government together with various stakeholders, constituted one of the pioneering standards developed covering social and environmental issues. The Roundtable on Sustainable Biofuels (RSB) released its Version 1 and Version 2 certification standards, covering all biofuel feedstocks for liquid biofuels. A more recent initiative, the Jatropha Alliance, is working steadily to develop standards for jatropha. The alliance considered many different certification schemes but decided on using the RSB and the British-based Renewable Transport Fuel Obligation (RTFO) as the basis for its own jatropha certification scheme (Froger et al., undated). The project "Towards Sustainability Certification of Jatropha Biofuels in Mozambique" is cofunded by the Dutch development organization NL Agency within the Global Sustainable Biomass Fund, and it has completed pilot studies in Mozambique (Mattias and Visser, 2010).

Not all sectors support the development of standards, with some civil society organizations suggesting that standards will allow for the further expansion of agrofuels (ATC, 2010). However, voluntary standards and codes of conduct are limited to assessing the impacts and activities of individual projects (this is their greatest weakness). While each individual project may not have a large impact, collectively in a region, projects may have a substantial impact. It is essential that regulatory mechanisms be strengthened in southern Africa so that impacts that can be felt far from the project can be managed effectively.

5.6. *Socioeconomic impacts of jatropha compared to fossil fuels*

The true impacts of jatropha growing are still emerging, so detailed comparison with fossil fuels is difficult. Some overarching differences are likely, and these are based on the assumption that an economically viable jatropha-based biofuel industry can be established. If that happens, then jatropha production might have both positive and negative impacts when compared to conventional fossil fuels:

- Biofuel is estimated to create at least 2 orders of magnitude more local jobs per unit of fuel than fossil fuels (see NBTT, 2006). Jatropha, with its high labor requirements and possible suitability for small-scale farming, could be expected to create even more job opportunities.
- Biofuel production could have a substantive impact on the national economy through impacts on the trade balances. Fossil fuel imports are the single biggest foreign exchange expenditure in most southern African countries, including South Africa (excluding Angola). A biofuel industry would also have a noticeable impact on GDP and especially on the contribution of agriculture to the GDP (Arndt et al., 2009; NBTT, 2006).
- Jatropha production creates pro-poor development as most jobs are located in underdeveloped rural areas. Income per job opportunity is, however, less when compared to jobs in the fossil fuel sector, which is predominantly urban based.
- There is a potential for food–fuel conflict from jatropha production, both at the level of the farmer and at the national level. The interactions between fuel and food are likely to be complex, and it is labor rather than land that might be the key limiting factor (see Chapter 10).
- Biofuels, especially when grown on a large scale, may displace current land users and hence lead to increased poverty in some sectors.

6. Conclusions

Our study suggests that jatropha exhibits mixed results in how it fares over a range of criteria and how it has fared in actual project implementations to date. An early concern regarding jatropha introduction was that it would have a negative impact on catchment hydrology. In this regard, jatropha seems to perform well, and for South Africa, the evidence would suggest that jatropha plantations would have very similar impacts to those of the natural vegetation. Jatropha does use substantially more water than fossil fuels, but the total environmental cost of either option needs to be considered. Though jatropha is drought hardy, this does not mean that the plant can yield sustainable high levels of fruit in low-rainfall areas. Irrigation may also be necessary, and this water impact needs to be considered.

The carbon balance for jatropha would seem relatively favorable when compared to other biofuels, though this depends on cropping practices and yields. If woodlands are cleared to plant jatropha, then the carbon payback period can be quite long, especially if high yields are not achieved. In practice, grasslands, abandoned agricultural land, or degraded woodlands tend to be favored over mature woodlands when plantations are established as some of the standing carbon stock is already lost, hence reducing the payback time. Despite suggestions that jatropha will be grown predominantly on degraded land, local experience is that yields are likely to benefit from good, arable land, and this is where the majority of plantations are located. This could then lead to competition with food crops and indirect land use change.

Jatropha growing will have a biodiversity impact unless the plant is restricted to already degraded areas. Jatropha grown on degraded areas may have slight beneficial impacts on biodiversity, helping the stabilization of soils and reducing erosion. If jatropha is grown on existing agricultural land, then there is a very real threat of causing indirect land use change. Larger biodiversity impacts are expected if jatropha is established in virgin land (forest or grassland). Current experience with jatropha indicates that it is more often than not grown on good land rather than on degraded land.

Although to date, deforestation from jatropha appears relatively rare in the region, if the rate of jatropha expansion remains the same, then direct or indirect deforestation from jatropha expansion will become a concern. Current evidence is that jatropha is not suitable for arid areas but requires about 800–1,500 mm of rainfall to produce meaningful amounts of fruit. Therefore it will grow best in woodland (miombo) or forested areas, and thus deforestation is likely if the location for planting does not have regulations preventing deforestation. This concern, however, depends on jatropha proving itself to be a financially viable crop. Current data suggest that this is unlikely, unless improved varieties are developed.

In theory, jatropha appears to be a perfect smallholder crop as it is easy to grow; there is a large potential market for the fruit, which is relatively easily transported; and if stored properly, the fruit does not perish rapidly. In practice, smallholders are currently gaining very limited benefits and in fact may be experiencing high opportunity costs from growing jatropha. This is due primarily to low yields and the intensive labor requirement. Lack of markets is also a constraint, considering that jatropha is not a high-value crop. This is because its value is limited by the price of fossil diesel, which is currently still relatively cheap compared to the costs of growing biofuels. Current smallholder projects are quite new, but it is expected that benefits will improve over time as markets develop. However, unless the yields improve, farmers will gain limited income. The high labor requirement is a key concern as it will reduce profitability and potentially take labor away from food production, without gaining sufficient income to justify the shift in labor allocation. Individual farmers will determine if the opportunity costs of jatropha are worthwhile and will move in and out of production as it suits them, provided they are not locked into unfavorable long-term contracts.

Jatropha, when grown on large-scale plantations, can provide numerous job opportunities, and these are welcomed by local community members even if they are paid relatively low wages. However, there is concern over the potential for the displacement of local community members from their land as a consequence of large plantations. There are also concerns over poor or limited compensation for losses of the use of land. The financial viability of large-scale plantations is threatened if yields remain low. A number of large projects have experienced financial constraints, which have led to their closure or sale. Some of this might be due to financial constraints resulting from the financial crisis of 2008, but lower than expected yields and longer time than expected between planting and yields also seem partly to blame.

Jatropha is functionally still a wild crop. Many plantations have been established with unproven genetic material and no reliable data on yields. Yields from these plantations are likely to fall short of original expectations. It seems that there is varied success with current projects, but some are clearly not performing as well as expected. As the plantations mature, yields are expected to improve, though it is still too early to know what the final yields will be. Low yield could very well jeopardize the financial viability of the projects. The current variability in yields from individual trees and different varieties lends hope to the fact that high-yielding varieties with proven and reliable yields may be available in the future.

Jatropha cultivation as a smallholder crop for biofuel production is as yet an unproven economic opportunity. Reports have been received of low yields and markets that either pay less than the farmer expected or that do not exist at all. Labor requirements are also higher than anticipated (German et al., 2010; Haywood et al., 2008). If biofuels fail as an energy crop, then there will be a number of socioeconomic consequences, and smallholder farmers who have been convinced to invest in jatropha will be particularly vulnerable in this regard as they will have borne the opportunity cost of investing in jatropha, without reward.

It is clear that a large amount of research is still required before a viable jatropha industry can be established. In addition to the need for improved high-yielding varieties, there is still extensive research needed on management practices, including spacing, pruning, fertilizing, pest control, and other agronomic issues. Further research is needed on many socioeconomic aspects, including advantages and disadvantages of small-scale versus plantation models of production. Mechanization is potentially an option in the future to reduce labor costs, but if this approach is taken, it might mitigate against one of the key reasons for choosing jatropha in the first place – its high employment potential.

Given the high degree of uncertainty surrounding jatropha, great caution is recommended when contemplating the establishment of new jatropha projects for biofuel production. New projects should be considered very carefully, unless proven jatropha varieties with proven yields for the area being considered are used. Even then, consideration must be given to ensuring that there is overall financial viability. Planning a jatropha project must take into consideration the local socioeconomic and environmental situation, and site-specific impacts need to be considered before project approval. Jatropha remains a high-risk crop with currently low proven returns for the southern African region.

Part Five

Synthesis

14

Biofuels in developing countries: A synthesis

ALEXANDROS GASPARATOS
Oxford University, Oxford, UK

PER STROMBERG
Institute of Advanced Studies, United Nations University, Yokohama, Japan[1]

1. Main observations

This edited volume provided a systematic study of the drivers, impacts, and trade-offs of different biofuel production (and use) practices in different developing nations. The evidence provided in the 13 chapters of this volume suggests that several biofuel practices exist, each having radically different impacts on the environment and on human well-being. In fact, the term *biofuels* includes an array of different practices that are carried out in different ecosystems, for different purposes, and which can complement or compete with other human activities and natural processes. For example, the environmental and socioeconomic impacts as well as the drivers of large-scale sugarcane bioethanol production in Brazil (Chapters 6–8) are certainly different from those of small-scale jatropha biodiesel production in rural sub-Saharan Africa (Chapters 1, 11, and 13). What is more, the drivers, impacts, and potential of the *same* biofuel production practice might vary significantly across different world regions, as is the case for sugarcane bioethanol (Chapters 6, 7, and 12) and jatropha biodiesel (Chapters 10 and 13).

With this in mind, it would be reckless to group biofuels under the same banner when discussing their impacts and potential as it is very important to understand the environmental and socioeconomic contexts within which current biofuel expansion is taking place. This can contribute to a better understanding of the trade-offs and factors shaping the acceptability of biofuels by stakeholders. However, despite this context specificity of drivers and impacts, some general trends have been observed in this edited volume, and are summarized below.

2. Drivers

Despite certain context-specific drivers biofuel expansion in developing nations is, in most cases, driven by energy security and rural development concerns. Environmental and climate change mitigation concerns play a negligible role, if any.

---

[1] Per Stromberg also works for the Swedish Environmental Protection Agency, Stockholm, Sweden.

2.1. Energy security

Energy security is the key, overarching motivation behind biofuel expansion for most developed and developing countries (Chapter 1). Energy security is identified as a driver of biofuel expansion in Brazil (Chapter 6), Indonesia (Chapter 9), Thailand (Chapter 9), the Philippines (Chapter 9), China (Chapter 10), and several sub-Saharan African nations (Chapters 12 and 13).

2.2. Rural development

Rural development and its associated ripple effects on poverty alleviation are a significant motivation for biofuel expansion, particularly for some African countries (Chapters 1 and 11–13). Rural development concerns have also driven biofuel policies in countries such as Indonesia (Chapter 9), Thailand (Chapter 9), and China (Chapter 10). Finally, rural development and the social inclusion of poor farmers has been one of the key, officially proclaimed objectives of the Brazilian biodiesel program (Chapter 6).

2.3. Climate change mitigation

In general, climate change mitigation does not seem to be a significant driver for biofuel expansion in developing countries (Chapter 1). This is most likely due to the fact that these countries are Annex A countries that, according to the Kyoto Protocol, have no binding commitments to reduce their greenhouse gas (GHG) emissions. Malaysia is the only developing country identified in this volume for which climate mitigation concerns have been expressed as one of the drivers of its biofuel program (Chapter 9). It is interesting to note that developing countries that are (or aspire to become) major biofuel exporters, such as Brazil, emphasize the significant GHG savings of their biofuels when lobbying other governments, for example, the European Union (Chapter 6). In this sense, climate change mitigation can be considered an emerging indirect driver for biofuel expansion in export-oriented developing nations.

2.4. Other drivers

Other drivers of biofuel expansion are country specific and are in several instances related to economic development. These economic concerns include foreign exchange savings (Chapters 1, 6, and 10–12), export potential (Chapters 6, 8, and 12), or maintaining the competitiveness of ailing or mature agroindustries (Chapters 1, 6, and 7). The Brazilian case is of particular interest. An emerging feature of sugarcane bioethanol expansion in the country is the Brazilian government's strategic effort to make ethanol an international commodity that would benefit the nation's ethanol exports in large economies that currently enforce tariff barriers against ethanol imports such as the economy of the United States (Chapter 6). This phenomenon might drive sugarcane bioethanol expansion in other developing countries, particularly those located in southern Africa (Chapter 12; see also next section).

3. Production practices and policy instruments

A large variety of different biofuel production practices are being pursued throughout the world. Industrial biofuel production in developing countries is currently limited to first-generation feedstocks (i.e., food crops and jatropha). Even though there are plans for producing second-generation biofuels, most of the biofuel expansion in developing countries in the near to medium term, will come from first-generation feedstocks. In fact, with the exception of Brazil, China, and India, the discussion on second-generation biofuels is almost nonexistent in developing countries (Chapters 1 and 6).

The most widely pursued first-generation biofuel practices in developing countries include bioethanol from sugarcane and sugarcane molasses (Chapters 6–8 and 12), soybean biodiesel (Chapter 6), palm oil biodiesel (Chapter 9), and jatropha biodiesel (Chapters 10 and 13). The adoption of other, less common biofuel types depends on the country's specific climatic and agroecological conditions and includes cassava bioethanol (Chapter 9), coconut oil biodiesel (Chapter 9), and biodiesel from different types of oil crops (Chapters 1 and 6).

Among the main policy instruments used to promote biofuel expansion are blending mandates (Chapter 1). These mandates essentially safeguard national biofuel demand but are not in themselves enough to ensure a biofuel program's viability, which usually requires additional policy instruments. For example, consistent and coordinated institutional support has been identified as one of the main reasons behind the success of the Brazilian ethanol program (Chapter 6). Other countries have put massive subsidies in place to make their biofuels economically competitive substitutes to fossil fuels, aiming not only to enhance energy security but also to advance other policy objectives such as rural development. However, it is not obvious to what extent subsidies contribute to meeting such objectives, particularly compared to other renewable fuels and ways of promoting rural development (Chapter 1).

Markets are, of course, important means for boosting biofuel production. Perhaps the most substantial international market initiative has been the Brazilian government's attempt to turn ethanol into an internationally traded commodity (Chapter 6). In fact, global biofuel demand is expected to increase partly due to this effort, and as a result, Brazil may benefit in the next decade thanks to significant increases in domestic and international bioethanol demand (Chapter 8). It can be argued that Brazil's diffusion of ethanol technology to developing countries serves to pave the way for Brazil's own ethanol exports by reducing potential importers' worries of becoming too dependent on one single ethanol supplier. This "ethanol diplomacy" has been seen particularly in sub-Saharan Africa through technology transfer and other incentives (Chapters 6 and 12). Increased global biofuel trade may also go in tandem with biofuel certification, which could have positive environmental and social effects (Chapters 7 and 9). It should also be mentioned that even though international biofuel trade (particularly trade with developed countries) can be a lucrative activity for developing countries, regional biofuel trade schemes can end up being just as economically significant for developing countries as a way of better matching market development to differing infrastructure and economic conditions (Chapter 12).

Finally, the existence of markets for biofuel coproducts can improve the viability of biofuel programs. For example, while the lack of any formal market for jatropha coproducts may hamper the long-term economic viability of jatropha production in China (Chapter 10), the existence of markets for the coproducts of the sugarcane sector is seen as a potential booster for sugarcane bioethanol production in southern Africa as it essentially reduces business risks for investors (Chapter 12).

4. Impacts

While biofuels' negative impacts have attracted the most attention, there are examples of biofuel practices that provide significant environmental and socioeconomic benefits.

4.1. Energy

Several first-generation biofuel practices are significant net energy providers (high energy return on investment, high percentage fossil improvement) and as such can provide a dependable and renewable source of energy that can improve energy security in the short to medium term at both national and local levels (Chapter 1). In some cases, the utilization of by-products, such as sugarcane bagasse for electricity cogeneration, can further boost final energy provision (Chapters 6 and 12). Conversely, the overreliance on fossil fuel–intensive commodities (e.g., fertilizers and agrochemicals) for feedstock production throws in doubt the long-term energy viability of biofuels as long as current production practices are pursued. Second-generation biofuels can produce even higher net energy gains, but the prospect of their large-scale production in developing nations is still unripe (Chapter 1). Furthermore, the use of fossil fuels for land preparation, transport, and distribution might still be significant for second-generation biofuels.

4.2. Climate

The impacts of first-generation biofuels on climate constitute one of the most contentious aspects of the biofuel debate. Biofuel GHG emissions have been the topic of several recent studies and meta-analyses. An important caveat when interpreting and comparing the results of such studies is that a wide range of methodologies exist. This means that differences in results are, essentially, as likely to be due to the different methodologies being employed as to the different biofuel production pathways being considered (Chapter 3). Most studies suggest that the choice of feedstock crop has a large impact on the anticipated life cycle GHG emissions due to the CO_2 fluxes associated with tillage and the N_2O fluxes associated with nitrogen fertilizer use. Biofuels produced from perennial crops (e.g., switchgrass, miscanthus), which are not tilled annually and typically use nitrogen more efficiently, tend to have lower GHG emissions than those from annual crops (e.g., corn) (Chapter 3). However, most life cycle analyses (LCAs) in the past have disregarded the effect of land use and cover change (LUCC) on overall GHG emissions. If direct and indirect LUCC is considered in biofuel LCAs, then several first-generation practices, particularly those that depend on feedstocks grown on forest or grassland, can incur significant carbon debts (Chapters 3, 6, and 13). For this reason, there is great uncertainty about whether food- and animal feed–based biofuels such as corn grain ethanol and soybean biodiesel can reduce GHG emissions relative to petroleum-derived fuels (Chapter 3). Nevertheless, it should be noted that certain first-generation biofuel practices show very high GHG savings, even when factoring in LUCC. The main example is Brazilian sugarcane ethanol, which consistently shows GHG emission savings of over 50 percent and is therefore considered an advanced biofuel (equivalent to second-generation biofuels) by the U.S. Environmental Protection Agency (Chapter 6).

4.3. Air quality

Biofuel production and use emit several types of atmospheric pollutants throughout their life cycle. The main pollutants emitted include particulate matter (PM), nitrogen oxides (NO_x), sulfur dioxide (SO_2), ammonia (NH_3), volatile organic compounds (VOCs), and ozone (O_3) (Chapter 3). These atmospheric pollutants can be particularly harmful to human health and ecosystems (Chapters 3 and 6). Given the complexity of biofuel chains, the wide variety of biofuel practices, and the numerous ambient air pollutants emitted, no general trends can be discerned regarding the effect of biofuel production and use on ambient air quality. Studies investigating potential air quality impacts of biofuel production and use have shown that displacing conventional fuels will likely lead to reduced emission for some pollutants (not all) and increases for others (Chapter 3). Such varied findings across pollutants and biofuel practices result partly from the differing experimental setups and modeling approaches employed in the different studies and partly from the differing effects of the fuels themselves as they are deployed in actual use in transportation.

4.4. Water

In general, biofuels consume much more water throughout their life cycle; that is, they have a higher water footprint (WF) than other energy carriers. This difference owes primarily to the high water consumption during feedstock production (agricultural phase) (Chapter 4). However, different biofuel production practices can have radically different water requirements, depending on the characteristics of the feedstock used, the region where it is produced, and the structure of the agricultural system itself (e.g., extensive irrigation) (Chapter 4). For example, jatropha exhibits a very high WF in countries such as India (Chapter 4), but in other contexts, it is considered a conservative water user when compared to natural vegetation (Chapter 13). Feedstock production and processing have also been identified as major sources of water pollution due to fertilizer and pesticide use and industrial effluent in, for example, Brazil (Chapter 6) and Southeast Asia (Chapter 9). Biofuel-related water pollution can be a significant human and ecosystem health hazard (Chapter 6).

4.5. Land use and deforestation

Biofuel production can be a major driver of direct and indirect LUCC (Chapters 2, 5, and 8), while it has been suggested that biofuel expansion is already contributing significantly to deforestation in the tropics (Chapters 5 and 9). It is estimated that global biofuel demand in the future will affect regional land use change in Brazil, although the extent of the incorporation of new land is dampened by the ability to intensify agricultural production (Chapter 8). Even though it is relatively straightforward to estimate the *amount* of land that must be converted to meet this added demand, it is very difficult to identify the exact locations where land will be converted for feedstock production and therefore the original use of the converted land. The multifunctional nature of biofuel feedstocks, the lack of a standard definition of deforestation, and the lack of updated data sets with sufficient spatial resolution and global coverage are only some of the reasons why it is extremely difficult to quantify, at a global level, the impact of biofuel expansion on deforestation (Chapter 5).

4.6. Soil erosion

Certain feedstocks, such as palm oil, have been shown to cause significant soil erosion (Chapter 9), which can lead to a loss of soil fertility and a deterioration of terrestrial and aquatic habitats, thus having an adverse effect on biodiversity. Conversely, other feedstocks, such as jatropha, can stabilize soil and reduce soil erosion if grown on degraded lands (Chapter 13).

4.7. Biodiversity

Biofuel expansion is becoming an important emerging threat to biodiversity in different parts of the world. Feedstock production is associated with extended monocultures, habitat loss, pollution, and other frontier-opening activities, which are important drivers of biodiversity loss (Chapter 9). Other biofuel feedstocks, such as jatropha and certain perennial crops, can become invasive (Chapter 13). The production of palm oil, which is becoming an important biodiesel feedstock, has already been associated with significant biodiversity loss in Southeast Asia (Chapter 9). For other feedstocks, the evidence is not conclusive. For example, there are fears that sugarcane and soybean expansion in Brazil might directly and indirectly affect the highly biodiverse Cerrado ecosystems and the Amazon (Chapter 6). Jatropha production in southern Africa can also potentially affect biodiversity due to land use change and jatropha's invasiveness, but so far very little research has been conducted (Chapter 13).

4.8. Rural development

Biofuel production and use can contribute to rural development by generating income and employment opportunities (Chapter 1). Biofuel feedstock production has been a major source of employment in Brazil (Chapter 6), while it has produced more mixed results in China (Chapter 10) and Africa (Chapters 12 and 13). Moreover, employment generation might be very uneven along the biofuel chain. Most jobs are created during the feedstock production stage (agricultural phase, including usually low-skill activities) and not during feedstock processing and biofuel production (higher-skill activities) (Chapters 1, 6, and 7). The impact of biofuel expansion on rural development also depends greatly on the type of production system adopted. Despite the high employment opportunities, economies of scale, and stable salaries that large plantations offer (Chapters 1, 6, and 13), smallholder/outgrower schemes might offer far greater benefits, particularly in poorer developing nations (Chapters 1 and 11–13). Strengthening institutions that regulate foreign investments and guarantee local citizens' rights and access to key resources might ensure that the welfare of smallholders is improved (Chapter 11). In some cases, the use of locally produced biofuel (small-scale biofuel production) could further contribute to rural development in a positive way (Chapters 1 and 6). Nevertheless, biofuel-related income and employment generation can be precarious, given the uncertainties and risks associated with biofuel production (Chapters 1 and 11), international trade (Chapter 11), and the mechanization of agricultural production (Chapter 6).

4.9. Food production

The competition of biofuels with food production is perhaps the most hotly debated topic in the biofuel polemic. The competition between food and biofuels is a very complex process that cannot be easily delineated. It can be direct (i.e., diversion of edible crops to biofuel production) or indirect (i.e., diversion of factor inputs such as labor, land, and water to biofuel production). Evidence suggests that biofuel expansion contributed to the high food prices observed during the 2007–2008 food crisis (Chapter 1). Several developing countries, such as China, India, and South Africa, have banned the use of (some) food crops for biofuel production to avoid direct competition between food and biofuel production (Chapters 1, 10, and 12). However, in some cases, indirect competition for inputs such as labor (Chapters 10 and 11), water (Chapter 4), and land (Chapters 2 and 8) may in the future affect food production in developing countries. Such an example is the case of jatropha, which might affect food security due to competition for production inputs (Chapter 10), including good-quality land (Chapter 13). It is interesting to note that the impact of the food–biofuel conflict is not felt equally across different segments of society. For example, net food consuming households, rural landless, and urban poor households are the most vulnerable to high food prices resulting from biofuel expansion, facing a greater risk of food insecurity than other segments of society (Chapters 1 and 11). While some developing, food-deficit countries have been affected by biofuel-induced increases in food prices (Chapter 1), others have not (Chapter 11). For example, in Brazil, the largest biofuel producer among developing nations, there are no signs that domestic food production and food security have declined (or will decline) due to biofuel expansion (Chapters 6 and 8). On the basis of the preceding, it is no surprise that global biofuel potential greatly depends on future diets. In general, the richer the diet is, the lower the bioenergy potential (Chapter 2). However, the world can afford to forgo some potential intensification of agricultural production without jeopardizing food supply and still yield substantial amounts of bioenergy (Chapter 2). However, scenario calculations that take into account food demand and land use intensity also indicate that the amount of biofuels that can be produced sustainably is likely to be much lower than previous estimates have assumed (Chapter 2).

4.10. Other social issues

Certain feedstock production practices (e.g., sugarcane) have been associated with harsh, degrading working conditions and concentration of power held by a few actors (Chapters 6 and 7). The latter point has raised concerns about threatening small farmers' land rights (Chapters 1 and 13). There are also concerns that biofuel impacts are gender differentiated, with women being more likely to face the negative socio-economic and environmental impacts associated with biofuel expansion (Chapter 1). Nevertheless, there are cases in which biofuel production and use can benefit women in developing countries, if planned properly (Chapters 1 and 13).

5. Biofuel sustainability

The policy instruments adopted during biofuel production, use, and trade are crucial in determining whether biofuel production can provide a net benefit on the economy, environment, and human well-being. However, the complexity of biofuel chains, the multiple uses of the land appropriated for feedstock production, and the many different incentives that drive the biofuel market contribute to the difficulty of designing and implementing policies for sustainable biofuels.

Various issues have been associated with biofuel sustainability. For example, Hill et al. (2006) suggest that biofuels need to be (1) net energy providers, (2) environmentally sustainable, (3) economically competitive, and (4) not in competition with food production. Lately, additional social criteria have been articulated (Borras et al., 2010; Rist et al., 2009). In our view, all the impacts discussed previously (and their interrelationships) need to be considered when aiming for sustainable biofuel production and use. Assessing only a subset of biofuels' impacts is probably worse than not addressing any of them at all because a piecemeal approach lends itself to inappropriate resource use decisions and is easily used to support narrow economic interests.

As a result, it is important to develop appropriate sustainability criteria that can be feasibly implemented in biofuel-related policies. Currently two main types of policy instruments are aiming to enhance the sustainability of biofuel practices, regulatory–policy initiatives, and voluntary standards (Kunen and Chalmers, 2010).

Biofuel mandates, the main type of biofuel policies in developing countries, fall into the first category as they essentially provide incentives and set boundaries for biofuel expansion (Kunen and Chalmers, 2010).[2] However, in most cases, such instruments target the economic viability of biofuel production and usually lack provisions to ensure the social and environmental sustainability of biofuel production and use (particularly in developing countries). For example, it is discussed throughout this edited volume that environmental considerations are virtually absent from developing countries' biofuel policies (Chapters 1, 6, 9, and 12). Social criteria are sometimes articulated – as in the Brazilian biodiesel program (Chapter 6) – but can also be blatantly neglected, as in the Brazilian bioethanol program (Chapters 6 and 7).

Voluntary standards are a second class of policy initiatives that seek to enhance biofuel sustainability. Such standards are promoted by multistakeholder alliances and can either target biofuels, for example, the Roundtable on Sustainable Biofuels (RSB, 2010), or specific biofuel feedstocks, for example, the Roundtable on Sustainable Palm Oil (RSPO, 2007) and the Roundtable on Responsible Soy (RTRS, 2010). Usually, such standards are comprehensive in the sense that they encompass an ample range of economic, environmental, and social criteria that have to be met if a biofuel–feedstock practice is to be considered sustainable. However, given their voluntary nature, such initiatives suffer from a lack of strong implementation and enforcement

---

[2] With respect to the environmental and socioeconomic impacts of biofuels, such mechanisms can articulate biofuel-specific sustainability criteria, e.g., the European Union Renewable Energy Directive (EU, 2009) and the UK Renewable Transport Fuel Obligation (RFA, 2008), or can target broader sustainability issues and have, as a result, a ripple effect on biofuel sustainability (Kunen and Chalmers, 2010).

mechanisms. The potential for increased biofuel exports to developed countries is seen as an opportunity to boost certification efforts in developing countries, which can have a positive effect on sustainable biofuel production (Chapters 7, 9, 12, and 13).[3]

An important characteristic of policy initiatives and voluntary standards is that they are performance based, meaning that they set specific sustainability targets (e.g., GHG savings) that need to be met if a biofuel or feedstock production practice is to be considered sustainable (Kunen and Chalmers, 2010). In the authors' opinion, both the main advantage and the main limitation of these frameworks is their simplicity. While the simplification and reduction of biofuel sustainability to a discrete list of measurable criteria/targets that have to be met (similar to an indicator list) provides clear guidance information, at the same time, the interrelations between these different criteria and the dynamics of the social–ecological systems that accommodate biofuel production are also, regrettably, lost.

6. Methods

The chapters of this edited volume have adopted highly diverse perspectives for assessing biofuel impacts. This is testified to by the variety of analytic methodologies used throughout this volume. Techniques have included material balances (Chapter 2), LCA (Chapters 3 and 6), WF analysis (Chapter 4), remote sensing (Chapter 5), sociological research (Chapter 7), partial equilibrium models (Chapter 8), interviews and questionnaires (Chapter 10), econometric models (Chapters 8 and 11), and a number of other field techniques (Chapters 9 and 13). However, the diversity of the environmental and socioeconomic impacts of biofuel production and use combined with the fact that these impacts can take place at multiple spatial and temporal scales mean that no single methodology can capture the multitude of biofuels' impacts. For example, powerful techniques, such as LCA, can capture the energy, GHG emissions, and pollution impacts of biofuels but fail to accommodate other important impacts such as biodiversity loss, competition with food production, and other socioeconomic issues. For this reason, the feasibility, constraints, and implications of biofuel expansion must be evaluated in a multidimensional manner. This implies that a coherent parallel multiscale analysis and multicriteria evaluation should be adopted (Borzoni, 2011). It must also be considered that in developing countries, the weights given to different sustainability criteria may be very different from those applied in developed countries. For instance, GHG savings may be of far lesser importance than rural development Section 2).

---

[3] The introduction of sustainability criteria for biofuels would also need to consider a system of implementation to address issues such as conformity assessment and how to accredit the organizations and institutes providing certification.

Additionally, while there are a great number of powerful techniques that can explain the full range of biofuels' environmental and socioeconomic impacts, there is a lack of a clear and multidisciplinary synthesis of the evidence surrounding biofuels' trade-offs. This is in part due to the lack of a comprehensive, consistent, and coherent conceptual framework that can structure the biofuel debate and put these diverse trade-offs into perspective. The absence of such a conceptual framework has resulted in a lack of appropriate assessment tools capable of integrating and evaluating the main impacts and trade-offs of biofuel production and use (Gasparatos et al., 2011; Robertson et al., 2008; Tilman et al., 2009) and, in our opinion, has curtailed the effectiveness of current biofuel policies.

In this respect, synthesizing biofuels' trade-offs in a cohesive manner is as important as ensuring a robust assessment of individual impacts. Two very promising frameworks that can be used to synthesize the evidence about biofuel impacts in a policy-relevant manner are sustainability science (Takeuchi et al., in press) and the ecosystem services approach (Gasparatos et al., 2011; Stromberg et al., 2010).

Such syntheses can complement and improve the quality of existing initiatives and frameworks that aim to enhance biofuel sustainability (Section 5). In particular, both sustainability science and the ecosystem services approach – and their application to tackle biofuel issues – can assist policy makers in obtaining a better grasp on the trade-offs associated with biofuel production and use. Understanding and communicating the dynamics and trade-offs of biofuel production is something that other assessment frameworks in their current format (similar to an indicator list) miss (Section 5). However, significant research is needed before such synthesis frameworks can be used effectively for assessing different biofuel practices. Furthermore, when developing such methods, particular effort should be made to ensure the ability to accommodate the various, often equally legitimate but mutually incompatible perspectives, framings, assumptions, and alternative types of knowledge relevant for the assessment of biofuel options.

7. Proposals

The preceding clearly suggest that biofuel chains are complex processes that involve several production steps and have different types of environmental and socioeconomic impacts that intersect vast temporal and spatial scales. Furthermore, biofuel supply chains are tightly linked with other economic activities and ecosystem processes. While these attributes are common within many other economic sectors, the mere speed and scale with which biofuel expansion is taking place in some developing nations is particularly impressive. It is our hope that the evidence presented in this collective volume makes the case that it is important to assess the full range of environmental and socioeconomic impacts of biofuel expansion as well as to articulate their trade-offs in a context-specific manner.

The main biofuel policies, such as those based on biofuel mandates, the free market, and voluntary standards, are inadequate in their current format to catalyze the minimization of the negative environmental and socioeconomic impacts due to biofuel production and use. The authors of this volume agree that concerted action should be taken by researchers, practitioners, and policy makers to ensure that the diverse biofuel production practices increase human welfare without inadvertently compromising ecosystem functioning. For this reason, we conclude this edited volume by offering specific research, practice, and policy proposals that emerge from its 14 chapters.

Research

- Need to develop integrated tools that adopt a systemic view of land systems when evaluating biofuel practices; such tools should be able to consider possible synergies and trade-offs between food, energy, and rural development (Chapters 1 and 2)
- Need to better understand the potential, impacts, and ways forward for second-generation biofuel production in developing countries (Chapter 1)
- Need for nation-specific studies of air pollutant emissions from biofuel production and use (Chapter 3)
- Need to understand better the impact of bioenergy crops on water resources (Chapter 4)
- Need for more accurate data regarding the location of feedstock cultivation and the amount of food crops used for biofuel production (as opposed to other uses) for assessing the impacts of biofuel expansion on deforestation (Chapter 5)
- Need for a systematic comparative evaluation of the impacts of the Brazilian bioethanol and biodiesel programs (Chapter 6)
- Need to assess the possibility of intensifying bioethanol production (and the limits to sustainable intensification) to lessen the pressure and competition for limited resources, land in particular (Chapter 8)
- Need for a comprehensive understanding of what constitutes degraded lands that could be targeted for future biofuel production (Chapter 9)
- Need for further exploring how to maintain agricultural activities that support rural livelihoods and long-term national food security in countries with rapid economic development such as China, India, and Brazil; in this regard, assessment of the economic viability of agricultural industries (biofuels or their by-products) should be coupled with an analysis of the interactions between industrial development and governance systems (land tenure system, etc.) as well as evaluation of impacts on ecosystem services (Chapter 10)
- Need to better understand the human welfare dimensions of biofuel production and use and particularly which types of households benefit the most from national biofuel schemes in developing countries (Chapter 11)
- Need to understand the economics of household energy and where biofuels might play a useful role (e.g., ethanol stoves for cooking vs. transport fuel). In some cases biofuels could contribute directly to better-quality energy services and improved energy access (e.g., in

some countries, only the very wealthy can afford cars); this is particularly pertinent for the African context (Chapter 12)

- Need to provide proven genetic material with a known production potential for feedstock production sites through plant breeding for jatropha and other new biofuel crops; once yields are known, additional research is needed to identify how biofuel development will bring broad-based socioeconomic benefits (Chapter 13)
- Need to develop (1) a cohesive and consistent conceptual framework that can frame the biofuel debate and (2) integrated assessment frameworks and tools that can assess biofuels' diverse impacts; sustainability science and the ecosystem services approach have the versatility and the acceptability by academics, practitioners, and policy makers to fulfill this task, but very limited research has been directed toward that end (Chapter 14)

Practice

- Small-scale biofuel production can contribute positively to local energy security, poverty alleviation, and rural development (Chapter 1).
- The bioenergy potential of by-products from food production are considerable, and the possibility to exploit this potential in a cost-effective and sustainable manner deserves more attention (Chapter 2).
- The evaluation of the air pollution impacts of biofuel production and use should be shifted from its current main focus on GHGs to also include air quality (Chapter 3).
- When comparing different biofuel practices, it is more water efficient to produce bioethanol than biodiesel (Chapter 4).
- National and local governments should require the detailed accounting of biofuel feedstock production from both producers and buyers and make this information freely available in the public domain (Chapter 5).
- For the energy security and social objectives of the Brazilian biodiesel program to be met, biodiesel feedstock should be diversified and move away from the near-monopoly of soybeans. Mechanization of sugarcane production should be well planned to ensure that the people who lose their jobs will have other employment opportunities (Chapter 6).
- Concerted effort should be made toward integrating social impacts, particularly those related to the historically shaped asymmetrical relations of power between the various stakeholders and participants (Chapter 7).
- Future biofuel expansion in the tropics requires a combination of wildlife-friendly farming and land-sparing approaches, which can be achieved by way of carefully designing landscapes (Chapter 9).
- Biofuel ventures should try to create positive spillovers when sourcing feedstock from local producers. This has been shown to have significant potential for improving livelihoods and food security in developing countries (Chapter 11).
- In developing nations (particularly the least developed countries), there will be significant differences between the expectations for new biofuel feedstocks and what they can deliver in particular locations, due to both biophysical differences and the complex economics of new market development. To support agroindustrial development in particular regions, crops

that are well proven (e.g., sugarcane, corn) should be prioritized for expanded commercial development before those crops with which there is limited experience (e.g., jatropha) (Chapter 12).

- The sustainability assessment of biofuel practices should go beyond the current performance-led paradigm (Chapter 14).

Policy Making

- Strategies that maximize synergies and benefits but avoid adverse social, economic, and ecological effects should be developed (Chapters 1 and 2)
- The full life cycle of biofuel air pollution emissions should be considered in policy making (Chapter 3).
- A shift toward greater biofuel production will bring a need for more water, diverting it from food production and thus causing competition over this scarce resource. This needs to be considered by policies promoting biofuel expansion (Chapter 4).
- Environmental and social criteria should be articulated and strengthened in the Brazilian bioethanol and biodiesel programs (Chapter 6).
- When designing international sustainability certification schemes, particular attention should be paid to the design of the process to ensure the equitable participation of the weakest and most marginalized groups as well as to mitigate the excessive power exercised by the established parties. This would be particularly important in view of the potentially highly unequal distribution of costs and benefits from the development of next-generation biofuels in the global South (Chapter 7).
- Mechanisms that provide incentives to intensify biofuel production (either increasing land reserves, facilitating the diffusion of new technologies, or implementing financial incentives) should be implemented (Chapter 8).
- Policy makers and decision makers need to have a clear understanding of the consequences and trade-offs of pursuing ambitious biofuel policies in Southeast Asia to avoid doing more harm than good to the environment. Emphasis on multistakeholder collaboration to develop more sustainable policies for expanding the biofuel industry is the best immediate remediation for ensuring the protection of the tropics' remaining natural habitats and biodiversity (Chapter 9).
- National governments should develop mechanisms that can ensure, or at least encourage, the private biofuel industry (or state-owned enterprises) to comply with the proposed sustainability criteria. Additionally, national governments should provide more support for biofuel R&D and assist local governments with the sustainability of their overall biofuel development agendas (Chapter 10).
- Foreign-led biofuel ventures in developing countries (such as Africa) need to be better designed at the policy level to reduce the risk of exposing poor households to biofuel project failures and missed opportunities for wage earning and revenue generation (Chapter 11).
- Policies must address the multilevel nature of biofuels markets – local, national, regional, and global. For example, where infrastructure is lacking and energy security is a key concern, such as in the case of southern Africa, regional mechanisms can address strategic economic issues in a way that national schemes cannot (Chapter 12).

- Jatropha and other new biofuel crops should not be promoted until there is clearly proven national benefit. At present, a lack of certainty around jatropha yields and jatropha's social benefits (in both small-scale and large-scale projects) means that a precautionary approach should be applied until there is better evidence that jatopha projects will meet their desired objectives (Chapter 13).

References

Abiove, 2010. Associação Brasileira das indústrias de óleos vegetais: Dados do complexo soja. Available at http://www.abiove.com.br/menu_br.html.

ABIQUIM, 2007. Anuário da indústria química brasileira. Associação Brasileira da Indústria Química, São Paulo.

Abramovay, R., 2008. *A political-cultural approach to the biofuels market in Brazil.* University of São Paulo, São Paulo.

Achard, F., DeFries, R., Eva, H., Mansen, M., Mayaux, P., Stibig, H.-J., 2007. Pan-tropical monitoring of deforestation. *Environment Research Letters* 2, 1–11.

Acharya, V.V., Richardson, M., 2009. *Restoring financial stability: How to repair a failed system.* John Wiley, Hoboken, NJ.

Achten, W.M.J., Mathijs, E., Verchot, L., Singh, V.P., Aerts, R., Muys, B., 2007. Jatropha biodiesel fueling sustainability? *Biofuels, Bioproducts, and Biorefining* 1, 283–291.

Achten, W.M.J., Verchot, L., Franken, Y.J., Mathijs, E., Singh, V.P., Aerts, R., Muys, B., 2008. Jatropha bio-diesel production and use. *Biomass and Bioenergy* 32, 1063–1084.

Achten, W.M.J., Almeida, J., Fobelets, V., Bolle, E., Mathijs, E., Singh, V.P., Tewari, D.N., Verchot, L.V., Muys, B., 2010a. Life cycle assessment of Jatropha biodiesel as transportation fuel in rural India. *Applied Energy* 87, 3652–3660.

Achten, W.M.J., Maes, W.H., Aerts, R., Verchot, L., Trabucco, A., Mathijs, E., Singh, V.P., Muys, B., 2010b. Jatropha: From global hype to local opportunity. *Journal of Arid Environments* 74, 164–165.

Achten, W.M.J., Maes, W.H., Reubens, B., Mathijs, E., Singh, V.P., Verchot, L., Muys, B., 2010c. Biomass production and allocation in *Jatropha curcas* L. seedlings under different levels of drought stress. *Biomass and Bioenergy* 34, 667–676.

Achten, W.M.J., Almeida, J., Fobelets, V., Bolle, E., Mathijs, E., Singh, V., Tewari, D., Verchot, L., Muys, B., 2010d. Life cycle assessment of jatropha biodiesel as transportation fuel in rural India. *Applied Energy* 87, 3652–3660.

Actionaid, 2010. *Meals per gallon: The impact of industrial biofuels on people and global hunger.* Actionaid, London.

Adam, S., Minnemeyer, S., Hansen, M., Potapov, P., Pittman, K., 2007. *Painting the global picture of tree cover change: Tree cover loss in the humid tropics.* World Resource Institute, Washington, DC.

AfDB, 2006. High oil prices and the African economy. Prepared for the 2006 African Development Bank Annual Meetings, Ouagadougou.

AGAMA Energy, 2003. Employment potential of renewable energy in South Africa. AGAMA Energy, Cape Town.

Agoramoorthy, G., Hsu, M.J., Chaudhary, S., Shieh, P., 2009. Can biofuel crops alleviate tribal poverty in India's drylands? *Applied Energy* 86, S118–S124.

Ahmed, A.U., Hill, R.V., Smith, L.C., Wiesmann, D.M., Frankenberger, T., Gulati, K., Quabili, W., Yohannes, Y., 2007. *The world's most deprived characteristics and causes of extreme poverty and hunger. 2020 discussion paper 43.* International Food Policy Research Institute, Washington, DC.

Ahmed, I., Decker, J., Morris, D., 1994. How much energy does it take to make a gallon of soybean biodiesel? Report prepared for National Soy Diesel Development Board, Institute for Local Self Reliance, Minneapolis, MN, and Washington, DC.

Ajanovic, A., 2011. Biofuels versus food production: Does biofuels production increase food prices? *Energy* 36, 2070–2076.

Alhajeri, N.S., McDonald-Buller, E.C., Allen, D.T., 2011. Comparisons of air quality impacts of fleet electrification and increased use of biofuels. *Environmental Research Letters* 6, doi:10.1088/1748-9326/6/2/024011.

Allen, A.G., Cardozo, A.A., da Rocha, G.O., 2004a. Influence of sugar cane burning on aerosol soluble ion composition in southeastern Brazil. *Atmospheric Environment* 38, 5025–5038.

Allen, R.G., Pereira, L.S., Raes, D., Smith, M., 2004b. Crop evapotranspiration: Guidelines for computing crop water requirements. FAO irrigation and drainage paper 56. Food and Agriculture Organization, Rome.

Almeida, W., 2009. Ethanol diplomacy: Brazil and U.S. in search of renewable energy. *Globalisation, Competitiveness, and Governability* 3, 114–124.

Alves, F., 2006. Porque Morrem os Cortadores de Cana? *Saúde e Sociedade* 15, 90–80.

Amaral, W.A.N., do, 2008. A sustainability analysis of the Brazilian biodiesel. Report to the UK Department for Environment, Food, and Rural Affairs, London.

Amaral, W.A.N., do, Marinho, J.P., Tarasantchi, R., Beber, A., Giuliani, E., 2008. Environmental sustainability of sugarcane ethanol in Brazil. In Zuurbier, P., van de Vooren, J. (Eds.), *Sugarcane ethanol: Contributions to climate change mitigation and the environment.* Wageningen Academic, Wageningen, pp. 113–138.

Anderson, L.G., 2009. Ethanol fuel use in Brazil: Air quality impacts. *Energy and Environmental Science* 2, 1015–1037.

Anderson-Teixeira, K.J., Snyder, P.K., DeLucia, E.H., 2011. Do biofuels life cycle analyses accurately quantify the climate impacts of biofuels-related land use change? *University of Illinois Law Review* 2011, 589–622.

Andrade, M.C., de, 1988. Area do Sistema Canavieiro. Superintendência de Desenvolvimento do Nordeste (SUDENE), Recife. *Estudos Regionais* 18.

Aneja, V.P., Schlesinger, W.H., Erisman, J.W., 2009. Effects of agriculture upon the air quality and climate: Research, policy and regulations. *Environmental Science and Technology* 43, 4234–4240.

Angelsen, A., Kaimowitz, D., 1999. Rethinking the causes of deforestation: Lessons from economic models. *World Bank Research Observer* 14, 73–98.

ANP, 2010a. Anuário Estatístico. Agência Nacional do Petróleo, Gás natural e Biocombustíveis. Available at http://www.anp.gov.br/.

ANP, 2010b. Boletim Mensal do Biodiesel. Agência Nacional do Petróleo, Gás natural e Biocombustíveis. Available at http://www.anp.gov.br/.

ANP, 2010c. Capacidade Autorizada Biodiesel. Agência Nacional do Petróleo, Gás natural e Biocombustíveis. Available at http://www.anp.gov.br/.

ANP, 2010d. Leilões de Biodiesel. Agência Nacional do Petróleo, Gás natural e Biocombustíveis. Available at http://www.anp.gov.br/.

ANP, 2010e. Produção de Biodiesel. Agência Nacional do Petróleo, Gás natural e Biocombustíveis. Available at http://www.anp.gov.br/.

Antonopoulos, R., 2008. Impact of employment guarantee schemes on gender equality and pro-poor economic development: Case study on South Africa. Levy Economics Institute of Bard College, Annandale-on-Hudson, NY.

APROSOJA, 2010. Safra de soja 2009/2010. Associação dos Produtores de Soja e Milho do Estado de Mato Grosso, Cuiabá.

Araujo, A., Quesada-Aguilar, A., 2007. *Gender and bioenergy*. World Conservation Union, Gland.

Aratrakorn, S., Thunhikorn, S., Donald, P.F., 2006. Changes in bird communities following conversion of lowland forest to oil palm and rubber plantations in southern Thailand. *Bird Conservation International* 16, 71–82.

Arbex, M.A., Zanobetti, A., Braga, A.L.F., 2006. The impact of sugar-cane burning emissions on the respiratory system of the children and the elderly. *Environmental Health Perspectives* 114, 725–729.

Arbex, M.A., Saldiva, P.H.N., Pereira, L.A.A., Braga, A.L.F., 2010. Impact of outdoor biomass air pollution on hypertension hospital admissions. *Journal of Epidemiological Community Health* 64, 573–579.

Ariza-Montobbio, P., Lele, S., Kallis, G., Martinez-Alier, J., 2010. The political ecology of jatropha plantations for biodiesel in Tamil Nadu, India. *Journal of Peasant Studies* 37, 875–879.

Arndt, C., Benfica, R., Tarp, F., Thurlow, J., Aiene, R., 2009. Biofuels, poverty, and growth: A computable general equilibrium analysis of Mozambiqu. Paper presented at International Association of Agricultural Economists Conference, Beijing.

Arndt, C., Benfica, R., Tarp, F., Thurlow, J., Uaiene, R., 2010a. Biofuels, poverty and growth: A computable general equilibrium analysis of Mozambique. *Environment and Development Economics* 15, 81–105.

Arndt, C., Pauw, K., Thurlow, J., 2010b. Biofuels and economic development in Tanzania. Discussion paper 966. International Food Policy Research Institute, Washington, DC.

Arndt, C., Benfica, R., Thurlow, J., 2010c. Gender implications of biofuels expansion: A CGE analysis of Mozambique. United Nations University World Institute for Development Economics Research, Helsinki.

Arndt, C., Msangi, S., Thurlow, J., 2010d. Are biofuels good for African development? An analytical framework with evidence from Mozambique and Tanzania. Working paper 2010/110. United Nations University World Institute for Development Economics Research, Helsinki.

Arrow, K., 1962. The economic implications of learning by doing. *Review of Economic Studies* 29, 155–173.

ATC, 2010. Agribusiness Transnational Corporations (TNCs) and UNFCCC process: Background document – La Via Campesina. Agribusiness Transnational Corporations (ATC) 30/11/2010. Available at http://www.viacampesina.org/en/index.php?option=com_content&view=article&id=979:agribusiness-transnational-corporations-tncs-and-unfccc-process&catid=48:-climate-change-and-agrofuels&Itemid=75.

Azevedo, L., 2009. Certificação sócio-ambiental do álcool. Pastoral do Migrante, São Pualo.

Bacoccina, D., 2007. Produtores querem "selo social" para etanol do Nordeste. Available at http://www.bbc.co.uk/portuguese/reporterbbc/story/2007/03/070306_etanolparaibadb.shtml.

Bahé, M., 2007. Nordeste fica de fora da "festa" do álcool. Folha de Pernambuco. Available at http://acertodecontas.blog.br/economia/nordeste-fica-de-fora-da-festa-do-alcool/.

Balsadi, O., 2007. O Mercado de Trabalho Assalariado na Agricultura Brasileira no Período 1992–2004 e suas Diferenciações Regionais. Doctoral thesis. Universidade Estadual de Campinas, Campinas.

Bancada do Nordeste, 2007. Nordeste e o endividamento rural. Bancada do Nordeste. Available at http://www.bancadadonordeste.com.br/noticia-imprimir.asp?secaoID= 5&NotID=573.

Barlow, J., Gardner, T.A., Araujo, I.S., vila-Pires, T.C.A., Bonaldo, A.B., Costa, J.E., Esposito, M.C., Ferreira, L.V., Hawes, J., Hernandez, M.I.M., Hoogmoed, M.S., Leite, R.N., Lo-Man-Hung, N.F., Malcolm, J.R., Martins, M.B., Mestre, L.A.M., Miranda-Santos, R., Nunes-Gutjahr, A.L., Overal, W.L., Parry, L., Peters, S.L., Ribeiro-Junior, M.A., da Silva, M.N.F., Motta, C., Peres, C.A., 2007. Quantifying the biodiversity value of tropical primary, secondary, and plantation forests. *Proceedings of the National Academy of Sciences of the United States of America* 104, 18555–18560.

Barney, J.N., DiTomaso, J.M., 2010. Bioclimatic predictions of habitat suitability for the biofuel switchgrass in North America under current and future climate scenarios. *Biomass and Bioenergy* 34, 124–133.

Barros, G.S.A. Alves, L.R.A., Osaki, M. 2009. Análise dos custoseconômicos do programa do biodiesel no Brasil. Paper presented at the 47th Sober Congress, Porto Alegre, Brazil.

Basri, M.W., Norman, K., Hamdan, A.B., 1995. Natural enemies of the bagworm, *Metisa plana* Walker (Lepidoptera: Psychidae) and their impact on host population regulation. *Crop Protection* 14, 637–645.

Batidzirai, B., 2002. Cogeneration in Zimbabwe: A utility perspective. AFREPREN occasional paper 19. AFREPREN/FWD, Nairobi.

Batidzirai, B., Wamukonya, N., 2009. Biofuels and their implications for Africa's development. Background paper. Forum of Energy Ministers of Africa, Maputo, Mozambique.

Batidzirai, B., Faaij, A., Smeets, E., 2006. Biomass and bioenergy supply from Mozambique. *Energy for Sustainable Development* 10, 54–81.

Bekunda, M., Palm, C.A., de Fraiture, C., Leadley, P., Maene, L., Martinelli, L.A., McNeely, J., Otto, M., Ravindranath, N.H., Victoria, R.L., Watson, H., Woods, J., 2009. Biofuels in developing countries. In Howarth, R.W., Bringezu, S. (Eds.), *Biofuels: Environmental consequences and implications of changing land use. Proceedings of the Scientific Committee on Problems of the Environment (SCOPE) International Biofuels Project Rapid Assessment.* Cornell University, Ithaca, NY, pp. 249–269.

Bell, M.L., Peng, R.D., Dominici, F., 2006. The exposure-response curve for ozone and risk of mortality and the adequacy of current ozone regulations. *Environmental Health Perspectives* 114, 532–536.

Benedick, S., 2005. Impacts of tropical forest fragmentation on fruit-feeding nymphalid butterflies in Sabah, Borneo. Doctoral thesis. Institute for Tropical Biology and Conservation. Universiti Malaysia Sabah, Kota Kinabalu, Malaysia.

Benge, M., 2006. Assessment of the potential of *Jatropha curcas* (biodiesel tree) for energy production and other uses in developing countries. Educational Concerns for Hunger Organization, Fort Myers, FL.

Bermann, C., Macedo Moreno, L., Soares Domingues, M., Rosenberg, R., 2008. Desafios e Perspectivas dos Agrocombustíveis no Brasil: A agricultura familiar face ao etanol da cana-de-açúcar e ao biodiesel da soja, mamona e dende. In *Agrocombustíveis and agricultura familiar e camponesa: Subsídios ao debate*. Federação de Órgãos para Assistência Social e Educacional, and Rede Brasileira pela Integração dos Povos, Rio de Janeiro, pp. 59–113.

Berndes, G., 2002. Bioenergy and water: The implications of large-scale bioenergy production for water use and supply. *Global Environmental Change* 12, 253–271.

Berndes, G., Hoogwijk, M., van den Broek, R., 2003. The contribution of biomass in the future global energy supply: A review of 17 studies. *Biomass and Bioenergy* 25, 1–28.

Bessou, C., Ferchaud, F., Gabrielle, B., Mary, B., 2010. Biofuels, greenhouse gases and climate change: A review. *Agronomy for Sustainable Development* 31, 1–79.

Bird, N., Cherubini, F., Pena, N., Zanchi, G., 2010. Greenhouse gas emissions and bioenergy. In Amezaga, J.M., von Maltitz, G.P., Boyes, S.L. (Eds.), *Assessing the sustainability of bioenergy projects in developing countries: A framework for policy evaluation.* Newcastle University, Newcastle upon Tyne, pp. 53–79.

Black, F., Tejada, S., Gurevich, M., 1998. Alternative fuel motor vehicles tailpipe and evaporative emissions composition and ozone potential. *Journal of the Air and Waste Management Association* 48, 578–591.

Blok, K., 2006. *Introduction to energy analysis.* Techne Press, Amsterdam.

Boddey, R.M., de B. Soares, L.H., Alves, B.J.R., Urquiaga, S., 2008. Bio-ethanol production in Brazil. In Pimentel, D. (Ed.), *Biofuels, solar and wind as renewable energy systems: Benefits and risks.* Springer, New York, pp. 321–356.

Borman, G., von Maltitz, G.P., Tiwari, S., Scholes, M.C., in press. Modeling the economic returns to labour for jatropha cultivation in southern Africa and India at differential local fuel prices. *Biomass and Bioenergy.*

Borras Jr., S.M., McMichael, P., Scoones, I. (Eds.), 2010. Biofuels, land and agrarian change: special issue. *Journal of Peasant Studies* 37, 575–962.

Borzoni, M. 2011. Multi-scale integrated assessment of soybean biodiesel in Brazil. *Ecological Economics* 70, 2028–2038.

Boserup, E., 1965. The conditions of agricultural growth: The economics of agrarian change under population pressure. Aldine/Earthscan, Chicago.

Botha, T., von Blottnitz, H., 2006. A comparison of the environmental benefits of bagasse-derived electricity and fuel ethanol on a life-cycle basis. *Energy Policy* 34, 2654–2661.

Bound, K., 2008. *Brazil: The natural knowledge economy.* Demos, London.

Boustead, I., Hancock, G.F., 1979. *Handbook of industrial energy analysis.* Horwood Limited, Ellis.

Bouwman, A.F., Boumans, L.J.M., Batjes, N.H., 2002. Estimation of global NH3 volatilization loss from synthetic fertilizers and animal manure applied to arable lands and grasslands. *Global Biogeochemical Cycles* 16, 1024–1038.

Bouwman, A.F., van Grinsven, J.J.M., Eickhout, B., 2010. Consequences of the cultivation of energy crops for the global nitrogen cycle. *Ecological Applications* 20, 101–109.

Bradford, G.E., Baldwin, R.L., 2003. Letter to the editor regarding: S. Wirsenius's "The biomass metabolism of the food system: A model-based survey of the global and regional turnover of the food system." *Journal of Industrial Ecology* 7, 47–80.

Braimoh, A.K., Subramanian, S.M., Elliott, W.S., Gasparatos, A., 2010. *Climate and human-related drivers of biodiversity decline in Southeast Asia.* United Nations University–Institute for Advanced Studies, Yokohama.

Branford, S., Freris, N., 2000. One great big hill of beans. *The Ecologist* 30, 46–47.

Brasil Energy, 2010. An internationalized business. *Brasil Energy 419*, April 2010.

Brilhante, O.M., 1997. Brazil's Alcohol Programme: From an attempt to reduce oil dependence in the seventies to the green arguments in the nineties. *Journal of Environmental Planning and Management* 40, 435–449.

Brinkmann Consultancy, 2009. Greenhouse gas emissions from palm oil production: Literature review and proposals from the RSPO Working Group on Greenhouse Gases. Final report. Brinkhamm Consultancy, Hoevelaken.

Brown, J.C., Koeppe, M., Coles, B., Price, K.P., 2005. Soybean production and conversion of tropical forest in the Brazilian Amazon: The case of Vilhena, Rondonia. *Royal Swedish Academy of Sciences* 34, 462–469.

Brown, L.R., 2006. Plan B 2.0: Rescuing a planet under stress and a civilization in trouble. Earth Policy Institute, Washington, DC.

Bruhl, C.A., 2001. Leaf litter ant communities in tropical lowland rain forests in Sabah, Malaysia: effects of forest disturbance and fragmentation. Doctoral thesis. University of Würzburg, Würzburg, Germany.

Bruinsma, J., 2003. World agriculture: Towards 2015/2030. An FAO perspective. Earthscan, London.

Buddenhagen, C.E., Chimera, C., Clifford, P., 2009. Assessing biofuel crop invasiveness: A case study. *PLoS ONE* 4: e5261.

Burgard, D.A., Bishop, G.A., Stedman, D.H., 2006. Remote sensing of ammonia and sulfur dioxide from on-road light duty vehicles. *Environmental Science and Technology* 40, 7018–7022.

Burgess, S.S.O., Adams, M.A., Turner, N.C., Beverly, C.R., Ong, C.K., Khan, A.A.H., Bleby, T.M., 2001. An improved heat pulse method to measure low and reverse rates of sap flow in woody plants. *Tree Physiology* 21, 589–598.

Butler, R.A., Koh, L.P., Ghazoul, J., 2009. REDD in the red: Palm oil could undermine carbon payment schemes. *Conservation Letter* 2, 67–73.

Buyx, A., Tait, J., 2011. Ethical framework for biofuels. *Science* 332, 540–541.

Cai, X., Zhang, X., Wang, D., 2011. Land availability for biofuel production. *Environmental Science and Technology* 45, 334–339.

Camp, K.D., 1999. *The bioresource groups of KwaZulu-Natal.* Cedara report N/A99/14. KwaZulu-Natal Department of Agriculture, Cedara.

Campbell, J.E., Lobell, D.B., Genova, R.C., Field, C.B., 2008. The global potential of bioenergy on abandoned agriculture lands. *Environmental Science and Technology* 42, 5791–5795.

Cançado, J.E.D., Saldiva, P.H.N., Pereira, L.A.A., Lara, L.B.L.S., Artaxo, P., Martinelli, L.A., Arbex, M.A., Zanobetti, A., Braga, A.L.F., 2006. The impact of sugar cane–burning emissions on the respiratory system of children and elderly. *Environmental Health Perspectives* 114, 725–729.

Cançado, J.E.D., Saldiva, P.H.N., Pereira, L.A.A., Lara, L.B.L.S., Artaxo, P., Martinelli, L.A., Cardona, C.A., Quintero, J.A., Paz, I.C., 2010. Production of bioethanol from sugar cane bagasse: Status and perspectives. *Bioresource Technology* 101, 4754–4766.

Cardona, C.A., Quintero, J.A., Paz, I.C., 2010. Production of bioethanol from sugarcane bagasse: Status and perspectives. *Bioresource Technology* 101, 4754–4766.

Carroll, M.L., DiMiceli, C.M., Townshend, J.R.G., Sohlberg, R.A., DeFries, R.S., 2006. Vegetation cover conversion MOD44A. Burned Vegetation, collection 4. University of Maryland, College Park.

Cassman, K.G., 1999. Ecological intensification of cereal production systems: Yield potential, soil quality, and precision agriculture. *Proceedings of the National Academy of Sciences of the United States of America* 96, 5952–5959.

Cassman, K.G., Liska, A.J., 2007. Food and fuel for all: Realistic or foolish? *Biofuels, Bioproducts, and Biorefining* 1, 18–23.

Casson, A., 2000. *The hesitant boom: Indonesia's oil palm sub-sector in an era of economic crisis and political change.* Center for International Forestry Research, Bogor.

Casson, A., 2003. Oil palm, soybeans and critical habitat loss. WWF Forest Conservation Initiative, Hohlstrasse.

Casson, A., Tacconi, L., Deddy, K., 2007. Strategies to reduce carbon emissions from the oil palm sector in Indonesia. Paper prepared for the Indonesian Forest Climate Alliance, Jakarta.

Castanho, A.D., Artaxo, P., 2001. Wintertime and summertime São Paulo aerosol source apportionment study. *Atmospheric Environment* 35, 4889–4902.

Cavalcanti, C., Dias, A., Lubambo, C., Barros, H., Cruz, L., de Araújo, M.L.C., Moreira, M., Galindo, O., 2002. Programa de apoio ao desenvolvimento sustentavel da zona da mata de Pernambuco – PROMATA. Trabalhos Para Discussão 135/2002. Fundação Joaquim Nabuco, Recife.

Cavalett, O., Ortega, E., 2010. Integrated environmental assessment of biodiesel production from soybean in Brazil. *Journal of Cleaner Production* 18, 55–70.

CETESP, 2009. Qualidade do ar do Estado de São Paulo. Companhia Ambiental do Estado de São Paulo, São Paulo.

Chang, M.S., Hii, J., Buttner, P., Mansoor, F., 1997. Changes in abundance and behaviour of vector mosquitoes induced by land use during the development of an oil palm plantation in Sarawak. *Transactions of the Royal Society of Tropical Medicine and Hygiene* 91, 382–386.

Chapagain, A.K., Hoekstra, A.Y., 2004. Water footprints of nations. Value of Water Research Report Series 16. United Nations Educational, Scientific, and Cultural Organization–International Institute for Infrastructural, Hydraulic, and Environmental Engineering, Delft.

Cheesman, O.D., 2004. *Environmental impacts of sugar production.* CABI, Wallingford.

Chey, V.K., 2006. Impacts of forest conversion on biodiversity as indicated by moths. *Malayan Nature Journal* 57, 383–418.

Chiu, Y., Walseth, B., Suh, S., 2009. Water embodied in bio-ethanol in the United States. *Environmental Science and Technology* 43, 2688–2692.

Chung, A.Y.C., Eggleton, P., Speight, M.R., Hammond, P.M., Chey, V.K., 2000. The diversity of beetle assemblages in different habitat types in Sabah, Malaysia. *Bulletin of Entomological Research* 90, 475–496.

Cleveland, C.J., 2005. Net energy from the extraction of oil and gas in the United States. *Energy* 30, 769–782.

Cleveland, C.J., Hall, C.A.S., Herendeen, R.A., 2006. Energy returns on ethanol production. *Science* 312, 1746–1746.

Cluver, F.H., Cooper, C.J., Kotzè, D.J., 1998. The role of energy in economic growth. Paper presented at 17th Congress of the World Energy Council, London.

Coate, S., Ravallion, M., 1993. Reciprocity without commitment: Characterization and performance of informal insurance arrangements. *Journal of Development Economics* 40, 1–24.

Coltro, L., Garcia, E.E.C., Queiroz, G., 2003. Life cycle inventory for electric energy system in Brazil. *International Journal of Life Cycle Assessment* 8, 290–296.

Compéan, R.G., Polenske, K.R., 2010. Antagonistic bioenergies: Technological divergence of the ethanol industry in Brazil. *Energy Policy* 39, 6951–6961.

CONAB, 2010. *Safras.* Companhia Nacional de Abastecimento, Brasilia.

Confalonieri, U., Menne, B., Akhtar, R., Ebi, K.L., Hauengue, M., Kovats, R.S., Revich, B., Woodward, A., 2007. Human health. In Parry, M.L., Canziani, O.F., Palutikof, J.P., van der Linden, P.J., Hanson, C.E. (Eds.), *Climate change 2007: Impacts, adaptation and vulnerability. Contribution of Working Group II to the fourth assessment report of the Intergovernmental Panel on Climate Change.* Cambridge University Press, Cambridge, pp. 391–431.

Connor, D.J., 2008. Organic agriculture cannot feed the world. *Field Crops Research* 106, 187–190.

Conservation International, 2010. Biodiversity hotspots. Available at http://www.biodiversity hotspots.org/.

Constanza, R., d'Arge, R., de Groots, R., Farber, S., Grasso, M., Hannon, B., Limbirg, K., Naeem, S., O'Neill, R.V., Paruelo, J., Raskin, R.G., Sutton, P., van den Belt, M., 1997. The value of the world's ecosystem services and natural capital. *Nature* 387, 253–260.

Cook, R., Phillips, S., Houyoux, M., Dolwick, P., Mason, R., Yanca, C., Zawacki, M., Davidson, K., Michaels, H., Harvey, C., Somers, J., Luecken, D., 2010. Air quality impacts of increased use of ethanol under the United States Energy Independence and Security Act. *Atmospheric Environment* 45, 7714–7724.

Cordeiro, A., 2008. Etanol para alimentar carros ou comida para alimentar gente? In *Impactos da industria canavieira no Brasil: Poluição atmosférica, ameaça a recursos hídricos, riscos para a produção de alimentos, relações de trabalho atrasadas e proteção insuficiente à saúde de trabalhadores*. Instituto Brasileiro de Análises Sociais e Econômicas, Rio de Janeiro, pp. 9–22.

Corley, R.H.V., Tinker, P.B., 2003. *The oil palm*. 4th ed. Blackwell Science, Oxford.

Cornland, D.W., Johnson, F.X., Yamba, F., Chidumayo, E.N., Morales, M.M., Kalumiana, O., Mtonga-Chidumayo, S.B., 2001. *Sugar cane resources for sustainable development: A case study in Luena, Zambia*. Stockholm Environment Institute, Stockholm.

Correa, S.M., Arbilla, G., 2008. Carbonyl emissions in diesel and biodiesel exhaust. *Atmospheric Environment* 42, 769–775.

Correa, S.M., Martins, E.M., Arbilla, G., 2003. Formaldehyde and acetaldehyde in a high traffic street of Rio de Janeiro, Brazil. *Atmospheric Environment* 37, 23–29.

Correa, S.M., Arbilla, G., Martins, E.M., Quiterio, S.L., Guimaraes, C.S., Gatti, L.V., 2010. Five years of formaldehyde and acetaldehyde monitoring in the Rio de Janeiro downtown area – Brazil. *Atmospheric Environment* 44, 2302–2308.

Cotula, L., Dyer, N., Vermeulen, S., 2008. Fuelling exclusion? The biofuels boom and poor people's access to land. International Institute for Environment and Development, London.

Cotula, L., Vermeulen, S., Leonard, R., Keeley, J., 2009. Land grab or development opportunity? Agricultural investment and international land deals in Africa. Joint publication of the Food and Agriculture Organization, the International Fund for Agricultural Development, Rome, and the International Institute for Environment and Development, London.

Crago, C.L., Khanna, M., Barton, J., Giuliani, E., Amaral, W., 2010. Competitiveness of Brazilian sugar cane ethanol compared to U.S. corn ethanol. *Energy Policy* 38, 7404–7415.

Crosti, R., Cascone, C., Cipollaro, S., 2010. Use of a weed risk assessment for the Mediterranean region of central Italy to prevent loss of functionality and biodiversity in agro-ecosystems. *Biological Invasions* 12, 1607–1616.

CTC, 1989. Cooperativa de produtores de cana, açúcar e álcool no Estado de São Paulo – COPERSUCAR. Copersúcar Technology Center, São Paulo.

da Silva Dias, G.L., 2007. Um desafio novo: O biodiesel. *Estudos Avançados* 21, 178–183.

Dabat, C.R., 2007. Moradores de engenho: Relações de trabalho e condições de vida dos trabalhadores rurais na zona canavieira de Pernambuco segundo a literatura, a academia e os próprios atores sociais. Editora Universitária da UFPE, Recife.

Dai, D., Hu, Z., Pu, G., Li, H., Wang, C.T., 2006. Energy efficiency and potential of cassava fuel ethanol in Guangxi region of China. *Energy Conversion and Management* 47, 1686–1699.

Danielsen, F., Beukema, H., Burgess, N.D., Parish, F., Bruhl, C.A., Donald, P.F., Murdiyarso, D., Phalan, B., Reijnders, L., Struebig, M., Fitzherbert, E.B., 2009. Biofuel plantations on forested lands: Double jeopardy for biodiversity and climate. *Conservation Biology* 23, 348–358.

Danielsen, F., Heegaard, M., 1995. Impact of logging and plantation development on species diversity: a case study from Sumatra. In Sandbukt, Ø. (Ed.), *Management of tropical forests: Towards an integrated perspective*. University of Oslo, Oslo, pp. 73–92.

Dar, W., 2007. Research needed to cut risks to biofuel farmers. IDRC/ICRISAT. SciDevNet. Available at http://www.scidev.net/en/climate-change-and-energy/biofuels/opinions/research-needed-to-cut-risks-to-biofuel-farmers.html.

Dauber, J., Jones, M.B., Stout, J.C., 2010. The impact of biomass crop cultivation on temperate biodiversity. *GCB Bioenergy* 2, 289–309.

Dawson, W., Burslem, D.F.R.P., Hulme, P.E., 2009. The suitability of weed risk assessment as a conservation tool to identify invasive plant threats in East African rainforests. *Biological Conservation* 142, 1018–1024.

de Abrantes, R., Assuncao, J.V., Pesquero, C.R., Bruns, R.E., Nobrega, R.P., 2009. Emissions of polycyclic aromatic hydrocarbons from gasohol and ethanol vehicles. *Atmospheric Environment* 43, 648–654.

de Camargo Barros, G.S.A., Alves, L.R.A., Osaki, M., 2009. Análise dos custos econômicos do programa do biodiesel no Brasil. Paper presented at 47th Sober Congress, Porto Alegre.

de Castro Santos, M.H., 1987. Fragmentacão e informalismo na tomada de decisão: O caso da política do álcool combustível no Brasil autoritario pos-64. *Dados – Revista de Ciências Sociais* 30, 73–93.

de Figueiredo, E.B., Panosso, A.R., Romao, R., La Scala, N., 2010. Greenhouse gas emission associated with sugar production in southern Brazil. *Carbon Balance and Management* 5, doi:10.1186/1750-0680-5-3.

de Fraiture, C., Berndes, G., 2009. Biofuels and water. In Howarth, R.W., Bringezu, S. (Eds.), *Biofuels: Environmental consequences and implications of changing land use.* Proceedings of the Scientific Committee on Problems of the Environment International Biofuels Project Rapid Assessment. Cornell University, Ithaca, pp. 139–152.

de Fraiture, C., Giordano, M., Liao, Y., 2008. Biofuels and implications for agricultural water use: Blue impacts of green energy. *Water Policy* 10, 67–81.

de Gorter, H., Just, D., 2008. "Water" in the U.S. ethanol tax credit and mandate: Implications for rectangular deadweight costs and the corn-oil price relationship. *Review of Agricultural Economics* 30, 397–410.

de Janvry, A., Fafchamps, M., Sadoulet, E., 1991. Peasant household behavior: Some paradoxes explained. *Economic Journal* 101, 1400–1417.

de Miranda, A.C., Moreira, J.C., de Carvalho, R., Peres, F., 2007. Neoliberalismo, uso de agrotóxicos e a crise da soberania alimentar no Brasil. *Ciência and Saúde Coletiva* 12, 7–14.

de Vries, S.C., van de Ven, G.W.J., Ittersum, M.K., Giller, K.E., 2010. Resource use efficiency and environmental performance of nine major biofuel crops, processed by first-generation conversion techniques. *Biomass and Bioenergy* 34, 588–601.

de Wit, C.T., 1992. Resource efficiency in agriculture. *Agricultural Systems* 40, 125–151.

Decreto, 2009. Decreto 6961/09, de 17 de setembro de 2009. Government of Brazil. Available at http://www.agricultura.gov.br/arq_editor/file/Desenvolvimento_Sustentavel/Agroenergia/DECRETO%206961%20%20-%20ZONEAMENTO%2026072010_0.pdf.

Dehue, B.W., Hettinga, W., 2008. The GHG performance of Jatropha biodiesel. *Commissioned by D1 Oils.* Ecofys bv, Utrecht.

Delucchi, M., 2003. A lifecycle emissions model (LEM): Lifecycle emissions from transportation fuels, motor vehicles, transportation modes, electricity use, heating and cooking fuels and materials. Institute of Transportation Studies, University of California, Davis.

Delucchi, M.A., 2006. Lifecycle analyses of biofuels. University of California Institute of Transportation Studies, Davis.

Delzeit, R., Holm-Müller, K., 2009. Steps to discern sustainability criteria for a certification scheme of bioethanol in Brazil: Approach and difficulties. *Energy* 34, 662–668.

Demetrius, F.J., 1990. *Brazil's national alcohol program: Technology and development in an authoritarian regime.* Praeger, New York.

Demirbas, A., 2009. Political, economic and environmental impacts of biofuels: A review. *Applied Energy* 86, S108–S117.

Dennis, R.A., Mayer, J., Applegate, G., Chokkalingam, U., Pierce, C.J., Kurniawan, I., Lachowski, H., Maus, P., Permana, R.P., Ruchiat, Y., Stolle, F., Suyanto, S., Tomich, T.P., 2005. Fire, people and pixels: Linking social science and remote sensing to understand underlying causes and impacts of fires in Indonesia. *Human Ecology* 33, 465–504.

Dessus, S., Herrera, S., de Hoyos, R.E., 2008. The impact of food inflation on urban poverty and its monetary cost: Some back-of-the-envelope calculations. *Agricultural Economics* 39, 417–429.

Devereux, S., 2009. Why does famine persist in Africa? *Food Security* 1, 25–35.

DIEESE, 2007. Desempenho do setor sucroalcooleiro brasileiro e os trabalhadores. Departamento Intersindical de Estatística e Estudos Socio-econômicos, São Paulo.

Dillon, H.S., Laan, T., Dillon, H.S., 2008. *Biofuels – at what cost? Government support for ethanol and biodiesel in Indonesia.* International Institute of Sustainable Development, Winnipeg.

Diop, A.M., 1999. Sustainable agriculture: New paradigms and old practices? Increased production with management of organic inputs in Senegal. *Environment, Development, and Sustainability* 1, 285–296.

DME, 2007. Biofuels industrial strategy of the Republic of South Africa. Department of Minerals and Energy, Pretoria.

Dockery, D.W., Pope, C.A., Xu, X., Spengler, J.D., Ware, J.H., Fay, M.E., Ferris, B.G., Speizer, F.E., 1993. An association between air pollution and mortality in six U.S. cities. *The New England Journal of Medicine* 329, 1753–1759.

DOE, 2010. *Energy efficiency and renewable energy. Alternative fuels and advanced vehicles data center.* Department of Energy (DOE), Washington, DC.

Domalsky, E.S., Jobe, T.L., Milne, T.A., 1986. *Thermodynamic data for biomass conversion and waste inceniration.* Solar Technical Information Program of the Solar Energy Research Institute.

Donald, P.F., 2004. Biodiversity impacts of some agricultural commodity production systems. *Conservation Biology* 18, 17–37.

Doornbosch, R., Steenblik, R., 2007. Biofuels: Is the cure worse than the disease? Organization for Economic Co-operation and Development, Paris.

Dornburg, V., Faaij, A.P.C., Verweij, P., Langeveld, H., van de Ven, G., van Keulen, H., van Diepen, K., Meeusen, M., Banse, M., Ros, J., van den Born, G.J., Oorschot, M., Smout, F., Van Vliet, J., Aiking, H., Londo, M., Mozaffarian, H., Smekens, K., Lysen, E., van Egmont, S., 2008. Biomass assessment: Assessment of global biomass potentials and their links to food, water, biodiversity, energy demand and economy. Netherlands Environmental Assessment Agency, Bilthoven.

Dorward, A., Kydd, J., Morrison, J., Urey, I., 2004. A policy agenda for pro-poor agricultural growth. *World Development* 32, 73–89.

Dorward, A., Poole, N., Morrison, J., Kydd, J., Urey, I., 2003. Markets, institutions and technology: Missing links in livelihoods analysis. *Development Policy Review* 21, 319–332.

Drigo, R., Lasserre, B., Marchetti, M., 2009. Patterns and trends in tropical forest cover. *Plant Biosystems* 143, 311–322.

Dufey, A. 2007. International trade in biofuels: Good for development? And good for environment? International Institute for Environment and Development, London.

Dufey, A., Baldock, D., Farmer, M., 2004. Impacts of changes in key E.U. policies on trade and production displacement of sugar and soy. Study commissioned by WWF. International Institute for Environment and Development and Institute for European Environmental Policy, London.

Dufey, A., Ferreira, M., Togeiro, L., 2007. Capacity building in trade and environment in the sugar/bioethanol sector in Brazil. UK Department for Environment, Food, and Rural Affairs, London.

Dumortier, J., Hayes, D.J., Carriquiry, M., Dong, F., Du, X., Elobeid, A., Fabiosa, J.F., Tokgoz, S., 2009. Sensitivity of carbon emission estimates from indirect land use change. Working paper 09-WP493. Center for Agriculture and Rural Development, Iowa State University, Ames.

Dunn, R.R., 2004. Managing the tropical landscape: A comparison of the effects of logging and forest conversion to agriculture on ants, birds, and lepidoptera. *Forest Ecology and Management* 191, 215–224.

Durbin, T.D., Miller, J.W., Younglove, T., Huai, T., Cocker, K., 2007. Effects of fuel ethanol content and volatility on regulated and unregulated emissions for the latest technology gasoline vehicles. *Environmental Science and Technology* 41, 4059–4064.

Durbin, T.D., Pisano, J.T., Younglove, T., Sauer, C.G., Rhee, S.H., Huai, T., Miller, J.W., MacKay, G.I., Hochhauser, A.M., Ingham, M.C., Gorse, R.A., Jr., Beard, L.K., DiCicco, D., Thompson, N., Stradling, R.J., Rutherford, J.A., Uihlein, J.P., 2004. The effect of fuel sulfur on NH3 and other emissions from 2000–2001 model year vehicles. *Atmospheric Environment* 38, 2699–2708.

Dye, P.J., 1996. Response of *Eucalyptus grandis* trees to soil water deficits. *Tree Physiology* 16, 233–238.

Dye, P.J., Versfeld, D., 2007. Managing the hydrological impacts of South African plantation forests: An overview. *Forest Ecology and Management* 251, 121–158.

Éboli, E., 2010. Minc diz que Pernambuco é "desastre do desastre." O Globo. Available at http://oglobo.globo.com/pais/mat/2008/07/01/minc_diz_que_pernambuco_desastre_do_desastre_usinas_sao_multadas_em_120_milhoes-547045643.asp.

EC, 2009. Directive 2009/28/EC on the promotion of the use of energy from renewable sources. *Official Journal of the European Union* L140, 16–62.

EcoEnergy, 2008. *Mozambique biofuel assessment. Final report.* Ministry of Agriculture and the Ministry of Energy of Mozambique, Maputo.

Edwards, R., Larivé, J.-F., Mahieu, V., Rouveirolles, P., 2007b. Well-to-wheels analysis of future automotive fuels and powertrains in the European context, Version 2c, March 2007. CONCAWE, European Council for Automotive R&D and European Commission Joint Research Centre, Brussels.

Edwards, S., Asmelash, A., Araya, H., Berhan, T., Egziabher, G., 2007a. Impact of compost use on crop yields in Tigray, Ethiopia. Food and Agriculture Organization, Rome.

Eide, A., 2009. The right to food and the impact of liquid biofuels (agrofuels). Right to Food Studies. Food and Agriculture Organization, Rome.

Eller, A.S.D., Sekimoto, K., Gilman, J.B., Kuster, W.C., de Gouw, J.A., Monson, R.K., Graus, M., Crespo, E., Warneke, C., Fall, R., 2011. Volatile organic compounds emissions from switchgrass cultivars used as biofuel crops. *Atmospheric Environment* 45, 3333–3337.

Ellis, F., 1988. *Peasant economics.* Cambridge University Press, Cambridge.

Energia, 2009. Biofuels for sustainable rural development and empowerment of women: Case studies from Africa and Asia. Energia, Leusden. Available at http://www.energia.org/nl/knowledge-centre/energia-publications/biofuels-case-studies/.

EPA, 2002. A comprehensive analysis of biodiesel impacts on exhaust emissions. Draft Technical Report EPA420-P-02-001. Environmental Protection Agency, Washington, DC.

EPA, 2007. Regulatory impact analysis: Renewable fuel standard program. Report EPA420-R-07-004. Environmental Protection Agency, Washington, DC.

EPA, 2009. EPA lifecycle analysis of greenhouse gas emissions from renewable fuels. Report EPA-420-F-09-024. Environmental Protection Agency, Washington, DC.

EPA, 2010. EPA lifecycle analysis of greenhouse gas emissions from renewable fuels. Environmental Protection Agency, Ann Arbor.

Erb, K.-H., Gaube, V., Krausmann, F., Plutzar, C., Bondeau, A., Haberl, H., 2007. A comprehensive global 5min resolution land-use dataset for the year 2000 consistent with national census data. *Journal of Land Use Science* 2, 191–224.

Erb, K.-H., Haberl, H., Krausmann, F., Lauk, C., Plutzar, C., Steinberger, J.K., Müller, C., Bondeau, A., Waha, K., Pollack, G., 2009. Eating the planet: Feeding and fuelling the world sustainably, fairly and humanely, a scoping study. Social Ecology Working Paper. Institute of Social Ecology, Vienna, and PIK Potsdam, Potsdam.

Esbert, R.M., Diaz-Pache, F., Grossi, C.M., Alonso, F.J., Ordaz, J., 2001. Airborne particulate matter around the Cathedral of Burgos (Castilla y Leon, Spain). *Atmospheric Environment* 35, 441–452.

EU, 2009. Directive 2009/28/EC on the promotion of the use of energy from renewable sources. *Official Journal of the European Union* L140, 16–62.

Everson, C.S., 2001. The water balance of a first order catchment in the montane grasslands of South Africa. *Journal of Hydrology* 241, 110–123.

Everson, C.S., Mengistu, M., Gush, M.B., in press. A field assessment of the agronomic performance and water use of Jatropha curcas in South Africa. *Biomass and Bioenergy*.

Ewing, M., Msangi, S., 2009. Biofuels production in developing countries: Assessing tradeoffs in welfare and food security. *Environmental Science and Policy* 12, 520–528.

Extra Alagoas, 2008. Firma responsável pelas operações do terminal açucareiro muda o código de atividade para sonegar impostos. Extra Alagoas. Available at http://www.extralagoas.com.br/noticia.kmf?noticia=7504353&canal=333.

Faay, A.P.C., 1997. *Energy from biomass and waste*. PhD thesis, University of Utrecht.

Fabiosa, J.F., Beghin, J.C., Dong, F., Elobeid, A., Fuller, F., Matthey, H., Tokgoz, S., Wailes, E., 2007. The impact of the European enlargement and CAP reforms on agricultural markets: Much ado about nothing? *Journal of International Agricultural Trade and Development* 3, 57–69.

Fabiosa, J.F., Beghin, J.C., Dong, F., Elobeid, A., Tokgoz, S., Yu, T.H., 2009. Land allocation effects of the global ethanol surge: Predictions from the international FAPRI model. Working paper 09-WP 488. Center for Agricultural and Rural Development, Iowa State University, Ames.

Fabiosa, J.F., Beghin, J.C., Dong, F., Elobeid, A., Tokgoz, S., Yu, T., 2010. Land allocation effects of the global ethanol surge: Predictions from the international FAPRI model. *Land Economics* 86, 687–706.

Fafchamps, M., 1992. Solidarity networks in pre-industrial societies: Rational peasants with a moral economy. *Economic Development and Cultural Change* 41, 147–174.

Fafchamps, M., 1993. Sequential labor decisions under uncertainty: An estimable household model of west-African farmers. *Econometrica* 6, 1173–1197.

Fairbanks, M., 2009. Glicerina: Crescimento do biodiesel provoca inundação no mercado de glicerina, incentivando a descobrir novas aplicações. *Revista Química e Derivados* 487. Available at http://www.quimica.com.br/revista/qd487/glicerina/glicerina01.htm.

Fairhurst, T., McLaughlin, D., 2009. Sustainable oil palm development on degraded land in Kalimantan. World Wildlife Fund for Nature, Washington, DC.

Fairless, D., 2007. Biofuel: The little shrub that could – maybe. *Nature* 449, 652–655.

Falkenmark, M., 1989. Comparative hydrology – a new concept. In Falkenmark, M., Chapman, T. (Eds.), *Comparative hydrology: An ecological approach to land and water resources*. United Nations Educational, Scientific, and Cultural Organization, Paris, pp. 10–42.

FAO, 1996. Declaration on world food security. World Food Summit. Food and Agriculture Organization, Rome.

FAO, 2002a. World agriculture: Towards 2015/2030 – Summary report. Food and Agriculture Organization, Rome.

FAO, 2002b. Small-scale palm oil processing in Africa. FAO Agricultural Services Bulletin 148. Food and Agriculture Organization, Rome.

FAO, 2004. Fertiliser use by crop in Brazil. Food and Agriculture Organisation, Rome.

FAO, 2005. FAO country profiles and mapping information system – Compendium of Food and Agriculture Indicators. Food and Agriculture Organization, Rome.

FAO, 2006a. World agriculture: Towards 2030/2050: Prospects for food, nutrition, agriculture and major commodity groups. Food and Agricultural Organization, Rome.

FAO, 2006b. Choosing forest definition for the clean development mechanism. Forest and climate change working paper 4. Food and Agriculture Organization, Rome.

FAO, 2008a. FAOSTAT databases. Food and Agriculture Organization, Rome.

FAO, 2008b. Biofuels: Prospects, risks and opportunities: The state of food and agriculture. Food and Agriculture Organization, Rome.

FAO, 2009a. Small-scale bioenergy initiatives: Brief description and preliminary lessons on livelihood impacts from case studies in Asia, Latin America and Africa. Food and Agricultural Organization, Rome.

FAO, 2009b. The state of food insecurity in the world, economic crises: Impacts and lessons learned. Food and Agriculture Organization, Rome.

FAO, 2010a. Jatropha: A smallholder's bioenergy crop. The potential for pro-poor development. Food and Agriculture Organization, Rome.

FAO, 2010b. Low-income food-deficit countries (LIFDC) – List for 2010. Food and Agricultural Organization, Rome.

FAO, 2010c. Bioenergy and food security: The BEFS analysis for Tanzania. Food and Agriculture Organization, Rome.

FAO, 2010d. The State of Food Insecurity in the World. Addressing food insecurity in protracted crises. Food and Agriculture Organization, Rome.

FAO, 2011. FAOSTAT databases. Food and Agriculture Organization, Rome.

FAPRI, 2004. Documentation of the FAPRI modeling system. FAPRI-UMC Report 12-04. Food and Agricultural Policy Research Institute, University of Missouri, Columbia.

FAPRI, 2008. Model of the U.S. ethanol market. FAPRI-UMC Report 07-08. Food and Agricultural Policy Research Institute, University of Missouri, Columbia.

FAPRI, 2010. FAPRI 2010 U.S. and World Agricultural Outlook. Iowa State University and University of Missouri Staff Report 1-09. Food and Agricultural Policy Research Institute, Ames.

FAPRI, 2011. FAPRI models. Food and Agriculture Policy Research Institute, Ames. Available at http://www.fapri.iastate/edu/models/.

Fargione, J., Hill, J., Tilman, D., Polasky, S., Hawthorne, P., 2008. Land clearing and the biofuel carbon debt. *Science* 319, 1235–1238.

Fargione, J., Plevin, R., Hill, J., 2010. The ecological impact of biofuels. *Annual Review of Ecology, Evolution and Systematics* 41, 351–377.

Farinelli, B., Carter, C.A., Cynthia Lin, C.Y., Sumner, D.A., 2009. Import demand for Brazilian ethanol: A cross-country analysis. *Journal of Cleaner Production* 17, S9–S17.

Fearnside, P.M., 2001. Soybean cultivation as a threat to the environment. *Environmental Conservation* 28, 23–38.

FEDEPALMA, 2009. Anuario estadístico 2009: Estadísticas de la palma de aceite en Colombia. Federación Nacional de Cultivadores de Palma de Aceite, Bogota.

Fernandes, B.M., Welch, C.A., Gonçalves, E.C., 2010. Agrofuel policies in Brazil: Paradigmatic and territorial disputes. *Journal of Peasant Studies* 37, 793–819.

FIAN, 2008. Agrofuels in Brazil. Food First Information and Action Network, Heidelberg.

Field, C.B., Campbell, J.E., Lobell, D.B., 2008. Biomass energy: The scale of the potential resource. *Trends in Ecology and Evolution* 23, 65–72.

Filho, G.M., Figueiredo, O., 2008. EUA acusam Brasil de explorar trabalho escravo. O Estado de S.Paulo. Available at http://www.espacopublico.blog.br/?p=3635.

Filoso, S., Martinelli, L.A., Williams, M.R., Lara, L.B., Krusche, A., Ballester, M.V., Victoria, R.L., Camargo, P.B., 2003. Land use and nitrogen export in the Piracicaba River basin, southeast Brazil. *Biogeochemistry* 65, 275–294.

Finco, M.V.A., Doppler, W., 2010. Bioenergy and sustainable development: The dilemma of food security and climate change in the Brazilian savannah. *Energy for Sustainable Development* 14, 194–199.

Finlayson-Pitts, B.J., Pitts, J.N., 2000. Chemistry of the upper and lower atmosphere: Theory, experiments and applications. Academic Press, San Diego.

FIPE, 1991. Avaliação do programa de agroindustria canavieira em Pernambuco, vol. 2, estudo exploratorio sobre a diversificação agropecuaria na zona da Mata de Pernambuco. Fundação Instituto Pernambuco, Recife.

Firbank, L., 2008. Assessing the ecological impacts of bioenergy projects. *Bioenergy Research* 1, 12–19.

Fischer, G., Heilig, G.K., 1997. Population momentum and the demand on land and water resources. *Philosophical Transactions of the Royal Society B: Biological Sciences* 352, 869–889.

Fischer, G., van Velthuizen, H., Nachtergaele, F.O., 2000. Global agro-ecological zones assessment: Methodology and results. IIASA interim report IR-00-064. International Institute for Applied Systems Analysis, Laxenburg.

Fischer, G., Hizsnyik, E., Prielder, S., Shah, M., van Velthuizen, H., 2009. Biofuels and food security. International Institute for Applied Systems Analysis, Vienna.

Fischer, J., 2009. A fair deal for forest people: Working to ensure that REDD forests bear fruit for local communities. Available at http://news.mongabay.com/2009/1127-redd_commentary_ffi.html.

Fischer, J., Brosi, B., Daily, G.C., Ehrlich, P.R., Goldman, R., Goldstein, J., Lindenmayer, D.B., Manning, A.D., Mooney, H.A., Pejchar, L., Ranganathan, J., Tallis, H., 2008. Should agricultural policies encourage land sparing or wild-life friendly farming? *Frontiers in Ecology and the Environment* 6, 380–385.

Fitzherbert, E.B., Struebig, M.J., Morel, A., Danielsen, F., Brühl, C.A., Donald, P.F., Phalan, B., 2008. How will oil palm expansion affect biodiversity? *Trends in Ecology and Evolution* 23, 538–545.

F.O. Licht, 2010. World ethanol and biofuels report. F.O. Licht, London. Available at http://www.agra-net.com/portal2/home.jsp?template=productpage&pubid=ag072.

F.O. Licht, n.d. Agra-net.com: Agra Informa web site. F.O. Licht, London. Available at http://www.agra-net.com/portal2/.

Foley, G., 1990. *Electricity for rural people*. Panos Institute, London.

Folha de Pernambuco, 2009. *Resultado não satisfaz Sindaçúcar*. Folha de Pernambuco. Available at http://host-1-14-127.hotlink.com.br/index.php/component/content/article/54-economia/530088-resultado-nao-satisfaz-sindacucar.

Folke, C., 2006. Resilience: The emergence of a perspective for social-ecological systems analyses. *Global Environmental Change* 16, 253–267.

Fontana, M., Wood, A., 2000. Modeling the effect of trade on women at work and at home. *World Development* 28, 1173–1190.

Forster, P., Ramaswamy, V., Artaxo, P., Berntsen, T., Betts, R., Fahey, D.W., Haywood, J., Lean, J., Lowe, D.C., Myhre, G., Nganga, J., Prinn, R., Raga, G., Schulz, M., Van Dorland, R., 2007. Changes in atmospheric constituents and in radiative forcing. In Solomon, S., Qin, D., Manning, M., Chen, Z., Marquis, M., Averyt, K.B., Tignor, M., Miller, H.L. (Eds.), *Climate change 2007: The physical science basis. Contribution of Working Group I to the fourth assessment report of the Intergovernmental Panel on Climate Change.* Cambridge University Press, Cambridge, pp. 129–234.

Franco, J., Levidow, L., Fig, D., Goldfarb, L., Honicke, M., Mendonca, M.L., 2010. Assumptions in the European Union biofuels policy: Frictions with experiences in Germany, Brazil and Mozambique. *Journal of Peasant Studies* 37, 611–698.

Friends of the Earth, 2008. Malaysian palm oil: Green gold or green wash? Social justice, forests, and agrofuels. Friends of the Earth International, Brussels.

Friends of the Earth, 2010. Africa up for grabs: The scale and impact of land-grabbing for agrofuels. Friends of the Earth International, Brussels.

Fritsche, U.R., 2008. *Impacts of biofuels on greenhouse gas emissions*. Food and Agriculture Organization, Rome.

Fritsche, U.R., Weigmann, K., 2008. Externe expertise für das WBGU-hauptgutachten "Welt im wandel: Zukunftsfähige bioenergie und nachhaltige landnutzung." German Advisory Council on Global Change, Berlin.

Fritz, T., 2008. Agroenergia na América Latina. In *Estudo de caso de quatro países: Brasil, Argentina, Paraguai e Colômbia*. Brot für die Welt Forschungs-und Dokumentationszentrum Chile-Lateinamerika, Berlin.

Frizzone, J.A., Santo Matioli, C., Rezende, R., Gonçalves, A.C.A. 2001. Viabilidade econômica da irrigação suplementar da cana-de-açúcar, Saccharum spp., para a região Norte do Estado de São Paulo. *Acta Scientiarum Maringá* 23, 1131–1137.

Fulton, L., Howes, T., Hardy, J., 2004. *Bio-fuels for transport: An international perspective*. International Energy Agency, Paris.

Furtado, A.T., Scandiffio, M.I.G., Cortez, L.A.B., 2011. The Brazilian sugarcane innovation system. *Energy Policy* 39, 156–166.

Galtung, J.A., 1971. Structural theory of imperialism. *Journal of Peace Research* 8, 81–117.

Gan, L., Yu, J., 2007. Bioenergy transition in rural China: Policy options and co-benfits. *Energy Policy* 36, 531–540.

Garcez, C.A.G., Vianna, J., N.d.S., 2009. Brazilian biodiesel policy: Social and environmental considerations of sustainability. *Energy* 34, 645–654.

Garcia, A.R., 1988. Libertos e sujeitos: Sobre a transição para trabalhadores livres do nordeste. *Revista Brasileira do Ciências Sociais* 3, 5–41.

Garcia, C.A., Fuentes, A., Hennecke, A., Riegelhaupt, E., Manzini, F., Masera, O., 2011. Life-cycle greenhouse gas emissions and energy balances of sugarcane ethanol production in Mexico. *Applied Energy* 6, 2088–2097.

Gasparatos, A., Stromberg, P., Takeuchi, K., 2011. Biofuels, ecosystem services and human wellbeing: Putting biofuels in the ecosystem services narrative. *Agriculture, Ecosystems, and Environment* 142, 111–128.

Gauder, M., Graeff-Hönninger, S., Claupein, W., 2011. The impact of a growing bioethanol industry on food production in Brazil. *Applied Energy* 88, 672–679.

GBO3, 2010. Global Biodiversity Outlook 3. Secretariat of the Convention on Biological Diversity, Montréal.

GEF-STAP, 2006. *Report of the GEF-STAP workshop on liquid biofuels*. Global Environment Facility's Scientific and Technical Advisory Panel, Washington, DC.

Geist, H.J., Lambin, E.F., 2002. Proximate causes and underlying driving forces of tropical deforestation. *Bioscience* 52, 143–150.

GEM, 2009. Commencement of commercial jatropha oil production. Company news release, November 25. GEM BioFuels, Douglas, Isle of Man.

Gerbens-Leenes, P.W., Hoekstra, A.Y., 2012. The water footprint of sweeteners and bio-ethanol. *Environment International* 40, 202–211.

Gerbens-Leenes, P.W., Hoekstra, A.Y., van der Meer, T.H., 2009a. The water footprint of bioenergy. *Proceedings of the National Academy of Sciences of the United States of America* 106, 10219–10223.

Gerbens-Leenes, P.W., Hoekstra, A.Y., van der Meer, T.H., 2009b. The water footprint of bioenergy and other primary energy carriers. *Ecological Economics* 68, 1052–1060.

Gerbens-Leenes, P.W., Hoekstra, A.Y., van der Meer, T.H., 2009c. Reply to: Maes et al., A global estimate of the water footprint of *Jatropha curcas* under limited data availability. *Proceedings of the National Academy of Sciences of the United States of America* 106, E113.

German, L., Schoneveld, G.C., Gumbo, D., 2010. The local social and environmental impacts of large-scale investments in biofuels in Zambia. Report prepared as part of the European Community Contribution Agreement EuropeAid/ENV/2007/143936/TPS. Center for International Forestry Research, Bogor.

Gerten, D., Rost, S., Bloh, W.V., Lucht, W., 2008. Causes of change in 20th century global river discharge. *Geophysical Research Letters* 35, L20405, doi:10.1029/2008GL035258.

GEXSI, 2008a. Global market study on jatropha, project inventory: Latin America. Global Exchange for Social Investment. Prepared for World Wide Fund for Nature, London and Berlin.

GEXSI, 2008b. Global market study on jatropha, project inventory: Asia. Global Exchange for Social Investment. Prepared for: World Wide Fund for Nature, London and Berlin.

GEXSI, 2008c. Global market study on jatropha, project inventory: Africa. Global Exchange for Social Investment. Prepared for: World Wide Fund for Nature, London and Berlin.

GEXSI, 2008d. Jatropha biofuel 2006–2008. Global Exchange for Social Investment, Berlin.

GFC, 2010. Open letter: Growing opposition to Round Table on Responsible Soy. Global Forest Coalition, Amsterdam. Available at http://www.globalforestcoalition.org/?p=396.

Ghazoul, J., Koh, L.P., Butler, R.A., 2010a. A REDD light for wildlife friendly farming. *Conservation Biology* 24, 644–645.

Ghazoul, J., Butler, R.A., Mateo-Vega, J., Koh, L.P., 2010b. REDD: A reckoning of environment and development implications. *Trends in Ecology and Evolution* 25, 396–402.

Ghezehei, S.B., Annandale, J.G., Everson, C.S., 2009. Shoot allometry of *Jatropha curcas*. *Southern Forests* 71, 279–286.

Giampietro, M., Ulgiati, S., 2005. Integrated assessment of large-scale biofuel production. *Critical Reviews in Plant Sciences* 24, 365–384.

Gibbs, H.K., Johnston, M., Foley, J.A., Holloway, T., Monfreda, C., Ramankutty, N., Zaks, D., 2008. Carbon payback times for crop-based biofuel expansion in the tropics: The effects of changing yield and technology. *Environmental Research Letters* 3, 1–10.

Gibbs, H., Ruesch, S., Achard, F., Clayton, M., Holmgren, P., Ramankutty, N., Foley, J., 2010. Tropical forests were the primary sources of new lands 1980–2000. *Proceedings of the National Academy of Sciences of the United States of America* 107, 16732–16737.

Gibbs Russell, G.E., 1986. Significance of different centres of diversity in sub-families of Poaceae in southern Africa. *Palaeoecology of Africa* 17, 183–191.

Ginnebaugh, D.L., Liang, J., Jacobson, M.Z., 2010. Examining the temperature dependence of ethanol (E85) versus gasoline emissions on air pollution with a largely-explicit chemical mechanism. *Atmospheric Environment* 44, 1192–1199.

GLCF, 2010. Global Land Cover Facility (GLCF). University of Maryland, College Park. Available at ftp://ftp.glcf.umiacs.umd.edu/modis/VCC.

Gleick, P.H., 1993. Water and energy. In Gleick, P.H. (Ed.), *Water in crisis: A guide to the world's freshwater resources*. Oxford University Press, New York, pp. 67–79.

Gleick, P.H., 1994. Water and energy. *Annual Review of Energy and the Environment* 19, 267–299.

Gmunder, S.M., Zah, R., Bhatacharjee, S., Classen, M., Mukherjee, P., Widmer, R., 2010. Life cycle assessment of village electrification based on straight jatropha oil in Chhattisgarh, India. *Biomass and Bioenergy* 34, 347–355.

Gnansounou, E., Dauriat, A., Villegas, J., Panichelli, L., 2009. Life cycle assessment of biofuels: Energy and greenhouse gas balances. *Bioresource Technology* 100, 4919–4930.

GNESD, 2007. *Reaching the Millennium Development Goals and beyond: Access to modern forms of energy as a prerequisite*. Global Network on Energy for Sustainable Development, Roskilde.

Godfray, H.C., Beddington, J.R., Crute, I.R., Haddad, L., Lawrence, D., Muir, J.F., Pretty, J., Robinson, S., Thomas, S.M., Toulmin, C., 2010. Food security: The challenge of feeding 9 billion people. *Science* 327, 812–818.

Goebes, M.D., Strader, R., Davidson, C., 2003. An ammonia emission inventory for fertilizer application in the United States. *Atmospheric Environment* 37, 2539–2550.

Goldemberg, J., 2007. Ethanol for a sustainable energy future. *Science* 315, 808–810.

Goldemberg, J., 2008. The Brazilian biofuels industry. *Biotechnology Biofuels* 1, doi: 10.1186/1754-6834-1-6.

Goldemberg, J., Guardabassi, P., 2009. Are biofuels a feasible option? *Energy Policy* 37, 10–14.

Goldemberg, J., Johansson, T.B., 2004. *World energy assessment – overview: 2004 update.* United Nations Development Programme, New York.

Goldemberg, J., Coelho, S.T., Mastari, P.M., Lucon, O., 2004a. Ethanol learning curve: The Brazilian experience. *Biomass and Bioenergy* 26, 301–304.

Goldemberg, J., Coelho, S.T., Lucon, O., 2004b. How adequate policies can push renewables. *Energy Policy* 32, 1141–1146.

Goldemberg, J., Coelho, S.T., Guardabassi, P., 2008. The sustainability of ethanol production from sugarcane. *Energy Policy* 36, 2086–2097.

Gonçalves, J.S., Castanho Filho, E.P., 2006. Defesa da reserva legal and a complexidade da agropecuária paulista. *Análise and Indicadores do Agronegócio* 1, Available at http://www.iea.sp.gov.br/out/verTexto.php?codTexto=6415.

Gonçalves, J.S., Souza, S.A.M., Ghobril, C.N., 2007. Agropecuária paulista: especialização regional and mudanças na composição de culturas de 1969–1971 a 2002–2006. Instituto de Economia Agrícola, São Paulo.

Gordon, D.R., Onderdonk, D.A., Fox, A.M., Stocker, R.K., 2008. Consistent accuracy of the Australian weed risk assessment system across varied geographies. *Diversity and Distributions* 14, 234–242.

Graham, L.A., Belisle, S.L., Baas, C., 2008. Emissions from light duty gasoline vehicles operating on low blend ethanol gasoline and E85. *Atmospheric Environment* 42, 4498–4516.

Grant, T., Beer, T., Campbell, P.K., Batten, D., 2008. Life cycle assessment of environmental outcomes and greenhouse gas emissions from biofuels production in western Australia. Department of Agriculture and Food, Government of Western Australia, Perth.

Grenier, P., 1985. The alcohol plan and the development of Northeast Brazil. *GeoJournal* 11, 61–68.

Grubler, A., Nakićenovic, N., Victor, D.G., 1999. Dynamics of energy technologies and global change. *Energy Policy* 7, 247–280.

GSI, 2008. Biofuels – at what cost? Government support for ethanol and biodiesel in China. Global Studies Initiative of the International Institute for Sustainable Development, Geneva.

GSI, 2011. Biofuel subsidies. Global Subsidies Initiative, Geneva.

GTZ, 2009. International survey of fuel prices. Deutsche Gesellschaft für Internationale Zusammenarbeit. Available at http://www.gtz.de/en/themen/29957.htm.

Guedes, L.F., 2010. Perspectivas para o setor sucroalcooleiro. Folha de Pernambuco. Available at http://www.folhape.com.br/index.php/luis-fernando-guedes-coluna/579066-luiz-fernando-guedes-11072010.

Gunkel, G., Kosmol, J., Sobral, M., Rohn, H., Montenegro, S., Aureliano, J., 2007. Sugar cane industry as a source of water pollution: Case study on the situation in Ipojuca River, Pernambuco, Brazil. *Water, Air, and Soil Pollution* 180, 261–269.

Gupta, R.B., Demirbas, A., 2010. *Gasoline, diesel and ethanol biofuels from grasses and plants.* Cambridge University Press, New York.

Gush, M.B., 2008. Measurement of water-use by *Jatropha curcas* L. using the heat-pulse velocity technique. *Water SA* 34, 579–584.

Gush, M.B., Hallowes, J., 2007. Assessing the likely water use impacts of large-scale *Jatropha curcas* production in South Africa. In *Proceedings of the 13th SANCIAHS Symposium,* Cape Town.

Gusmão, M., 1998. Êxodo nordestino: Fim de empréstimos baratos e incentivos leva usineiros a investir em outras regiões. Veja. Available at http://veja.abril.com.br/280198/p_082.html.

Gutmann, A., Thompson, D., 2004. *Why deliberative democracy?* Princeton University Press, Princeton.

Guyon, A., Simorangkir, D., 2002. The economics of fire use in agriculture and forestry: A preliminary review for Indonesia. International Union for Conservation of Nature Project Fire Fight South East Asia, Jakarta.

Haberl, H., Beringer, T., Bhattacharya, S.C., Erb, K.-H., Hoogwijk, M., 2010. The global technical potential of bio-energy in 2050 considering sustainability constraints. *Current Opinion in Environmental Sustainability* 2, 394–403.

Haberl, H., Erb, K.-H., Krausmann, F., Adensam, H., Schulz, N.B., 2003. Land-use change and socioeconomic metabolism in Austria. Part II: Land-use scenarios for 2020. *Land Use Policy* 20, 21–39.

Haberl, H., Erb, K.-H., Krausmann, F., Bondeau, A., Lauk, C., Müller, C., Plutzar, C., Steinberger, J.K., 2011. Global bioenergy potentials from agricultural land in 2050: Sensitivity to climate change, diets and yields. *Biomass and Bioenergy* 35, 4753–4769.

Haberl, H., Erb, K.-H., Krausmann, F., Gaube, V., Bondeau, A., Plutzar, C., Gingrich, S., Lucht, W., Fischer-Kowalski, M., 2007. Quantifying and mapping the human appropriation of net primary production in earth's terrestrial ecosystems. *Proceedings of the National Academy of Sciences of the United States of America* 104, 12942–12947.

Haberl, H., Geissler, S., 2000. Cascade utilisation of biomass: How to cope with ecological limits to biomass use. *Ecological Engineering* 16, S111–S121.

Habitat, 1993. Application of biomass-energy technologies. United Nations Centre for Human Settlements (Habitat), Nairobi.

Hagens, N.R., Costanza, R., Mulder, K., 2006. Energy returns on ethanol production. *Science* 312, 1746–1746.

Haggblade, S., Hazell, P.B.R., Gabre-Madhin, E., 2010. Challenges for African agriculture. In Haggblade, S., Hazell, P.B.R. (Eds.), *Successes in African agriculture: Lessons for the future*. Published for the International Food Policy Research Institute. Johns Hopkins University Press, Baltimore, pp. 3–26.

Haguiuda, C., Veneziani, R., 2006. Eficiência energética no saneamento básic. 3o Congresso Brasileiro de Eficiência Energética e Cogeração de Energia da ABESCO, São Paulo.

Halberg, N., Sulser, T.B., Høgh-Jensen, H., Rosegrant, M.W., Knudsen, M.T., 2006. The impact of organic farming on food security in a regional and global perspective. In Halberg, N., Alrøe, H.F., Knudsen, M.T., Kristensen, E.S. (Eds.), *Global development of organic agriculture: Challenges and prospects*. CABI, Wallingford, pp. 277–322.

Hall, C.A.S., Cleveland, C.J., Kaufmann, R., 1986. *Energy and resource quality*. John Wiley, New York.

Hall, J., Matos, S., Severino, L., Beltrao, N., 2009. Brazilian biofuels and social exclusion: Established and concentrated ethanol versus emerging and dispersed biodiesel. *Journal of Cleaner Production* 17, 577–585.

Hallowes, J., 2007. Bio-physical potential of *Jatropha curcas*. In Holl, M., Gush, M.B., Hallowes, J., Versfeld, D.B. (Eds.), *Jatropha curcas in South Africa: An assessment of its water use and bio-physical potential*. Water Research Commission, Pretoria, pp. 104–120.

Halonen, J.I., Lanki, T., Tiittanen, P., Niemi, J.V., Loh, M., Pekkanen, J., 2010. Ozone and case-specific cardiorespiratory morbidity and mortality. *Journal of Epidemiology and Community Health* 64, 814–820.

Hannah, L., Midgley, G.F., Millar, D., 2000. Climate change-integrated conservation strategies. *Global Ecology and Biogeography* 11, 485–495.

Hansen, M.C., Townshend, J.R.G., DeFries, R.S., Carroll, M., 2005. Estimation of tree cover using MODIS data at global, continental and regional/local scales. *International Journal of Remote Sensing* 26, 4359–4380.

Hansen, M.C., Stehman, S.V., Potapov, P.V., Thomas, R., Loveland, T.R., John, R.G., Townshend, J.R.G., DeFries, R.S., Pittman, K.W., Arunarwati, B., Stolle, F., Steininger, M.K.,

Carroll, M., DiMiceli, C., 2008. Humid tropical forest clearing from 2000 to 2005 quantified by using multitemporal and multiresolution remotely sensed data. *Proceedings of the National Academy of Sciences of the United States of America* 105, 9439–9442.

Hartemink, A.E., 2006. Soil erosion: Perennial crop plantations. *In Encyclopedia of soil science*. Taylor and Francis, New York, pp. 1613–1617.

Harvey, M., McMeekin, A., 2010. Political shaping of transitions to biofuels in Europe, Brazil and the USA. Paper presented at Sussex Energy Group conference "Energy transitions in an interdependent world: What and where are the future social science research agendas?" University of Sussex, Brighton.

Hassall, M., Jones, D.T., Taiti, S., Latipi, Z., Sutton, S.L., Mohammed, M., 2006. Biodiversity and abundance of terrestrial isopods along a gradient of disturbance in Sabah, East Malaysia. *European Journal of Soil Biology* 42, S197–S207.

Havlík, P., Schneider, U.A., Schmid, E., Böttcher, H., Fritz, S., Skalský, R., Aoko, K., de Cara, S., Kindermann, G., Kraxner, F., Leduc, S., McCallum, I., Mosnier, A., Sauer, T., Obersteiner, M., 2011. Global land-use implications of first and second generation biofuel targets. *Energy Policy* 39, 5690–5702.

Hayes, D.J., Babcock, B.A., Fabiosa, J., Tokgoz, S., Elobeid, A., Yu, T., Dong, F., Hart, C., Thompson, W., Meyer, S., Chavez, E., Pan, S., 2009. Biofuels: Potential production capacity, effects on grain and livestock sectors, and implications for food prices and consumers. *Journal of Agriculture and Applied Economics* 41, 465–491.

Haywood, L., von Maltitz, G.P., Setzkorn, K., Ngepah, N., 2008. Biofuel production in South Africa, Mozambique, Malawi and Zambia: A status quo analysis of the social, economic and biophysical elements of the biofuel industry in Southern Africa. Council for Scientific and Industrial Research, Pretoria.

Heller, J., 1996. Physic nut – *Jatropha curcas* L. International Plant Genetic Resources Institute, Rome.

Hellmann, F., Verburg, P.H., 2010. Impact assessment of the European biofuel directive on land use and biodiversity. *Journal of Environmental Management* 91, 1389–1396.

Henning, R.K., 2006. *Jatropha curcas* L. in Africa: Assessment of the impact of the dissemination of "the Jatropha System" on the ecology of the rural area and the social and economic situation of the rural population (target group) in selected countries in Africa. Global Facilitation Unit for Underutilized Species, Rome.

Herrero, M., Thornton, P.K., Gerber, P., Reid, R.S., 2009. Livestock, livelihoods and the environment: Understanding the trade-offs. *Current Opinion in Environmental Sustainability* 1, 111–120.

Hertel, T.W., 2008. Implications of US biofeuls production for global land use. Presentation to California Air Resources Board, Sacramento.

Hertel, T.W., Golub, A.A., Jones, A.D., O'Hare, M., Plevin, R.J., Kammen, D.M., 2010. Effects of US maize ethanol on global land use and greenhouse gas emissions: Estimating market-mediated responses. *BioScience* 60, 223–231.

Hess, P., Johnston, M., Brown-Steiner, B., Holloway, T., de Andrade, J.B., Artaxo, P., 2009. Air quality issues associated with biofuel production and use. In Howarth, R.W., Bringezu, S. (Eds.), *Biofuels: Environmental consequences and interactions of changing land use. Proceedings of the Scientific Committee on Problems of the Environment (SCOPE) International Biofuels Project Rapid Assessment*. Cornell University, Ithaca, pp. 169–194.

Hewitt, C.N., MacKenzie, A.R., Di Carlo, P., Di Marco, C.F., Dorsey, J.R., Evans, M., Fowler, D., Gallagher, M.W., Hopkins, J.R., Jones, C.E., Landford, B., Lee, J.D., Lewis, A.C., Lim, S.F., McQuaid, J., Misztal, P., Moller, S.J., Monks, P.S., Nemitz, E., Oram, D.E., Owen, S.M., Phillips, G.J., Pugh, T.A.M., Pyle, J.A., Reeves, C.E., Ryder, J., Siong,

J., Skiba, U., Stewart, D.J., 2009. Nitrogen management is essential to prevent tropical oil palm plantations from causing ground-level ozone pollution. *Proceedings of the National Academy of Sciences of the United States of America* 106, 18447–18451.

Hill, J., Nelson, E., Tilman, D., Polasky, S., Tiffany, D., 2006. Environmental, economic, and energetic costs and benefits of biodiesel and ethanol biofuels. *Proceedings of the National Academy of Sciences of the United States of America* 103, 11206–11210.

Hill, J., Polasky, S., Nelson, E., Tilman, D., Huo, H., Ludwig, L., Neumann, J., Zheng, H., Bonta, D., 2009. Climate change and health costs of air emissions from biofuels and gasoline. *Proceedings of the National Academy of Sciences of the United States of America* 106, 2077–2082.

Hira, A., 2011. Sugar rush: Prospects for a global ethanol market. *Energy Policy* 39, 6925–6935.

Hoefnagels, R., Smeets, E., Faaij, A., 2010. Greenhouse gas footprints of different biofuel production systems. *Renewable and Sustainable Energy Reviews* 14, 1661–1694.

Hoekstra, A.Y., Chapagain, A.K., 2007. Water footprints of nations: Water use by people as a function of their consumption pattern. *Water Resource Management* 21, 35–48.

Hoekstra, A.Y., Gerbens-Leenes, P.W., van der Meer, T.H., 2009. Reply to: Jongschaap et al., The water footprint of *Jatropha curcas* under poor growing conditions. *Proceedings of the National Academy of Sciences of the United States of America* 106, E119–E119.

Hoekstra, A.Y., Chapagain, A.K., Aldaya, M.M., Mekonnen, M.M., 2011. *The water footprint assessment manual: Setting the global standard*. Earthscan, London.

Hoffler, H., Owour Ochieng, B., 2008. High commodity prices: Who gets the money? A case sudy on the impact of high food and factor prices on Kenyan farmers. Heinrich Boell Foundation, Berlin.

Hollander, G., 2010. Power is sweet: Sugarcane in the global ethanol assemblage. *Journal of Peasant Studies* 37, 699–721.

Holt, M.T., 1999. A linear approximate acreage allocation model. *Journal of Agricultural and Resource Economics* 24, 383–397.

Hoogwijk, M., Faaij, A., Eickhout, B., de Vries, B., Turkenburg, W., 2005. Potential of biomass energy out to 2100, for four IPCC SRES land-use scenarios. *Biomass and Bioenergy* 29, 225–257.

Hoogwijk, M., Faaij, A., van den Broek, R., Berndes, G., Gielen, D., Turkenburg, W., 2003. Exploration of the ranges of the global potential of biomass for energy. *Biomass and Bioenergy* 25, 119–133.

Hu, Z., Pu, G., Fang, F., Wang, C., 2004. Economics, environment, and energy life cycle assessment of automobiles fueled by bio-ethanol blends in China. *Renewable Energy* 29, 2183–2192.

Huang, J., Qiu, H., Yang, J., Zhang, Y., Zhang, Y., 2008. Strategies and options for integrating biofuel and rural renewable energy production into rural agriculture for poverty reduction in the greater Mekong subregion: A case study of China. Asian Development Bank, Manila.

Hubert, B., Rosegrant, M., van Boekel, M.A.J.S., Ortiz, R., 2010. The future of food: Scenarios for 2050. *Crop Science* 50, 33–50.

Huertas, D.A., Berndes, G., Holmén, M., Sparovek, G., 2010. Sustainability certification of bioethanol: How is it perceived by Brazilian stakeholders? *Biofuels, Bioproducts, and Biorefining* 4, 369–384.

Hufstadter, C., 2010. The costs of biofuels. Oxfam America. Available at http://www.oxfamamerica.org/articles/the-costs-of-biofuel.

Huo, H., Wu, Y., Wang, M., 2009. Total versus urban: Well-to-wheels assessment of criteria pollutants emissions from various vehicle/fuel systems. *Atmospheric Environment* 43, 1796–1804.

IAASTD, 2009. *Agriculture at a Crossroads. International Assessment of Agricultural Knowledge, Science, and Technology for Development (IAASTD), Global Report.* Island Press, Washington, DC.

IATP, 2008. Biofuel and global biodiversity. Institute for Agriculture and Trade Policy, Minneapolis.

IBGE, 2009a. Censo Agropecuário. Instituto Brasileiro de Geografia e Estatística, Rio de Janeiro.

IBGE, 2009b. Produção Agrícola Municipal. Instituto Brasileiro de Geografia e Estatística, Rio de Janeiro.

IBGE, 2009c. Comunicação Social – Agricultura Familiar 2006. Instituto Brasileiro de Geografia e Estatística, Rio de Janeiro.

ICONE, 2009. Report to the U.S. Environmental Protection Agency regarding the proposed changes to the renewable fuel standard program: Impacts on land use and GHG emissions from a shock on Brazilian sugarcane ethanol exports to the United States using the Brazilian Land Use Model (BLUM). Available at http://www.iconebrasil.org.br/arquivos/noticia/1873.pdf.

ICRAF China, 2007. Biofuels in China: An analysis of the opportunities and challenges of *Jatropha curcas* in southwest China. World Agroforestry Centre China, Beijing.

IEA, 2004a. Biofuels for transport: An international perspective. International Energy Agency, Paris.

IEA, 2004b. World energy outlook 2004. International Energy Agency, Paris.

IEA, 2006. World energy outlook 2006. International Energy Agency, Paris.

IEA, 2010a. Sustainable production of second-generation biofuels: Potential and perspectives in major economies and developing countries. International Energy Agency, Paris.

IEA. 2010b. Energy Balances for Non-OECD Countries: 2007–2008. International Energy Agency, Paris.

IEA, 2010c. World energy outlook 2010. International Energy Agency, Paris.

IEA, 2011a. Extended energy balances of OECD and non-OECD countries. International Energy Agency, Paris.

IEA, 2011b. Area e producao Agricola: Estado de São Paulo. Instituto de Economica Agricola, São Paulo.

IGBP-IHDP, 1995. The Miombo network. Projection of land use and land cover change in southern Africa's Miombo region. Available at http://miombo.gecp.virginia.edu/.

IIASA and FAO, 2000. Global Agro-Ecological Zones 2000. International Institute for Applied Systems Analysis and Food and Agriculture Organization, Vienna and Rome.

Illovo, 2009. Group profile. Illovo, Durban. Available at http://www.illovo.co.za/About_Us/Group_Information/Group_Profile.aspx.

IMF, 2005. Zimbabwe: Selected issues and statistical appendix. IMF Country Report 05/359. International Monetary Fund, Washington, DC.

IMF, 2006a. Botswana: Statistical appendix. IMF Country Report 06/65. International Monetary Fund, Washington, DC.

IMF, 2006b. Mauritius: Selected issues and statistical appendix. IMF Country Report 06/224. International Monetary Fund, Washington, DC.

IMF, 2006c. Namibia: Selected issues and statistical appendix. IMF Country Report 06/153. International Monetary Fund, Washington, DC.

IMF, 2006d. The Kingdom of Swaziland: Statistical appendix. IMF Country Report 06/109. IMF Country Report 06/224. International Monetary Fund, Washington, DC.

Innes, J., 2010. Sugar in Africa: Status and prospects. International Sugar Journal 112, 280–286.

IPCC, 2007. *Climate change 2007: The physical science basis. Summary for policymakers. Intergovernmental Panel on Climate Change.* Cambridge University Press, Cambridge.

ISO, 2010. World trade in raw and white sugar: Recent trends and prospects. International Sugar Organization, London.

IUCN, 2009. Guidelines on biofuels and invasive species. International Union for Conservation of Nature, Gland.

Ivanic, M., Martin, W., 2008. Implications of higher global food prices for poverty in low-income countries. *Agricultural Economics* 39, 405–416.

Jacobson, M.Z., 2002. *Atmospheric pollution: History, science and regulation.* Cambridge University Press, Cambridge.

Jacobson, M.Z., 2007. Effects of ethanol (E85) versus gasoline vehicles on cancer and mortality in the United States. *Environmental Science and Technology* 41, 4150–4157.

Jacoby, H.G., Li, G., Rozelle, S., 2002. Hazards of expropriation: Tenure insecurity and investment in rural China. *American Economic Review* 92, 1420–1447.

Jakarta Post, 2011. Indonesian palm oil industry calls for export tax cut. Available at http://www.asianewsnet.net/home/news.php?id=17036.

Jarmain, C., Everson, C.S., Savage, M.J., Mengistu, M.G., Clulow, A.D., Gush, M.B. 2008. Refining tools for evaporation monitoring in support of water resources management. WRC Report 1567/1/08. Water Research Commission, Pretoria.

Jatobá, J., 1986. The labour market in a recession-hit region: The northeast of Brazil. *International Labour Review* 125, 227–242.

Javier, E.Q., 2008. The Philippine Biofuel Program. Paper presented at Philippine Energy Summit, Manila, January 29–February 5.

Jepma, C.J., 1995. *Tropical deforestation, a social-economic approach.* Earthscan, London.

Jerrett, M., Burnett, R.T., Pope, C.A., III, Ito, K., Thurston, G., Krewski, D., Shi, Y., Calle, E., Thun, M., 2009. Long-term ozone exposure and mortality. *New England Journal of Medicine* 360, 1085–1095.

Jewitt, G.P.W., Wen, H.W., Kunz, R.P., van Rooyen, A.M., 2009. Scoping study on water use of crops/trees for biofuels in South Africa. Report to the water research commission. University of KwaZulu-Natal, Pietermaritzburg.

Johnson, F.I., 1983. Sugar in Brazil: Policy and production. *Journal of Developing Areas* 17, 243–256.

Johnson, F.X., Matsika, E., 2006. Bio-energy trade and regional development: The case study of bio-ethanol in southern Africa. *Energy for Sustainable Development* 10, 42–54.

Johnson, F.X., Rosillo-Calle, F., 2007. Biomass, livelihoods and international trade: Challenges and opportunities for the EU and southern Africa. Stockholm Environment Institute, Stockholm.

Johnson, F.X., Seebaluck, V., Watson, H., Woods, J., 2008. Renewable resources for industrial development and export diversification: The case of bioenergy from sugar cane in southern Africa. In *African Development Perspectives Yearbook*. University of Bremen Institute for World Economics and International Management. Lit-Verlag, Muenster, pp. 599–610.

Johnston, M., Foley, J.A., Holloway, T., Kucharik, C., Monfreda, C., 2009. Resetting global expectations from agricultural biofuels. *Environmental Research Letters* 4, doi: 10.1088/1748-9326/4/1/014004.

Jolly, L., 2011. Sugar reforms, ethanol demand and market restructuring. In Johnson, F.X., Seebaluck, V. (Eds.), *Bioenergy for sustainable development and international competitiveness: The role of sugar cane in Africa.* Earthscan, London, pp. 183–211.

Jones, D.L., 2010. Potential air emission impacts of cellulosic ethanol production at seven demonstration refineries in the United States. *Journal of the Air and Waste Management Association* 60, 1118–1143.

Jones, N., Miller, J.H., 1992. *Jatropha curcas: A multipurpose species for problematic sites.* World Bank, Washington, DC.

Jongschaap, R.E.E., Corré, W.J., Bindraban, P.S., Brandenburg, W.A., 2007. *Claims and facts on* Jatropha curcas *L.* Plant Research International, Wageningen.

Jongschaap, R.E.E., Blesgraaf, R.A.R., Bogaard, T.A., van Loo, E.N., Savenije, H.H.G., 2009. The water footprint of bioenergy from *Jatropha curcas* L. *Proceedings of the National Academy of Sciences of the United States of America* 106, E92–E92.

Jornal do Commercio, 2007a. Cana-de-açúcar: Estado terá pequenas destilarias. *Jornal do Commercio*. Available at http://www.fetape.org.br/index.php?secao=noticiaunica &codnot=112.

Jornal do Commercio, 2007b. Multinacionais não preocupam fornecedor. *Jornal do Commercio*. Available at http://www.fetape.org.br/index.php?secao=noticiaunica&codnot=112.

JornalCana, 2007. Canal do Sertão abre caminhos para crescimento nordestino. *JornalCana Norte/Nordeste* 15. Available at http://www.canaweb.com.br/pdf/159/%5Cjcnone.pdf.

Jumbe, C.B.L., Msiska, F.B.M., Madjera, M., 2009. Biofuel development in sub-Saharan Africa: Are policies conducive? *Energy Policy* 37, 4980–4986.

Junginger, M., Faaij, A., Schouwenberg, P.P., Arthers, C., Bradley, D., Best, G., 2006. Sustainable international bioenergy trade: Securing supply and demand. Technology report: Opportunities and barriers for sustainable international bioenergy trade. International Energy Agency, Paris.

Junginger, M., Bolkesjø, T., Bradley, D., Dolzan, P., Faaij, A., Heinimö, J., 2008. Developments in international bioenergy trade. *Biomass and Bioenergy* 32, 717–729.

Junior, P., Albuquerque, R., 2010. A Mãe Terra é quem nos culpa. EcoDebate. Available at http://www.ecodebate.com.br/2010/07/09/a-mae-terra-e-quem-nos-culpa-artigo-de-placido-junior-e-renata-albuquerque/.

Kadam, K.L., 2002. Environmental benefits on a life cycle basis of using bagasse-derived ethanol as a gasoline oxygenate in India. *Energy Policy* 30, 371–384.

Kahn, B.M., Zaks, D., Fulton, M., Dominik, M., Soong, E., Baker, J., Cotter, L., Reilly, J., Foley, J.A., Schneider, G., Barford, C., Licker, R., Johnston, M., Monfreda, C., Ramankutty, N., Fortenberry, T.R., Kratz, O., Oberbannscheidt, R., Miltner, S., Just, T., Frey, H., 2009. Investing in agriculture: Far-reaching challenge, significant opportunity, an asset management perspective. DB Climate Change Advisors, Deutsche Bank Group, Frankfurt.

Kaltner, F.J., Azevedo, G.F.P., Campos, I.A., Mondim, A.O.F., 2005. Liquid biofuels for transportation in Brazil: Potential and implications for sustainable agriculture and energy in the 21st century. Study commissioned by the German Technical Cooperation, Berlin.

Kamahara, H., Hasanudin, U., Widiyanto, A., Tachibana, R., Atsuta, Y., Goto, N., Daimon, H., Fujie, K., 2010. Improvement potential for net energy balance of biodiesel derived from palm oil: A case study from Indonesian practice. *Biomass and Bioenergy* 34, 1818–1824.

Kammen, D.K., Farrel, A.E., Plevin, R.J., Jones, A.D., Nemet, G.F., Delucchi, M.A., 2007. Energy and greenhouse impacts of biofuels: A framework for analysis. Joint Transportation Research Center Discussion Paper 2007-2. Organization for Economic Co-operation and Development and International Transport Forum, Paris.

Kammen, D.M., Bailis, R., Herzog, A.V., 2003. Clean energy for development and economic growth: Biomass and other renewable energy options to meet energy and development needs in poor nations. United Nations Development Programme, New York.

Kean, A.J., Harley, R.A., Littlejohn, D., Kendall, G.R., 2000. On-road measurement of ammonia and other motor vehicle exhaust emissions. *Environmental Science and Technology* 34, 3535–3539.

Kelly, K.J., Bailey, B.K., Coburn, T.C., Clark, W., Lissiuk, P., 1996. Federal test procedure emissions test results from ethanol variable-fuel vehicle Chevrolet Luminas. Presented at Society of Automotive Engineers International Spring Fuels and Lubricants Meeting, Dearborn, May 6–8.

Kendall, A., Chang, B., 2009. Estimating life cycle greenhouse gas emissions from corn-ethanol: A critical review of current U.S. practices. *Journal of Cleaner Production* 17, 1175–1182.

Khan, S.R., Yusuf, M., Khan, S.A., Abbasy, R., 2010. Biofuels trade and sustainable development: The case of sugarcane bioethanol in Pakistan. In Dufey, A., Grieg-Gran, M. (Eds.), *Biofuels production, trade and sustainable development*. International Institute for Environment and Development, London, pp. 5–34.

Khanna, M., Chen, X., Huang, H., Onal, H., 2010. Supply of cellulosic biofuel feedstocks and regional production patterns. Paper presented at the annual meeting of the American Applied Economics Association, Denver, July 25–27.

Khatiwada, D., Silveira, S., 2009. Net energy balance of molasses based ethanol: The case of Nepal. *Renewable and Sustainable Energy Reviews* 13, 2515–2524.

Kidane, W., Maetz, M., Dardell, P., 2006. Food security and agricultural development in sub-Saharan Africa. Food and Agriculture Organization, Rome.

Kim, H., Kim, S., Dale, B.E., 2009. Biofuels, land use change and greenhouse gas emissions: Some unexplored variables. *Environmental Science and Technology* 43, 961–967.

Kim, S., Dale, B.E., 2008. Life cycle assessment of fuel ethanol derived from corn grain via dry milling. *Bioresource Technology* 99, 5250–5260.

Klemm, R.J., Mason, R.M., Heilig, C.M., Neas, L.M., Dockery, D.W., 2000. Is daily mortality associated specifically with fine particles? Data reconstruction and replication analysis. *Journal of the Air and Waste Management Association* 50, 1215–1222.

Knauf, G., Maier, J., Skuce, N., Sugrue, A., 2007. The challenge of sustainable bioenergy: Balancing climate protection, biodiversity and development policy. A discussion paper. CURES, Bonn.

Koh, L.P., 2007. Potential habitat and biodiversity losses from intensified biodiesel feedstock production. *Conservation Biology* 21, 1373–1375.

Koh, L.P., 2008a. Can oil palm plantations be made more hospitable for forest butterflies and birds? *Journal of Applied Ecology* 45, 1002–1009.

Koh, L.P., 2008b. Birds defend oil palms from herbivorous insects. *Ecological Applications* 18, 821–825.

Koh, L.P., 2009. Calling Indonesia's US$13 billion bluff. *Conservation Biology* 23, 789–789.

Koh, L.P., Ghazoul, J., 2008. Biofuels, biodiversity and people: Understanding the conflicts and finding opportunities. *Biological Conservation* 141, 2450–2460.

Koh, L.P., Ghazoul, J., 2010. Spatially explicit scenario analysis for reconciling agricultural expansion, forest protection, and carbon conservation in Indonesia. *Proceedings of the National Academy of Sciences of the United States of America* 107, 11140–11144.

Koh, L.P., Wilcove, D.S., 2007. Cashing in palm oil for conservation. *Nature* 448, 993–994.

Koh, L.P., Wilcove, D.S., 2008. Is oil palm agriculture really destroying tropical biodiversity? *Conservation Letters* 1, 60–64.

Koh, L.P., Wilcove, D.S., 2009. Oil palm: Disinformation enables deforestation. *Trends in Ecology and Evolution* 24, 67–68.

Koh, L.P., Butler, R.A., Bradshaw, C.J.A., 2009a. Conversion of Indonesia's peatlands. *Frontiers in Ecology and the Environment* 7, 238–238.

Koh, L.P., Levang, P., Ghazoul, J., 2009b. Designer landscapes for sustainable biofuels. *Trends in Ecology and Evolution* 24, 431–438.

Koizumi, T., 2009. Impact of the Chinese bioethanol import on the world sugar markets: An econometric simulation approach. *International Sugar Journal* 111, 38–49.

Koizumi, T., Ohga, K., 2009. Impact of the expansion of Brazilian FFV utilization and U.S. biofuel policy amendment on the world sugar and corn markets: An econometric simulation approach. *Japanese Journal of Rural Economy* 11, 9–32.

Kojima, M., Johnson, T., 2006. Potential for biofuels for transport in developing countries. World Bank, Washington, DC.

Koning, N., van Ittersum, M.K., 2009. Will the world have enough to eat? *Current Opinion in Environmental Sustainability* 1, 77–82.

Krausmann, F., Erb, K.-H., Gingrich, S., Lauk, C., Haberl, H., 2008. Global patterns of socioeconomic biomass flows in the year 2000: A comprehensive assessment of supply, consumption and constraints. *Ecological Economics* 65, 471–487.

Kretschmer, B., Narita, D., Peterson, S., 2009. The economic effects of the E.U. biofuel target. *Energy Economics* 31, S285–S294.

Křivánek, M., Pyšek, P., Jarošík, V., 2006. Planting history and propagule pressure as predictors of invasion by woody species in a temperate region. *Conservation Biology* 20, 1487–1498.

Krivonos, E., Olarreaga, M., 2006. Sugar prices, labor income, and poverty in Brazil. *Economía* 9, 95–123.

Kunen, E., Chalmers, J., 2010. Sustainable biofuel development policies, programs and practices in APEC economies. Winrock International, Arlington, VA.

Kusdiana, D., Saptono, A., 2008. Implementation of rural energy by renewable energy in Indonesia. Paper presented at Workshop on Rural Energization, Paris, May 28–29.

Kusiima, J.M., Powers, S.E., 2010. Monetary value of environmental and health externalities associated with production of ethanol from biomass feedstocks. *Energy Policy* 38, 2785–2796.

Lacey, T., 2009. RI biofuel development: An Asian dilemma. The Jakarta Post. Available at http://www.thejakartapost.com/news/2009/02/04/ri-biofuel-development-an-asian-dilemma.html.

Lal, R., 2005. World crop residues production and implications of its use as a biofuel. *Environment International* 31, 575–584.

Lal, R., Kimble, J., Stewart, B.A., 1995. World soils as a source or sink for radiatively active gases. In Lal, R., Kimble, J., Levine, E., Stewart, B.A. (Eds.), *Advances in soil science: Soil management and greenhouse effect*. CRC Press, Boca Raton, pp. 1–7.

Lamers, P., 2006. Emerging liquid biofuel markets – A dónde va la Argentina? Master's thesis, Lund University, Lund.

Lapola, D.M., Schaldach, R., Alcamo, J., Bondeau, A., Koch, J., Koelking, C., Priess, J.A., 2010. Indirect land-use changes can overcome carbon savings from biofuels in Brazil. *Proceedings of the National Academy of Sciences of the United States of America* 107, 3388–3393.

Lara, L.L., Artaxo, P., Martinelli, L.A., Victoria, R.L., Camargo, P.B., Krusche, A., Ayers, G.P., Ferraz, E.S.B., Ballester, M.V., 2001. Chemical composition of rainwater and anthropogenic influences in the Piracicaba River Basin, southeast Brazil. *Atmospheric Environment* 35, 4937–4945.

Lara, L.L., Artaxo, P., Martinelli, L.A., Camargo, P.B., Victoria, R.L., Ferraz, E.S.B., 2005. Properties of aerosols from sugar-cane burning emissions in southeastern Brazil. *Atmospheric Environment* 39, 4627–4637.

Laschefski, K., 2008. The agro fuel debate: Conflicts between diverse environmentalisms. Paper presented at the 7th Global Conference Making Sense of Health, Illness and Disease, Oxford, July 9–12.

Laurance, W.F., Lovejoy, T.E., Prance, G., Ehrlich, P.R., Mace, G., Raven, P.H., Cheyne, S.M., Bradshaw, C.J.A., Masera, O.R., Fredriksson, G., Brook, B.W., Koh, L.P., 2010a. An open letter about scientific credibility and the conservation of tropical forests. Available at http://blogs.nature.com/news/thegreatbeyond/Scientists-Letter-ITS-WGI-Oxley.pdf.

Laurance, W.F., Koh, L.P., Butler, R.A., Sodhi, N.S., Bradshaw, C.J.A., Neidel, J.D., Consunji, H., Mateo-Vega, J., 2010b. Improving the performance of the Roundtable on Sustainable Palm Oil for nature conservation. *Conservation Biology* 24, 377–381.

Le Maitre, D.C., van Wilgen, B.W., Gelderblom, C.M., Bailey, C., Chapman, R.A., Nel, J.A., 2002. Invasive alien trees and water resources in South Africa: Case studies of the costs and benefits of management. *Forest Ecology and Management* 160, 143–159.

Lehtonen, M., 1993. Proálcool, ympäristö ja valta: Tilanteellinen analyysi Brasilian alkoholipolttoaineohjelman arvioinnissa. (Proálcool, environment and power: The application of the situational analysis to the assessment of the Brazilian alcohol fuel programme). Master's thesis. University of Helsinki.

Lehtonen, M., 2011. Social sustainability of the Brazilian bioethanol: Power relations in a centre-periphery perspective. *Biomass and Bioenergy* 35, 2425–2434.

Lehtonen, M., 2010. Status report on sugarcane agrochemicals management. Ethical Sugar, Oullins.

Lemaux, P.G., 2008. Genetically engineered plants and foods: A scientist's analysis of the issues (Part I). *Annual Review of Plant Biology* 59, 771–812.

Leng, R., Wang, C., Zhang, C., Dai, D., Pu, G., 2008. Life cycle inventory and energy analysis of cassava-based fuel ethanol in China. *Journal of Cleaner Production* 16, 374–384.

Lerner, A., Matupa, O., Mothlathledi, F., Stiles, G., Brown, R. 2010. *SADC Biofuel State of Play*. Southern African Development Community, Gabarone.

Li, S.Z., Chan-Halbrendt, C., 2009. Ethanol production in (the) People's Republic of China: Potential and technologies. *Applied Energy* 86, S162–S169.

Ligon, E., Thomas, J.P., Worrall, T., 2001. Informal insurance arrangements in village economies. *Review of Economic Studies* 69, 209–244.

Lima, J.P.R., Sicsú, A.B., 2001. Revisitando o setor sucro-alcooleiro do Nordeste: O novo contexto e a reestruturação possível. Estudos infosucro no. 4. Universidade Federal do Rio de Janeiro, Rio de Janeiro.

Liow, L.H., Sodhi, N.S., Elmqvist, T., 2001. Bee diversity along a disturbance gradient in tropical lowland forests of South East Asia. *Journal of Applied Ecology* 38, 180–192.

Liska, A.J., Cassman, K.G., 2008. Towards standardization of life-cycle metrics for biofuels: Greenhouse gas emissions mitigation and net energy yield. *Journal of Biobased Materials and Bioenergy* 2, 187–203.

Liska, A., Perrin, R.K., 2009. Indirect land use emissions in the life cycle of biofuels: Regulations vs science. *Biofuels, Bioproducts, and Biorefining* 3, 318–328.

Liu, P., Andersen, M., Pazderka, C., 2004. Voluntary standards and certification for environmentally and socially responsible agricultural production and trade. Report 5. Food and Agriculture Organization, Rome.

Loos, T.G., 2009. Socio-economic impact of a Jatropha project on smallholder farmers in Mpanda, Tanzania. Master's thesis, University of Hohenheim, Stuttgart.

Lovett, J.C., Hards, S., Clancy, J., Snell, C., 2011. Multiple objectives in biofuels sustainability policy. *Energy and Environmental Science* 4, 261–268.

Low, T., Booth, C., 2007. The weedy truth about biofuels. Invasive Species Council, Melbourne.

Luo, L., van der Voet, E., Huppes, G., 2009. Life cycle assessment and life cycle costing of bioethanol from sugarcane in Brazil. *Renewable and Sustainable Energy Balances* 13, 1613–1619.

MA, 2005. Ecosystems and human well-being: Our human planet. Summary for decision makers. World Resources Institute Millennium Ecosystem Assessment, Washington, DC.

Macedo, I.C., 2005. Sugar cane's energy: Twelve studies on Brazilian sugar cane agribusiness and its sustainability. União da Indústria de Cana-de-açúcar and Berlendis and Vertecchia, São Paulo.

Macedo, I.C., Leal, M.R.L.V., Da Silva, J.E.A.R., 2004. Assessment of greenhouse gas emissions in the production and use of fuel ethanol in Brazil. Secretariat of the Environment of the State of São Paulo.

Macedo, I.C., Seabra, J.E.A., Silva, J.E.A.R., 2008. Green house gases emissions in the production and use of ethanol from sugarcane in Brazil: The 2005/2006 averages and a prediction for 2020. *Biomass and Bioenergy* 32, 582–595.

Maddox, T., Priatna, D., Gemita, E., Salampessy, A., 2007. The conservation of tigers and other wildlife in oil palm plantations. Jambi Province, Sumatra, Indonesia. ZSL Conservation Report 7. Zoological Society of London.

Maes, W.H., Achten, W.M.J., Muys, B., 2009. Use of inadequate data and methodological errors lead to an overestimation of the water footprint of *Jatropha curcas*. *Proceedings of the National Academy of Sciences of the United States of America* 106, E91–E91.

Magnusson, R., Nilsson, C., Andersson, B., 2002. Emissions of aldehydes and ketones from a two-stroke engine using ethanol and ethanol-blended gasoline as fuel. *Environmental Science and Technology* 36, 1656–1664.

Maltsoglou, I., Khwaja, Y., 2010. Bioenergy and food security. The BEFS analysis of Tanzania. Food and Agriculture Organization, Rome.

Manders, P., Godfrey, L., Hobbs, P., 2009. Acid mine drainage in South Africa. Briefing Note 2009/02. Council for Scientific and Industrial Research Natural Resources and Environment, Pretoria.

Marandu, E., Kayo, D., 2004. The regulation of the power sector in Africa: Attracting investment and protecting the poor. Zed Books, London.

Marinoni, N., Birelli, M.P., Rostagno, C., Pavese, A., 2003. The effects of atmospheric multi-pollutants on modern concrete. *Atmospheric Environment* 37, 4701–4712.

Martinelli, L.A., Filoso, S., 2008. Expansion of sugar cane ethanol production in Brazil: Environmental and social challenges. *Ecological Applications*, 18, 885–898.

Martinelli, L.A., Camargo, P.B., Lara, L.B.L.S., Victoria, R.L., Artaxo, P., 2002. Stable carbon and nitrogen isotope composition of bulk aerosol particles in a C4 plant landscape of southeast Brazil. *Atmospheric Environment* 36, 2427–2432.

Mathews, J.A., 2007. Biofuels: What a biopact between North and South could achieve. *Energy Policy* 35, 3550–3570.

Mathews, J.A., Tan, H., 2009. Biofuels and indirect land use change effects: The debate continues. *Biofuels, Bioproducts, and Biorefining* 2, 97–99.

Mattias, S., Visser, P., 2010. Towards sustainability certification of jatropha biofuels in Mozambique. Jatropha Alliance, Berlin.

Mbohwa, C.T., Fukuda, S., 2003. Electricity from bagasse in Zimbabwe. *Biomass and Bioenergy* 25, 197–207.

McCormick, R.L., 2007. The impact of biodiesel on pollutant emissions and public health. *Inhalation Toxicology* 19, 1033–1039.

MCT, 2009. Inventario brasileiro das emissões e remoções antrópicas de gases de efeito estufa. Informações gerais e valores preliminares. Ministério da Ciência e Tecnologia, Brasilia.

MDICE, 2011. Destino das exportacoes Brasileiras de alcool etilico. Ministério do Desenvolvimento, Indústria e Comércio Exterior, Brasília.

Mekonnen, M.M., Hoekstra, A.Y., 2010. The green, blue and grey water footprint of crops and derived crop products. Value of water research report series 47. United Nations Educational, Scientific, and Cultural Organization–International Institute for Infrastructural, Hydraulic, and Environmental Engineering, Delft.

Mekonnen, M.M., Hoekstra, A.Y., 2011. The water footprint of electricity from hydropower. Value of water research report series No. 51 United Nations Educational, Scientific, and Cultural Organization–International Institute for Infrastructural, Hydraulic, and Environmental Engineering, Delft.

Melillo, J.M., Gurgel, A.C., Kicklighter, D.W., Reilly, J.M., Cronin, T.W., Felzer, B.S., Paltsev, S., Schlosser, C.A., Sokolov, A.P., Wang, X., 2009. *Unintended environmental consequences of a global biofuels program*. MIT Joint Program on the Science and Policy of Global Change, Cambridge.

Mellko, H., 2008. Water footprint for biofuels for transport: Finland and the E.U. in the year 2010. MS thesis, Helsinki University of Technology.

Meloni, A., Rudorff, B.F.T., Antoniazzi, L.B., Alves de Aguiar, D., Piedade, M.R., Adami, M., 2008. *Prospects of sugarcane expansion in Brazil: Impacts on direct and indirect land use change*. Wageningen Academic, Wageningen.

Mendonça, M.L., 2006. A OMC e os efeitos destrutivos da indústria da cana no Brasil. Cadernos de formação 2. Rede Social de Justiça e Direitos Humanos, São Paulo.

Mendonça, M.L., 2008. Os impactos da produção de cana no Cerrado e Amazônia. Rede Social de Justiça e Direitos Humanos. São Paulo and Commissão Pastoral da Terra, Recife.

Menichetti, E., Otto, M., 2009. Energy balance and greenhouse gas emissions of biofuels from a product life-cycle perspective. In Howarth, R.W., Bringezu, S. (Eds.), *Biofuels: Environmental consequences and interactions with changing land use. Proceedings of the Scientific Committee on Problems of the Environment (SCOPE) International Biofuels Project Rapid Assessment*. Cornell University, Ithaca, pp. 81–109.

MEPU, 2010. Energy sector overview. Ministry of Energy and Public Utilities of Mauritius, Port Louis.

Mercado Etico, 2009. Certificação social favorece produtores de etanol e não muda práticas trabalhistas. Mercado Etico. Available at http://mercadoetico.terra.com.br/arquivo/certificacao-social-favorece-produtores-de-etanol-e-nao-muda-praticas-trabalhistas/.

Miller, S.A., Landis, A.E., Theis, T.L., 2006. Use of Monte Carlo analysis to characterize nitrogen fluxes in agroecosystems. *Environmental Science and Technology* 40, 2324–2332.

Milt, A., Milano, A., Garivait, S., Kamens, R., 2009. Effects of 10% biofuel substitution on ground level ozone formation in Bangkok, Thailand. *Atmospheric Environment* 43, 5962–5970.

Mitchell, D., 2005. Sugar policies: An opportunity for change. In Aksoy, A., Beghin, J. (Eds.), *Global agricultural trade and developing countries*. World Bank, Washington, DC, pp. 141–160.

Mitchell, D., 2008. *A note on rising food prices*. World Bank, Washington, DC.

Mitchell, D., 2011. Biofuels in Africa: Opportunities, prospects and challenges. World Bank, Washington, DC.

Mittermeier, R.A., Gil, P.R., Hoffman, M., Pilgrim, J., Brooks, T., Mittermeier, C.G., Lamoreux, J., da Fonseca, G.A.B., 2004. Hotspots revisited. Cemex, Conservation International and Agrupacion Sierra Madre, Monterrey.

MME, 2010. Balanço energético nacional. Ministério de Minas e Energia, Brasilia.

MME, 2011. National energy balances. Ministério de Minas e Energia, Brasilia.

MNRE, 2008. National policy on biofuels. Government of India, New Delhi.

Montobbio, P.A., Sharachchandra, L., Kallis, G., Martinez-Alier, J., 2010. The political ecology of jatropha plantations for biodiesel in Tamil Nadu, India. *Journal of Peasant Studies* 37, 875–897.

Montzka, S.A., Dlugokencky, E.J., Butler, J.H., 2011. Non-CO2 greenhouse gases and climate change. *Nature*, 476, 43–50.

Moreira, J.R., 2006. Brazil's experience with bioenergy. In Hazell, P., Pachauri, K. (Eds.), *Bioenergy and agriculture: Promises and challenges*. International Food Policy Research Institute, Washington, DC, pp. 17–18.

Morris, R.E., Pollack, A.K., Mansell, G.E., Lindhjem, C., Jia, Y., Wilson, G., 2003. Impact of biodiesel fuels on air quality and human health. Report NREL/SR-540-33793. National Renewable Energy Laboratory, Golden.

Morton, D.C., DeFries, R.S., Shimabukuro, E., Anderson, L.O., Arai, E., Espirito-Santo, F.d.B., Freita, R., Morisette, J., 2006. Cropland expansion changes deforestation dynamics in the southern Brazilian Amazon. *National Academy of Sciences* 103, 14637–14641.

Moura, E.P., Nascimento Mélo, M.A., do, Medeiros, D.D., de, 2004. Um estudo sobre o desempenho da agroindústria canavieira no Estado de Pernambuco no período de 1987 a 1996. *Revista Produção* 14, 78–91.

Mourad, A.L., 2008. *Avaliação da cadeia produtiva de biodiesel obtido a partir da soja*. PhD thesis, Universidade Estadual de Campinas, Campinas.

MPOB, 2010. Overview of the Malaysian oil palm industry 2009. Malaysia Palm Oil Board, Kuala Lumpur.

Msangi, S., Sulser, T., Rosegrant, M.W., Valmonte-Santos, R., 2007. Global scenarios for biofuels: Impacts and implications. *Farm Policy Journal* 4, 1–18.

Msangi, S., Sulser, T., Rosegrant, M., Valmonte-Santos, R., Ringler, C., 2008. Global scenarios for biofuels: Impacts and implications. International Food Policy Research Institute, Washington, DC.

Mucina, L., Rutherford, M.C. (Eds.), 2006. The vegetation of South Africa, Lesotho and Swaziland. South African National Biodiversity Institute, Pretoria.

Mueller, S.A., Anderson, J.E., Wallington, T.J., 2011. Impact of biofuel production and other supply and demand factors on food price increases in 2008. *Biomass and Bioenergy* 35, 1623–1632.

Muller, A.J., Schmidhuber, J., Hoogeveen, J., Steduto, P. 2008. Some insights in the effect of growing bioenergy demand on global food security and natural resources. *Water Policy* 10, 83–94.

Mullins, K.A., Griffin, W.M., Matthews, H.S., 2011. Policy implications of uncertainty in modeled life-cycle greenhouse gas emissions of biofuels. *Environmental Science and Technology* 45, 132–138.

Mulyoutami, E., Martini, E., Wulan, Y.C., Riswandi, K., Nasution, A., Susetyo, P.J., Sianturi, P., 2010. Chapter 2, component A: Land use and human livelihoods. In Tata, H.L., van Noordwijk, M. (Eds.), *Human livelihoods, ecosystem services and the habitat of the Sumatran orangutan: Rapid assessment in Batang Toru and Tripa*. World Agroforestry Centre, Bogor, pp. 9–34.

Myers, N., Mittermeier, R.A., Mittermeier, C.G., Fonseca, G.A.B., Kent, J., 2000. Biodiversity hotspots for conservation priorities. *Nature* 403, 853–858.

NAE, 2004. Biocombustíveis. Núcleo de Assuntos Estratégico da Presidência da República, Brasilia.

Nassar, A.M., Rudorff, B.F.T., Barcellos-Antoniazzi, L., Alves de Aguirar, D., Rumenos-Piedade-Bianchi, M., Adami, M., 2008. Prospects of the sugarcane expansion in Brazil: Impacts on direct and indirect land use change. In Zuurbier, P., van de Vooren, J. (Eds.), *Contributions to climate change mitigation and the environment*. Wageningen Academic, Wageningen, pp. 63–93.

Nastari, P.M., 2006. A expansão anunciada: A expansão industrial no setor sucroalcooleiro. Revista Opiniões. Available at http://www.revistaopinioes.com.br/aa/materia.php?id=477.

NAT, 2007. Construindo a soberania energética e alimentar: Experiências autônomas de produção de combustíveis renováveis na agricultura familiar e de enfrentamento do agronegócio da energia. Núcleo Amigos da Terra Brasil, Porto Alegre.

NBS, 2007. *China rural statistical yearbook*. National Bureau of Statistics of China. China Statistics Press, Beijing.

NBS, 2008. *China statistical yearbook 2008*. National Bureau of Statistics of China. China Statistics Press, Beijing.

NBTT, 2006. An investigation into the feasibility of establishing a biofuels industry in the Republic of South Africa: Prepared to assist the development of an industrial strategy. National Biofuel Task Team, Pretoria. Available at http://www.cityenergy.org.za/files/transport/resources/biofuels/bio_feasible_study.pdf.

Ndong, R., Montrejaud-Vignoles, M., Saint Girones, O., Gabrielle, B., Pirot, R., Domergue, D., Sablayrolles, C., 2009. Life cycle assessment of biofuels from *Jatropha curcas* in West Africa: A field study. *GCB Bioenergy* 1, 197–210.

NDRC, 2007. Medium and long-term development plan for renewable energy in China (English draft). National Development and Reform Commission, Beijing.

Nepstad, D., Stickler, C.M., Filho, B.S., Merry, F., 2008. Interactions among Amazon land use, forests and climate: Prospects for a near-term forest tipping point. *Philosophical Transactions of the Royal Society B: Biological Sciences* 363, 1737–1746.

Nepstad, D., Soares, B.S., Merry, F.D., Lima, A., Moutinho, P., Carter, J., Bowman, M., Cattaneo, A., Rodrigues, H., Schwartzman, S., McGrath, D., Stickler, C., Lubowski, R., Piris-Cabezas, R., Rivero, S., Alencar, A., Almeida, O., Stella, O., 2009. The end of deforestation in the Brazilian Amazon. *Science* 326, 1350–1351.

Nguyen, H.T.H., Takenaka, N., Bandow, H., Maeda, Y., de Oliva, S.T., Batelho, M.M.f., Tavares, T. M., 2001. Atmospheric alcohols and aldehydes concentrations measured in Osaka, Japan and in São Paulo, Brazil. *Atmospheric Environment* 35, 3075–3083.

Nguyen, T.L.T., Gheewala, S.H., Garivait, S., 2007. Full chain energy analysis of fuel ethanol from cassava in Thailand. *Environmental Science and Technology* 41, 4135–4142.

Nielsen, P.H., Wenzel, H., 2005. Environmental assessment of ethanol produced from corn starch and used as an alternative to conventional gasoline for car driving. Technical University of Denmark, Roskilde.

Noronha, S., Ortiz, L., Schlesinger, S., 2006. *Agribusiness and biofuels: An explosive mixture.* Friends of the Earth Brazil, Rio de Janeiro.

Novo, A., Jansen, K., Slingerland, M., Giller, K., 2010. Biofuel, dairy production and beef in Brazil: Competing claims on land use in São Paulo state. *Journal of Peasant Studies* 37, 769–792.

NRC, 1999. *Ozone-forming potential of reformulated gasoline.* National Research Council. National Academy Press, Washington, DC.

NREL, 1999. Fact sheet: Ford Taurus ethanol-fueled sedan. Report NREL/FS-540-26578. National Renewable Energy Laboratory, Golden.

Nyoka, B.I., 2003. State of forest and tree genetic resources in dry zone southern Africa development community countries. Forest Genetic Resources Working Paper FGR/41E. Food and Agriculture Organization, Rome.

O'Connor, T.G., Bredenkamp, G.J., 1997. Grassland. In Cowling, R.M., Richardson, D.M., Pierce, S.M. (Eds.), *Vegetation of South Africa.* Cambridge University Press, Cambridge, pp. 215–257.

OECD, 2008. Biofuel support policies: An economic assessment. Organization for Economic Co-operation and Development, Paris.

OECD-FAO, 2008. Agricultural outlook 2008–2017. Organisation for Economic Cooperation and Development, Paris, and Food and Agriculture Organization, Rome.

OECD-FAO, 2010. Agricultural Outlook 2010–201. Organisation for Economic Co-operation and Development, Paris, and Food and Agriculture Organization, Rome.

Oliveira, E.D.D., Seixas, F., 2006. Análise energética de dois sistemas mecanizados na colheita do eucalipto. *Scientia Florestalis* 70, 49–57.

Oliveira, M.E.D., Vaughan, B.E., Rykiel, E.J., 2005. Ethanol as fuel: Energy, carbon dioxide balances and ecological footprint. *Bioscience* 55, 593–602.

Openshaw, K., 2000. A review of *Jatropha curcas*: An oil plant of unfulfilled promise. *Biomass and Bioenergy* 19, 1–15.

Oppenheimer, C., Tsanev, V.I., Allen, A.G., McGonigle, A.J.S., Cardoso, A.A., Wiatr, A., Paterlini, W., de Mello Dias, C., 2004. NO_2 emissions from agricultural burning in São Paulo, Brazil. *Environmental Science and Technology* 38, 4557–4561.

Oxfam, 2008. Another inconvenient trouth. How biofuel policies are deepening poverty and accelerating climate change. Oxfam, Oxford.

PAC, 2009. Small-scale bioenergy initiatives: Brief description and preliminary lessons on livelihood impacts from case studies in Asia, Latin America and Africa. Practical Action Consulting, Rugby.

Panichelli, L., Dauriat, A., Gnansounou, E., 2009. Life cycle assessment of soybean-based biodiesel in Argentina for export. *International Journal of Life Cycle Assessment* 14, 144–159.

Parkinson, J., 2003. Legitimacy problems in deliberative democracy. *Political Studies* 51, 180–196.

Parton, W.J., Schimel, D.S., Cole, C.V., Ojima, D.S., 1987. Analysis of factors controlling soil organic matter levels in Great Plains grasslands. *Soil Science Society of America Journal* 51, 1173–1179.

Patzek, T., Pimentel, D., 2005. Thermodynamics of energy production from biomass. *Critical Reviews in Plant Sciences* 24, 327–364.

Peduzzi, P., 2009. Norte e Nordeste serão os mais beneficiados pelo etanol da mandioca, dizem especialistas. Agência Brasil. Available at http://www.noticiasdaamazonia.com.br/7620-norte-e-nordeste-serao-os-mais-beneficiados-pelo-etanol-da-mandioca-dizem-especialistas/.

Peh, K.S-H., De Jong, J., Sodhi, N.S., Lim, S.L.-H., Yap, C.A.-M., 2005. Lowland rainforest avifauna and human disturbance: persistence of primary forest birds in selectively logged forests and mixed rural habitats of southern Peninsular Malaysia. *Biological Conservation* 123, 489–505.

Peh, K.S.-H., Sodhi, N.S., De Jong, J., Sekercioglu, C.H., Yap, C.A.-M., Lim, S.L.-H., 2006. Conservation value of degraded habitats for forest birds in southern Peninsular Malaysia. *Diversity and Distributions* 12, 572–581.

Pellegrini, L.P., de Oliveira, S., Jr., 2011. Combined production of sugar, ethanol and electricity: Thermoeconomic and environmental analysis and optimisation. *Energy* 36, 3704–3715.

Pereira, C.L.F., Ortega, E., 2010. Sustainability assessment of large scale ethanol production from sugarcane. *Journal of Cleaner Production* 18, 77–82.

Peres, S., 2009. *Panorama dos biocombustíveis no Brasil: Norte/Nordeste*. Paper presented at Bioenergia: Desafios e Oportunidades de Negócios, August 26–27, Universidade de São Paulo, São Paulo.

Peskett, L., Slater, R., Stevens, C., Dufey, A., 2007. Biofuels, agriculture and poverty reduction. Overseas Development Institute Natural Resource Perspectives, London.

Pew Charitable Trusts, 2010. G-20 Clean energy factbook. Pew Charitable Trusts, Washington, DC.

Pickett, J., Anderson, D., Bowles, D., Bridgwater, T., Jarvis, P., Mortimer, N., 2008. Sustainable biofuels: Prospects and challenges. Royal Society, London.

Pilinis, C., 1989. Numerical-simulation of visibility degradation due to particulate matter – Model development and evaluation. *Journal of Geophysical Research* 94, 9937–9946.

Pimentel, D., Patzek, T.W., 2005. Ethanol production using corn, switch grass, and wood: Biodiesel production using soybean and sunflower. *Natural Resources Research* 14, 65–76.

Pinstrup-Andersen, P., 2009. Food security: Definition and measurement. *Food Security* 1, 5–7.

Pinto, E., Melo, M., Mendonça, M.L., 2007. O mito dos biocombustíveis. Brasil de Fato. Available at http://resistir.info/energia/mito_biocombustiveis.html.

Plataforma BNDES, 2008. Impactos da industria canavieira no Brasil: Poluição atmosférica, ameaça a recursos hídricos, riscos para a produção de alimentos, relações de trabalho atrasadas e proteção insuficiente à saúde de trabalhadores. Instituto Brasileiro de Análises Sociais e Econômicas, Rio de Janeiro.

Pleanjai, S., Gheewala, S.H., 2009. Full chain energy analysis of biodiesel production from palm oil in Thailand. *Applied Energy* 86, S209–S214.

Plevin, R.J., O'Hare, M., Jones, A.D., Torn, M.S., Gibbs, H., 2010. Greenhouse gas emissions from biofuels' indirect land use change are uncertain but may be much greater than previously estimated. *Environmental Science and Technology* 44, 8015–8021.

Porter, G., Dabat, C.R., de Souza, H.R., 2001. Local labour markets and the reconfiguration of the sugar industry in northeast Brazil. *Antipode* 33, 826–854.

Postel, S.L., Daily, G.C., Ehrlich, P.R., 1996. Human appropriation of renewable freshwater. *Science* 271, 785–788.

Potter, L., 2008. The oil palm question in Borneo. In Persson, G.A., Osseijer, M. (Eds.), *Reflections on the heart of Borneo*. Tropenbos International, Wageningen, pp. 69–90.

Prueksakorn, K., Gheewala, S.H., 2008. Full chain energy analysis of biodiesel from *Jatropha curcas* L. in *Thailand*. *Environmental Science and Technology* 42, 3388–3393.

Puppim de Oliveira, J., 2002. The policy making process for creating competitive assets for the use of biomass energy: The Brazilian alcohol programme. *Renewable and Sustainable Energy Reviews* 6, 129–140.

Qiu, H., Huang, J., Yang, J., Rozelle, S., Zhang, Y., 2010. Bioethanol development in China and the potential impacts on its agricultural economy. *Applied Energy* 87, 76–83.

Quintero, J.A., Montoya, M.I., Sanchez, O.J., Giraldo, O.H., Cardona, C.A., 2008. Fuel ethanol production from sugarcane and corn: Comparative analysis for a Colombian case. *Energy* 33, 385–399.

Raboni, A., 2009. *Etanol aparece em relatório de trabalho escravo*. Folha de São Paulo. Available at http://www1.folha.uol.com.br/folha/brasil/ult96u582224.shtml.

Raghu, S., Anderson, R.C., Daehler, C.C., Davis, A.S., Wiedenmann, R.N., Simberloff, D., Mack, R.N., 2006. Adding biofuels to the invasive species fire? *Science* 313, 1742–1742.

Rajagopal, D., 2008. Implications of India's biofuel policies for food, water and the poor. *Water Policy* 10, 95–106.

Ramachandran Nair, P.K., Mohan Kumar, B., Nair, V. D., 2009. Agroforestry as a strategy for carbon sequestration. *Journal of Plant Nutrition and Soil Science* 172, 10–23.

Ramalho, A., 2010. An internationalized business. Brasil Energy. Available at http://www.energiahoje.com.br/brasilenergy/2010/04/16/408597/an-internationalized-business.html.

Ramankutty, N., Foley, J.A., Norman, J., McSweeney, K., 2002. The global distribution of cultivable lands: Current patterns and sensitivity to possible climate change. *Global Ecology and Biogeography* 11, 377–392.

Ramos, P., 2006. O arrendamento nos lotes dos projetos de assentamento de trabalhadores rurais: Uma possibilidade a considerar? Paper presented at the 44th Sober Congress, Fortaleza.

Rao, P.J.M., 1997. Industrial utilisation of sugar cane and its co-products. ISPCK, New Delhi.

Ravindranath, N.H., Mauvie, R., Fargione, J., Canadell, J.G., Berndes, G., Woods, J., Watson, H., Sathaye, J., 2009. GHG implications of land use and land conversion to biofuel crops. In Howarth, R.W., Bringezu, S. (Eds.), *Biofuels: Environmental consequences and interactions with changing land use. Proceedings of the Scientific Committee on Problems of the Environment (SCOPE) International Biofuels Project Rapid Assessment*. Cornell University, Ithaca, pp. 111–125.

Regalbuto, J.R., 2009. Cellulosic biofuels – Got gasoline? *Science* 325, 822–824.

Reinhardt, G., Gärtner, S., Rettenmaier, N., Munch, J., von Falkenstein, E., 2007a. Screening life cycle assessment of jatropha biodiesel. Institute for Energy and Environmental Research, Heidelberg.

Reinhardt, G.A., Gärtner, S., Patyk, A., Rettenmaier, N., 2007b. Ökobilanzen zu BTL: Eine ökologische Einschätzung. Institute for Energy and Environmental Research, Heidelberg.

Repórter Brasil, 2009. Brazil of biofuels: Impact of crops over land, environment and society – sugarcane. *Repórter Brazil*, São Paulo.

Repórter Brasil, 2010a. Brazil of biofuels: Impact of crops over land, environment and society – sugarcane. *Repórter Brazil*, São Paulo.

Repórter Brasil, 2010b. A agricultura familiar e o programa nacional de biodiesel – retrato do presente, perspectivas de futuro. Centro de Monitoramento de Agrocombustíveis. Available at http://www.reporterbrasil.org.br/agrocombustiveis/relatorio.php.

Repórter Brasil, 2010c. O Brasil dos agrocombustíveis: Impactos das lavouras sobre a terra, o meio e a sociedade – Cana 2009. *Repórter Brasil*, São Paulo.

Rettenmaier, N., Koppen, S., Gartner, S.O., Reinhardt, G.A., 2010. Life cycle assessment of selected future energy crops for Europe. *Biofuels, Bioproducts, and Biorefining* 4, 620–636.

RFA, 2008. *The Gallagher review of the indirect effects of biofuels production*. Renewable Fuels Agency, East Sussex.

Ribeiro, D., Matavel, N., 2009. Jatropha! A Socio-economic pitfall for Mozambique. Justica Ambiental/Uniao Nacional de Camponesese, Mozambique.

Ribeiro, D., Matavel, N., 2010. The jatropha trap? The realities of farming jatropha in Mozambique. Friends of the Earth International, Amsterdam.

Richardson, B., 2010. Big sugar in southern Africa: Rural development and the perverted potential of sugar/ethanol exports. *Journal of Peasant Studies* 37, 917–938.

Richardson, D.M., Blanchard, R., 2011. Learning from our mistakes: Minimizing problems with invasive biofuel plants. *Current Opinion in Environmental Sustainability* 3, 36–42.

Rist, J., Lee, J.S.H., Koh, L.P., 2009. Biofuels: social benefits. *Science* 326, 1344–1344.

RIVM, 2008. Option for sustainable bioenergy: A jatropha case study. Report 607034001/ 2008. National Institute of Public Health and Environment, Bilthoven.

Robertson, B., Pinstrup-Andersen, P., 2010. Global land acquisition: Neo-colonialism or development opportunity? *Food Security* 2, 271–283.

Robertson, G.P., Dale, V.H., Doering, O.C., Hamburg, S.P., Melillo, J.M., Wander, M.M., Parton, W.J., Adler, P.R., Barney, J.N., Cruse, R.M., Duke, C.S., Fearnside, P.M., Follett, R.F., Gibbs, H.K., Goldemberg, J., Mladenoff, D.J., Ojima, D., Palmer, M.W., Sharpley, A., Wallace, L., Weathers, K.C., Wiens, J.A., Wilhelm, W.W., 2008. Sustainable biofuels redux. *Science* 322, 49–50.

Robles, M., Torero, M., von Braun, J., 2009. When speculation matters. International Food Policy Research Institute, Washington, DC.

Rodrigues, A.S.L., Ewers, R.M., Parry, L., Souza, C., Jr., Verissimo, A., Balmford, A., 2009. Boom-and-bust development patterns across the Amazon deforestation frontier. *Science* 324, 1435–1437.

Rogers, T.D., 2005. The deepest wounds: The laboring landscapes of sugar in northeastern brazil. PhD dissertation, Duke University, Durham.

Romijn, H.A., 2011. Land clearing and greenhouse gas emissions from Jatropha biofuels on African Miombo Woodlands. *Energy Policy* 39, 5751–5762.

Rosegrant, M.W., 2007. *Biofuels and grain prices: Impacts and policy responses*. International Food Policy Research Institute, Washington, DC.

Rosegrant, M.W., Paisner, M.S., Meijer, S., Witcover, J., 2001. Global food projections to 2020: Emerging trends and alternative futures. International Food Policy Research Institute, Washington, DC.

Rosegrant, M.W., Zhu, T.S., Msangi, S., Sulser, T., 2008. Global scenarios for biofuels: Impacts and implications. *Review of Agricultural Economics* 30, 495–505.

Rosenberg, N., 1982. *Inside the black box: Technology and economics*. Cambridge University Press, Cambridge.

Rosenthal, E., 2008. Biofuels deemed a greenhouse threat. *New York Times*. Available at http://www.nytimes.com/2008/02/08/science/earth/08wbiofuels.html.

Rosillo-Calle, F., Walter, A., 2006. Global market for bioethanol: Historical trends and future prospects. *Energy for Sustainable Development* 10, 20–32.

Rossi, A., Lambrou, Y., 2008. Gender and equity issues in liquid biofuels production: Minimizing the risks and maximizing the opportunities. Food and Agriculture Organization, Rome.

Rouget, M., Richardson, D.M., Nel, J.A., van Wilgen, B.W., 2002. Commercially important trees as invasive aliens: Towards spatially explicit risk assessment at a national scale. *Biological Invasions* 4, 397–412.

Roussef, D., 2004. Biodiesel o novo combustível do Brasil. Programa Nacional de Produção e Uso do Biodiesel. Ministra de Minas e Energia, Brasilia.

Royal Society, 2008. *Sustainable biofuels: Prospects and challenges.* Royal Society, London.

RSB, 2010. Global principles and criteria for sustainable biofuels production. Version One. Roundtable on Sustainable Biofuels, São Paulo.

RSPO, 2007. RSPO Certification Systems. Roundtable on Sustainable Palm Oil, Selangor.

RTRS, 2010. RTRS Standard for Responsible Soy Production: Version 1.0. Roundtable for Sustainable Soy Production, Buenos Ayres.

Rudel, T.K., Coomes, O.T., Moran, E., Achard, F., Angelsen, A., Xu, J., Lambin, E., 2005. Forest transitions: Toward a global understanding of land use change. *Global Environmental Change* 15, 23–31.

Runge C.F., Senauer, B., 2007. *How biofuels could starve the poor.* Foreign Affairs. Available at http://www.foreignaffairs.org/20070501faessay86305/c-ford-runge-benjamin-senauer/how-biofuels-could-starve-the-poor.html.

Rutz, D., Janssen, R., 2008. *Biofuel technology handbook.* WIP Renewable Energies, München.

Sachs, I., 2007. The biofuel controversy. United Nations Conference on Trade and Development, New York.

SADC, 2010. SADC framework for sustainable biofuel use and production. Approved by SADC Energy Ministers Meeting on April 29. South African Development Community, Gaborone.

Saghir, J., 2006. World Bank perspective on global energy security. Paper presented at G8 Energy Ministerial Meeting, Moscow, March 16.

Sala, O.E., van Vuuren, D., Pereira, H., Lodge, D., Alder, J., Cumming, G.S., Dobson, D., Wolters, V., Xenopoulos, M., 2005. Biodiversity across scenarios. In Carpenter, S.R., Pingali, P.L., Bennett, E.M., Zurek, M. (Eds.), *Ecosystems and human well-being: Scenarios.* Island Press, Washington, DC, pp. 375–410.

Saldanha, A., 2005. Alagoas: A "açúcarada" sucessão e a volta "delle." Fundação Joaquim Nabuco, Recife.

Sandalow, D., 2006. Ethanol: Lessons from Brazil. In *A high growth strategy for ethanol.* Aspen Institute, Washington, DC, pp. 67–74.

Sano, D., Romero, J., 2010. Sustainable biofuels in China: Lessons from jatropha production in Yunnan. Paper presented at 7th International Biofuels Conference Proceedings, New Delhi, February 11–12.

Sargeant, H.J., 2001. Vegetation fires in Sumatra Indonesia. Oil palm agriculture in the wetlands of Sumatra: Destruction or development? Forest fire prevention and control project. European Union, Brussels, and Ministry of Forestry, Jakarta.

Sasol, 2006. Our environmental performance. Sasol, Johannesburg. Available at http://www.sasolsdr.investoreports.com/sasol_sr_2006/downloads/segmented/sasol_sustainable_development_report%202006_our_environmental_performance.pdf.

Saturnino, M., Borras, J.R., McMichael, P., Scoones, I., 2010. The politics of biofuels, land use and agrarian change: Editors' introduction. *Journal of Peasant Studies* 37, 575–592.

Sawaya, M., Nappo, M., 2009. Etanol de cana-de-açucar: Uma solução energética global sob ataque. In Abramovay, R. (Ed.), *Biocomustíveis: A energia da controvérsia.* Senac, São Paolo, pp. 19–58.

Sawyer, D., 2008. Climate change, biofuels and eco-social impacts in the Brazilian Amazon and Cerrado. *Philosophical Transactions of the Royal Society B: Biological Sciences* 363, 1747–1752.

Schaffel, S.B., La Rovere, E., 2010. The quest for eco-social efficiency in biofuels production in Brazil. *Journal of Cleaner Production* 18, 1663–1670.

Scheper-Hughes, N., 1992. *Death without weeping: The violence of everyday life in Brazil.* University of California Press, Berkeley.

Schneider, U.A., McCarl, B.A., 2003. Economic potential of biomass based fuels for greenhouse gas emission mitigation. *Environmental and Resource Economics* 24, 291–312.

Scholes, R.J., Biggs, R., 2005. A biodiversity intactness index. *Nature* 434, 45–49.

Scholes, R.J., Walker, B.H., 1993. *An African savanna: Synthesis of the Nylsvley study.* Cambridge University Press, Cambridge.

Scholz, V., Berg, W., Kafulß, P., 1998. Energy balance of solid biofuels. *Journal of Agricultural Engineering Research* 71, 263–272.

Schoneveld, G.C., German, L., Nutakor, E., in press. Towards sustainable biofuel development in Ghana: Assessing the effectiveness of the Ghanaian legal and institutional framework. Center for International Forestry Research, Bogor.

Schubert, R., Schellnhuber, H.J., Buchmann, N., Epiney, A., Griebhammer, R., Kulessa, M., Messner, D., Rahmstorf, S., Schmid, J., 2008. *Future bioenergy and sustainable land use.* Earthscan, London.

Schut, M., Slingerland, M., Locke, A., 2010. Biofuel developments in Mozambique: Update and analysis of policy potential and reality. *Energy Policy* 38, 5151–5165.

Scott, D.M., Gemita, E., Maddox, T.M., 2004. Small cats in human modified landscapes in Sumatra. *Cat News* 40, 23–25.

Scurlock, J., Rosenschein, A., Hall, D.O., 1991. Fuelling the future: Power alcohol in Zimbabwe. Biomass Users Network and Acts Press, Nairobi.

SDE, 2011. Câmara setorial da cana-de-açúcar. Secretaria de Desenvolvimento Econômico, Pernambuco, Recife.

Searchinger, T., Heimlich, R., Houghton, R.A., Dong, F., Elobeid, A., Fabio, J., Tokgoz, S., Hayes, D., Yu, T.H., 2008. Use of U.S. croplands for biofuels increases greenhouse gases through emissions from land-use change. *Science* 319, 1238–1240.

Seebaluck, V., Mohee, R., Sobhanbabu, P.R.K., Rosillo-Calle, F., Leal, M.R.L.V., Johnson, F.X., 2008. Bioenergy for sustainable development and global competitiveness: The case of sugar cane in Southern Africa. Thematic Report 2: Industry. Stockholm Environment Institute, Stockholm.

SEI, 2009. Household energy in developing countries: A burning issue. Stockholm Environment Institute, Stockholm.

Seinfeld, J.H., Pandis, S.N., 2006. *Atmospheric chemistry and physics.* 2nd ed. John Wiley, New York.

Sen, A., 1983. *Poverty and famines: An essay on entitlement and deprivation.* Oxford University Press, New York.

SenterNovem, 2005. Participative LCA on biofuels. SenterNovem, The Hague.

Shackleton, C.M., 1997. *The prediction of woody productivity in the savanna biome.* PhD thesis, University of the Witwatersrand, Johannesburg.

Shah, T., Burke, J., Villholth, K., 2007. Groundwater: A global assessment of scale and significance. In Molden, D. (Ed.), *Water for food, water for life: A comprehensive assessment of water management in agriculture.* Earthscan, London, and International Water Management Institute, Colombo, pp. 395–424.

Shapouri, H., Duffield, J., McAloon, A., Wang, M., 2004. *The 2001 net energy balance of corn ethanol.* National Corn Growers Association, Chesterfield.

Sheehan, J., Aden, A., Paustian, K., Killian, K., Brenner, J., Walsh, M., Nelson, R., 2004. Energy and environmental aspects of using corn stover for fuel ethanol. *Journal of Industrial Ecology* 7, 117–146.

Sheehan, J., Camobreco, V., Duffield, J., Graboski, M., Shapouri, H., 1998. Life cycle inventory of biodiesel and petroleum diesel for use in an urban bus. National Renewable Energy Laboratory, Golden and Washington, DC.

Sheil, D., Casson, A., Meijaard, E., van Noordwijk, M., Gaskell, J., Sunderland-Groves, J., Wertz, K., Kanninen, M., 2009. The impacts and opportunities of oil palm in Southeast Asia: What do we know and what do we need to know? Center for International Forestry Research, Bogor.

Shi, A.Z., Koh, L.P., Tan, H.T.W., 2009. The biofuel potential of municipal solid waste. *Global Change Biology–Bioenergy* 1, 317–320.

Shiklomanov, I.A., 1997. Assessment of water resources and availability in the world. In Kjellén, M., Mcgranahan, G. (Eds.), *Comprehensive assessment of the freshwater resources of the world*. Environment Institute, Stockholm, pp. 88.

Shiklomanov, I.A., 2000. Appraisal and assessment of world water resources. *Water International* 25, 11–32.

Sicsú, A.B., Silva, K.S., 2001. Desenvolvimento rural na zona da mata canavieira do nordeste Brasileiro: Uma visão recente. Paper presented at the conference "Dilemas e perspectivas para o desenvolvimento regional com ênfases agrícola e rural no Brasil na primeira década do século XXI," Santiago, December 11–13.

Sierra, K., 2006. Meeting energy security and environment challenges: The crucial role of renewable energy policy. Keynote speech at International Grid-Connected Renewable Energy Policy Forum, Mexico City, February 1–3.

Silalertruska, T., Gheewala, S.H., 2009. Environmental sustainability assessment of bioethanol production in Thailand. *Energy* 34, 1933–1946.

Silva, J.M., de, Novato-Silva, E., Faria, H.P., Pinheiro, T.M.M., 2005. Agrotóxico e trabalho: Uma combinação perigosa para a saúde do trabalhador rural. *Ciência and Saúde Coletiva* 10, 891–903.

Simões Lopes, L.I., 2009. Private equity and the sugar ethanol and agrobusiness sector. Paper presented at the 2009 Ethanol Summit, São Paulo, June 1. Available at http://2009.ethanolsummit.com.br/upload/palestrante/20090615041736671–701210055.pdf.

Simpson, T.W., Martinelli, L.A., Sharpley, A.N., Howarth, R.W., 2009. Impact of ethanol production on nutrient cycles and water quality: The United States and Brazil as case studies. In Howarth, R.W., Bringezu, S. (Eds.), *Biofuels: Environmental consequences and interactions with changing land use. Proceedings of the Scientific Committee on Problems of the Environment (SCOPE) International Biofuels Project Rapid Assessment*. Cornell University, Ithaca, pp. 153–167.

Sims, R.E.H., Hastings, A., Schlamadinger, B., Taylor, G., Smith, P., 2006. Energy crops: Current status and future prospects. *Global Change Biology* 12, 2054–2076.

Sims, R.E.H., Mabee, W., Saddler, J.N., Taylor, M., 2010. An overview of second generation biofuel technologies. *Bioresource Technology* 101, 1570–1580.

Singh, I., Squire, L., Strauss, J. (Eds.), 1986. *Agricultural household models*. Johns Hopkins University Press, Baltimore.

Smeets, E., Junginger, M., Faaij, A., Walter, A., Dolzan, P., 2006. Sustainability of Brazilian bio-ethanol. Department of Science, Technology and Society Report NWS-E-2006-110, Copernicus Institute, Utrecht.

Smeets, E.M.W., Faaij, A.P.C., Lewandowski, I.M., Turkenburg, W.C., 2007. A bottom-up assessment and review of global bio-energy potentials to 2050. *Progress in Energy and Combustion Science* 33, 56–106.

Smeets, E., Junginger, M., Faaij, A., Walter, A., Dolzan, P., Turkenburg, W., 2008. The sustainability of Brazilian ethanol: An assessment of the possibilities of certified production. *Biomass and Bioenergy* 32, 781–813.

Smeets, E., Bouwman, L., Stehfest, E., van Vuuren, D., Posthuma, A., 2009. Contribution of N_2O to the greenhouse gas balance of first generation biofuels. *Global Change Biology* 15, 1–23.

Soccol, C.R., de Souza Vandenberghe, L.P., Medeiros, A.B.P., Karp, S.G., Buckeridge, M., Ramos, L.P., Pitarelo, A.P., Ferreira-Leitao, V., Gottschalk, L.M.F., Ferrara, M.A., da Silva Bon, E.P., de Moraes, L.M.P., de Amorim Araujo, J., Torres, F.A.G., 2010. Bioethanol from lignocelluloses: Status and perspectives in Brazil. *Bioresource Technology* 101, 4820–4825.

Souza, L., 2007. BB dá perdão bilionário para usineiros. Folha de São Paulo. Available at http://infoener.iee.usp.br/infoener/hemeroteca/imagens/100106.htm.

Sparovek, G., Berndes, G., Egeskog, A., de Freitas, F.L.M., Gustafsson, S., Hansson, J., 2007. Sugarcane ethanol expansion in Brazil: An expansion model sensitive to socioeconomic and environmental concerns. *Biofuels, Bioproducts, and Biorefining* 1, 270–282.

Spatari, S., MacLean, H.L., 2010. Characterizing model uncertainties in the life cycle of lignocelluloses-bases ethanol fuels. *Environmental Science and Technology* 44, 8773–8780.

Spatari, S., Bagley, D.M., MacLean, H.L., 2010. Life cycle evaluation of emerging lignocellulosic ethanol conversion technologies. *Bioresource Technology* 101, 654–667.

Sperling, D., Yeh, S., 2010. Toward a global low carbon fuel standard. *Transport Policy* 17, 47–49.

Sto, E., Standbacken, P., Scheer, D., Rubik, F., 2005. Background: Theoretical contributions, eco-labels and environmental policy. In Rubik, F., Frankl, P. (Eds.), *The future of eco-labeling: Making environmental product information systems effective*. Greenleaf, Sheffield, pp. 16–45.

Stokstad, E., 2001. U.N. report suggests slowed forest losses. *Science* 291, 2294–2294.

Stratton, R.W., Wong, H.M., Hileman, J.I., 2011. Quantifying variability in life cycle greenhouse gas inventories of alternative middle distillate transportation fuels. *Environmental Science and Technology* 45, 4637–4644.

Stromberg, P., Esteban, M., Thompson-Pomeroy, D., 2009. Interlinkages in climate change-vulnerability of a mitigation strategy: Implications of increased typhoon intensity on biofuel production. United Nations University–Institute of Advanced Studies, Yokohama.

Stromberg, P.M., Gasparatos, A., Lee, J.S.H., Garcia-Ulloa, J., Koh, L.P., Takeuchi, K., 2010. Impacts of liquid biofuels on ecosystem services and biodiversity. United Nations University–Institute of Advanced Studies, Yokohama.

Stromberg, P., Esteban, M., Gasparatos, A., 2011. Climate change effects on mitigation measures: The case of extreme wind events and Philippines' biofuel plan. *Environmental Science and Policy* 14, 1079–1090.

Sugrue, A., Andrew, M., 2010. SADC bioenergy policy development. GTZ, Hatfield and ProBEC, Pretoria.

Sulle, E., Nelson, F., 2009. Biofuels, land access, and rural livelihoods in Tanzania. International Institute for Environment and Development, London.

Swyngedouw, E., 2005. Governance innovation and the citizen: The janus face of governance beyond the state. *Urban Studies* 42, 1991–2006.

Tacconi, L., 2003. Fires in Indonesia: Causes, costs and policy implications. CIFOR Occasional Paper 38. Center for International Forestry Research, Bogor.

Taheripour, F., Tyner, W., 2007. Ethanol subsidies: Who gets the benefits? *Paper presented at Bio-Fuels, Food, and Feed Tradeoffs Conference*, St. Louis, April 12–13.

Taheripour, F., Hertel, T.W., Tyner, W.E., Beckman, J.F., Birur, D.K., 2010. Biofuels and their by-products: Global economic and environmental implications. *Biomass and Bioenergy* 34, 278–289.

Takama, T., Lambe, F., Johnson, F.X., Arvidson, A., Atanassov, B., Debebe, M., Nilsson, L., Tella, P., Tsephel, S., 2011. Will African consumers buy cleaner fuels and stoves? A household energy economic analysis model for the market introduction of bio-ethanol cooking stoves in Ethiopia, Tanzania, and Mozambique. Stockholm Environment Institute, Stockholm.

Takavarasha, T., Uppal, J., Hongo, H., 2006. Feasibility study for the production and use of biofuel in the SADC region. Southern African Development Community, Gaborone.

Takeuchi, K., Shiroyama, H., Matsuura, M., Saitoh, O. (Eds.), *Biofuels and sustainability*. United Nations University Press, Tokyo, in press.

Tata, H.L., van Noordwijk, M. (Eds.), 2010. Human livelihoods, ecosystem services and the habitat of the Sumatran orangutan: Rapid assessment in Batang Toru and Tripa. World Agroforestry Centre, Bogor.

Thamsiriroj, T., Murphy, J.D., 2009. Is it better to import palm oil from Thailand to produce biodiesel in Ireland than to produce biodiesel from indigenous Irish rapeseed? *Applied Energy* 86, 595–604.

Thomas, C.D., Cameron, A., Green, R.E., Bakkenes, M., Beaumont, L.J., Collingham, Y.C., Erasmus, B.F.N., de Siqueira, M.F., Grainger, A., Hannah, L., Hughes, L., Huntley, B., van Jaarsveld, A.S., Midgley, G.F., Miles, L., Ortega-Huerta, M.A., Peterson, A.T., Phillips, O.L., Williams, S.E., 2004. Extinction risk from climate change. *Nature* 427, 145–148.

Tilman, D., Fargione, J., Wolff, B., D'Antonio, C., Dobson, A., Howarth, R., Schindler, D., Schlesinger, W.H., Simberloff, D., Swackhamer, D., 2001. Forecasting agriculturally driven global environmental change. *Science* 292, 281–284.

Tilman, D., Cassman, K.G., Matson, P.A., Naylor, R., Polasky, S., 2002. Agricultural sustainability and intensive production practices. *Nature* 418, 671–677.

Tilman, D., Socolow, R., Foley, J.A., Hill, J., Larson, E., Lynd, L., Pacala, S., Reilly, J., Searchinger, T., Somerville, C., Williams, R., 2009. Beneficial biofuels: The food, energy, and environment trilemma. *Science* 325, 270–271.

Timilsina, G.R., Shrestha, A., 2010. Biofuels, markets, targets and impacts. *Policy Research Working Paper 5364*. World Bank, Washington, DC.

Timilsina, G.R., Beghin, J.C., van der Mensbrugghe, D., Mevel, S., 2011. The impacts of biofuels targets on land-use change and food supply: A global CGE assessment. World Bank Policy Research Working Paper 5513. World Bank, Washington, DC.

Toasa, J., 2009. Colombia: A new ethanol producer on the rise. U.S. Department of Agriculture Economic Research Service, Washington, DC.

Tokgoz, S., Elobeid, A., Fabiosa, J., Hayes, D.J., Babcock, B.A., Yu, T., Dong, F., Hart, C.E., 2008. Bottlenecks, drought and oil price spikes: Impact on U.S. ethanol and agricultural sectors. *Review of Agricultural Economics* 30, 604–622.

Tol, R., 2009. The economic effects of climate change. *Journal of Economic Perspectives* 23, 29–51.

Trentini, F., Saes, M.S.M., 2010. Sustentabilidade: O desafio dos biocombustíveis. Vol. 1. Annablume Editora, São Paulo.

Tsephel, S., Takama, T., Lambe, F., Johnson, F.X., 2009. Why perfect stoves are not always chosen: A new approach for understanding stove and fuel choice at the household level. *Boiling Point* 57, 6–8.

Turpie, J., Heydenrych, B., 2000. Economic consequences of alien infestation of the Cape Floral Kingdom's Fynbos vegetation. In Perrings, C., Williamson, M., Dalmazzone, S. (Eds.), *The economics of biological invasions*. Edward Elgar, Cheltenham, pp. 152–182.

Turrio-Baldassarri, L., Battistelli, C.L., Conti, L., Crebelli, R., de Berardis, B., Iamiceli, A.L., Gambino, M., Iannaccone, S., 2004. Emission comparison of urban bus engine fueled with diesel oil and "biodiesel" blend. *Science of the Total Environment* 327, 147–162.

Twigg, M.V., 2005. Controlling automotive exhaust emissions: Successes and underlying science. *Philosophical Transactions of the Royal Society A* 363, 1013–1033.

UEMOA, 2008. Sustainable bioenergy development in UEMOA member countries. Report written by United Nations Hub for Rural Development in West and Central Africa and the West African Economic and Monetary Union. United Nations Foundation, Washington, DC.

Uherek, E., Halenk, T., Borken-Kleefeld, J., Balkanski, Y., Berntsen, T., Borrego, C., Gauss, M., Hoor, P., Juda-Rezler, K., Lelieveld, J., Melas, D., Rypdal, K., Schmid, S., 2010. Transport impacts on atmosphere and climate: Land transport. *Atmospheric Environment* 44, 4772–4816.

UN Energy/Africa, 2007. Energy for sustainable development: Policy options for Africa. United Nations, New York.

UN, 2007a. World population prospects: The 2006 revision. United Nations Department of Economic and Social Affairs, New York.

UN, 2007b. The right to food. Interim report of the Special Rapporteur on the right to food, Jean Ziegler, submitted in accordance with General Assembly resolution 61/163, 22 August. United Nations, New York.

UN-DESA, 2007. Small-scale production and use of liquid biofuels in sub-Saharan Africa: Perspectives for sustainable development. United Nations Department of Economic and Social Affairs, New York.

UNDP, 2005. Energizing the Millenium Development Goals: A guide to energy's role in reducing poverty. United Nations Development Programme, New York.

UNECE, 2007. Secure and sustainable energy supplies. Annual report 2007. United Nations Economic Commission for Europe, New York and Geneva.

UNEP, 2009. Towards sustainable production and use of resources: Assessing biofuels. United Nations Environment Programme, Nairobi.

UNICA, 2011a. Venda de automóveis e veículos leves no Brasil. Uniao da Industria de Cana-de-Açucar, São Paulo.

UNICA, 2011b. Produção de etanol do Brasil. Uniao da Industria de Cana-de-Acusar, São Paulo.

UNICA, 2011c. Produção de cana-de-açúcar do Brasil. Uniao da Industria de Cana-de-Acusar, São Paulo.

Unisinos, 2010. Nunca um governo fez tanto por nosso setor, diz usineiro. Unisinos. Available at http://www.ihu.unisinos.br/index.php?option=com_noticias&Itemid=18&task=detalhe&id=31164.

Uriarte, M., Yackulic, C.B., Cooper, T., Flynn, D., Cortes, M., Crk, T., Cullman, G., McGinty, M., Sircely, J., 2009. Expansion of sugarcane production in São Paulo, Brazil: Implications for fire occurrence and respiratory health. *Agriculture, Ecosystems, and Environment* 132, 48–56.

U.S. House of Representatives, 2007. Public Law 110-140 2007: Energy Independence and Security Act. Washington, DC.

Valor Econômico, 2010. Etanol glorifica Lula, mas hesita na sucessão. Valor Econômico. Available at http://www.valoronline.com.br/impresso/politica/100/120412/etanol-glorifica-lula-mas-hesita-na-sucessao.

van Dam, J., Junginger, M., Faaij, A.P.C., 2010. From the global efforts on certification of bioenergy towards an integrated approach based on sustainable land use planning. *Renewable and Sustainable Energy Reviews* 14, 2445–2472.

van Dam, J., Junginger, M., Faaij, A., Jürgens, I., Best, G., Fritsche, U., 2008. Overview of recent developments in sustainable biomass certification. *Biomass and Bioenergy* 32, 749–780.

van der Voet, E., van Oers, L., Davis, C., Nelis, R., Cok, B., Heijungs, R., Chappin, E., Guinée, J.B., 2008. *Greenhouse gas calculator for electricity and heat from biomass*. Institute of Environmental Sciences, Leiden.

van der Voet, E., Lifset, R.J., Luo, L., 2010. Life-cycle assessment of biofuels, convergence and divergence. *Biofuels* 1, 435–449.

van Dingenen, R., Dentener, F.J., Raes, F., Krol, M.C., Emberson, L., Cofala, J., 2009. The global impact of ozone on agricultural crop yields under current and future air quality legislation. *Atmospheric Environment* 43, 604–618.

van Lienden, A.R., Gerbens-Leenes, P.W., Hoekstra, A.Y., van der Meer, T.H., 2010. Biofuel scenarios in a water perspective. Changes in water footprints due to changes in the transport sector. Value of Water Research Report Series 43. United Nations Educational, Scientific, and Cultural Organization–International Institute for Infrastructural, Hydraulic, and Environmental Engineering, Delft.

van Vuuren, D.P., van Vliet, J., Stehfest, E., 2009. Future bio-energy potential under various natural constraints. *Energy Policy* 37, 4220–4230.

Veiga Filho, A. de A., Ramos, P., 2006. Proálcool e evidências de concentração na produção e processamento de cana-de-açúcar. *Informações Econômicas* 36, 48–61.

Vermeulen, S., Cotula, L., 2010. Over the heads of local people: Consultation, consent, and recompense in large-scale land deals for biofuels projects in Africa. *Journal of Peasant Studies* 37, 899–916.

Volckaert, V., 2009. *Jatropha curcas*: Beyond the myth of the miracle crop. D1 Oils Plant Science, London. Available at http://www.ascension-publishing.com/BIZ/4ABVolckaert.pdf.

von Braun, J., 2008. Rising food prices: What should be done? International Food Policy Research Institute, Washington, DC.

von Braun, J., Pachuri, R.K., 2006. The promises and challenges of biofuels for the poor in developing countries. International Food Policy Research Institute, Washington, DC.

von Braun, J., Ahmed, A., Asenso-Okyere, K., Fan, S., Gulati, A., Hoddinott, J., Pandya-Lorch, R., Rosegrant, M.W., Ruel, M., Torero, M., van Rheenen, T., von Grebmer, K., 2008. High food prices: The what, who, and how of proposed policy actions. Policy Brief, May. International Food Policy Research Institute, Washington, DC.

von Fragstein und Niemsdorff, P., Kristiansen, P., 2006. Crop agronomy in organic agriculture. In Halberg, N., Alrøe, H.F., Knudsen, M.T., Kristensen, E.S. (Eds.), *Global development of organic agriculture: Challenges and prospects.* CABI, Wallingford.

von Maltitz, G.P., Brent, A., 2008. *Assessing the biofuel options for Southern Africa.* Council for Scientific and Industrial Research, Pretoria.

von Maltitz, G.P., Setzkorn, K. 2012. Potential impacts of biofuels on deforestation in Southern Africa. *Journal of Sustainable Forestry* 31, 80–97.

von Maltitz, G.P., Setzkorn, K., in press. A typology of southern African biofuel feedstock production projects. *Biomass and Bioenergy.*

von Maltitz, G.P., Scholes, R.J., Erasmus, B., Letsoalo, A., 2007. Adapting conservation strategies to accommodate climate change impacts in southern Africa. In Leary, N., Adejuwon, J., Barros, V., Burton, I., Kulkarni, J., Lasco, R. (Eds.), *Climate change and adaptation.* Earthscan, London, pp. 28–34.

von Maltitz, G.P., Haywood, L., Mapako, M., Brent, A., 2009. Analysis of opportunities for biofuel production in sub-Saharan Africa. Center for International Forestry Research, Bogor.

von Maltitz, G.P., Nickless, A., Blanchard, R., 2010. Maintaining biodiversity during biofuel development. In Amezaga, J.M., von Maltitz, G.P., Boyes, S.L. (Eds.), *Projects in developing countries: A framework for policy evaluation.* Newcastle University, Newcastle upon Tyne.

Waddington, S.R., Xiaoyun, L., Dixon, J., Hyman, G., de Vicente, C., 2010. Getting the focus right: Production constraints for six major food crops in Asian and African farming systems. *Food Security* 2, 27–48.

Wahl, N., Jamnadass, R., Baur, H., Munster, C., Liyama, M., 2009. Economic viability of Jatropha curcas L. plantations in northern Tanzania: Assessing farmers' prospects via cost-benefit analysis. Working Paper 97. World Agroforestry Centre, Nairobi.

Wang, M., Huo, H., Arora, S., 2011. Methods of dealing with co-products of biofuels in life-cycle analysis and consequent results within the U.S. context. *Energy Policy* 39, 5726–5736.

Wang, M.Q., Han, J., Haq, Z., Tyner, W.E., Wu, M., Elgowainy, A., 2011. Energy and greenhouse gas emission effects of corn and cellulosic ethanol with technology improvements and land use changes. *Biomass and Bioenergy* 35, 1885–1896.

Wang, X., 2007. Labor market behavior of Chinese rural households during transition. Institute of Agricultural Development in Central and Eastern Europe, Halle.

Watson, H.K., 2011. Potential to expand sustainable bioenergy from sugarcane in southern Africa. *Energy Policy* 39, 5746–5750.

Watson, H.K., Garland, G.G., Purchase, B., Dercas, N., Griffee, P., Johnson, F.X., 2008. Bioenergy for sustainable development and global competitiveness: The case of sugar cane in southern Africa. Thematic report 1: Agriculture. Stockholm Environment Institute, Stockholm.

WBGU, 2008. Welt im wandel: Zukunftsfähige bioenergie und nachhaltige landnutzung. Wissenschaftlicher Beirat der Bundesregierung Globale Umweltveränderungen, Berlin.

WCED, 1987. *Our common future*. World Commission on Environment and Development. Oxford University Press, Oxford.

West, T.O., Marland, G., 2002. A synthesis of carbon sequestration, carbon emissions, and net carbon flux in agriculture: Comparing tillage practices in the United States. *Agriculture Ecosystems and Environment* 91, 217–232.

Whitaker, M., Heath, G., 2009. Life cycle assessment of the use of jatropha biodiesel in Indian locomotives. National Renewable Energy Laboratory, Golden.

Wicke, B., Smeets, E., Watson, H., Faaij, A., 2011. The sustainable bioenergy production potential of semi-arid and arid regions in sub-Saharan Africa. *Biomass and Bioenergy* 35, 2773–2786.

Wiens, J., Fargione, J, Hill, J., 2011. Biofuels and biodiversity. *Ecological Applications* 21, 1085–1095.

Wilcove, D.S., Koh, L.P., 2010. Addressing the threats to biodiversity from oil palm agriculture. *Biodiversity and Conservation* 19, 999–1007.

Wilkinson, J., Herrera, S., 2008a. Subsídios para a discussão dos agrocombustíveis no Brasil. In Rede Brasileira pela integração dos povos agrocombustíveis e agricultura familiar e camponesa. Subsídios ao debate, Rio de Janeiro, pp. 22–53.

Wilkinson, J., Herrera, S., 2008b. *Agrofuels in Brazil: What outlook for its farming sector?* Oxfam Brazil, Brasília.

Wilkinson, J., Herrera, S., 2010. Biofuels in Brazil: Debates and impacts. *Journal of Peasant Studies* 37, 749–768.

Williamson, O., 1979. The governance of contractual relations. *Journal of Law and Economics* 22, 233–262.

Williamson, O., 1985. *The economic institutions of capitalism: Firms, markets, relational contracting*. Collier Macmillan, London.

Winebrake, J.J., Wang, M.Q., He, D., 2001. Toxic emissions from mobile sources: A total fuel cycle analysis from conventional and alternative fuel vehicles. *Journal of the Air and Waste Management Association* 51, 1073–1086.

Winrock, 2009. Implications of biofuel sustainability standards for Indonesia. Winrock International, Arlington VA.

Witt, A.B.R., 2010. Biofuels and invasive species from an African perspective: A review. *GCB Bioenergy* 2, 321–329.

Wodon, Q., Zaman, H., 2008. Rising food price in sub-Saharan Africa: Poverty impact and policy responses. World Bank, Washington, DC.

Wolford, W., 2004. Of land and labor: Agrarian reform on the sugarcane plantations of northeast Brazil. *Latin American Perspectives* 31, 147–170.

Woods, J., 2001. The potential for energy production using sweet sorghum in southern Africa. *Energy for Sustainable Development* 10, 31–38.

Woods, J., 2006. Science and technology options for harnessing bioenergy's potential. In Hazell, P., Pachauri, R.K. (Eds.), *Bioenergy and agriculture: Promises and challenges.* International Food Policy Research Institute, Washington, DC, pp. 13–14.

Woods, J., Mapako, M., Farioli, F., Bocci, E., Zuccari, F., Diaz-Chavez, R., Johnson, F.X., 2008. The impacts of exploiting the sugar industry bioenergy potential in southern Africa. Thematic report 4. Cane Resources Network for Southern Africa and Stockholm Environment Institute, Stockholm.

World Bank, 2006. Global economic prospects: Economic implications of remittances and migration. International Bank for Reconstruction and Development, Washington, DC.

World Bank, 2010. Africa development indicators. Available at http://data.worldbank.org/country.

World Growth, 2009. Palm oil: The sustainable oil. Palm Oil Green Development Campaign. World Growth Insitute, Arlington.

Worldwatch Institute, 2007. *Biofuels for transport: Global potential and implications for sustainable energy and agriculture.* Earthscan, London.

Wright, B., 2009. International grain reserves and other instruments to address volatility in grain markets. Policy Research Working Paper Series 5028. World Bank, Washington, DC.

Wu, W., Huang, J., Deng, X., 2010. Potential land for plantation of *Jatropha curcas* as feedstocks for biodiesel in China. *Science China Earth Sciences* 53, 120–27.

WWAP, 2009. The United Nations world water development report 3: Water in a changing world. United Nations Educational, Scientific, and Cultural Organization–World Water Assessment Programme, Paris, and Earthscan, London.

WWF Brasil, 2009. O impacto do mercado mundial de biocombustíveis na expansão da agricultura brasileira e suas consequências para as mudanças climáticas. Documento para consulta e debate. Estudo elaborado para Allianz Brasil. Programa de Agricultura e Meio Ambiente, WWF Brasil, Brasilia.

WWF-SNV, 2008. WWF and SNV's statement on bioenergy, CBD COP 9. Available at http://www.snvworld.org/.../SNV-WWF%20FSC%20Poster%2025-10-07.pdf.

Xunmin, O., Xiliang, Z., Shiyan, C., Qingfang, G., 2009. Energy consumption and GHG emissions of six biofuels pathways by LCA in (the) People's Republic of China. *Applied Energy* 86, S197–S208.

Yamba, F.D., Brown, G., Johnson, F.X., Jolly, L., Woods, J., 2008. Markets, technologies, investment, economics, implementation strategies, barriers and policy issues. Thematic report 3. Stockholm Environment Institute, Stockholm.

Yan, J., Lin, T., 2009. Biofuels in Asia. *Applied Energy* 86, S1–S10.

Yang, H., Zhou, Y., Liu, J., 2009. Land and water requirements of biofuels and implications for food supply and the environment in China. *Energy Policy* 37, 1876–1885.

Yanowitz, J., McCormick, R.L., 2009. Effect of E85 on tailpipe emissions from light-duty vehicles. *Journal of the Air and Waste Management Association* 59, 172–182.

Young, A., 1998. *Land resources: Now and for the future.* Cambridge University Press, Cambridge.

Young, A., 1999. Is there really spare land? A critique of estimates of available cultivable land in developing countries. *Environment, Development, and Sustainability* 1, 3–18.

Zah, R., Böni, H., Gauch, M., Hischier, R., Lehmann, M., Wäger, P., 2007. Life cycle assessment of energy products: Environmental impact assessment of biofuels. ETH-EMPA, St. Gallen.

Zahabu, E., 2008. Sinks and sources: A strategy to involve forest communities in Tanzania in global climate policy. PhD thesis, University of Twente, Enschede.

Zhan, X., Sohlberg, R.A., Townshend, J.R.G., DiMiceli, C., Carroll, M.L., Eastman, J.C., Hansen, M.C., and DeFries, R.S., 2002. Detection of land cover changes using MODIS 250m data. *Remote Sensing of Environment* 83, 336–350.

Zhang, C., Xie, G., Li, S., Ge, L., He, T., 2010. The productive potentials of sweet sorghum ethanol in China. *Applied Energy* 87, 2360–2368.

Zhong, C., Cao, Y., Li, B., Yuan, Y., 2010. Biofuels in China: Past, present and future. *Biofuels, Bioproducts, and Biorefining* 4, 326–342.

Zhou, A., Thomson, E., 2009. The development of biofuels in Asia. *Applied Energy* 86, S11–S20.

Zuurbier, P., van de Vooren, J. (Eds.), 2008. *Sugarcane ethanol: Contributions to climate change mitigation and the environment*. Wageningen Academic, Wageningen.

Zuzarte, F., 2007. Ethanol for cooking: Feasibility of small-scale ethanol supply and its demand as a cooking fuel: Tanzania case study. MS thesis, KTH School of Energy and Environmental Technology, Stockholm.

Zylbersztajn, D., 2010. Reshaping the global agricultural landscape: Perspectives from Brazil. *Agricultural Economics* 41, 57–64.

Index